NANOTECHNOLOGY

PROFESSOR MICK WILSON is Dean of the College of Science, Technology and Environment at the University of Western Sydney. He is a former Head of the Department of Chemistry, Materials and Forensic Science and Director of the Centre of Materials Technology at the University of Technology Sydney and has worked as a Chief Research Scientist at CSIRO. He has over 300 refereed scientific publications in international journals and over 100 on nanotechnology, of which seven have appeared in *Nature*.

DR KAMALI KANNANGARA joined the University of Technology Sydney in 1995 and is now the Director of the Nuclear Magnetic Resonance (NMR) facility at that university.

PROFESSOR GEOFF SMITH is Director of the Centre for Materials Technology, and Professor of Applied Physics at the University of Technology Sydney. He is a world leader in materials for solar energy, daylighting and windows.

ASSOCIATE PROFESSOR MICHELLE SIMMONS is a Queen Elizabeth II Research Fellow and Director of the Atomic Fabrication Facility, part of the Special Research Centre for Quantum Computer Technology at the University of New South Wales. She has over ten years' experience in all aspects of semiconductor crystal growth, device fabrication and electrical and optical characterisation of quantum electronic devices.

DR BURKHARD RAGUSE is Principal Research Scientist in the Biomimetric Engineering project at CSIRO Telecommunications and Industrial Physics. He is co-holder of over 50 patents on applications of nanotechnology, including the 'Ion Channel Switch' (ICS) biosensor technology.

NANOTECHNOLOGY

BASIC SCIENCE AND EMERGING TECHNOLOGIES

Mick Wilson

College of Science, Technology and Environment, University of Western Sydney

Kamali Kannangara Geoff Smith

Faculty of Science, University of Technology Sydney

Michelle Simmons

School of Physics, University of New South Wales

Burkhard Raguse

Biomimetric Engineering, CSIRO Telecommunications and Industrial Physics

UNSW PRESS

A UNSW Press book

Published by
University of New South Wales Press Ltd
University of New South Wales
UNSW Sydney NSW 2052
AUSTRALIA
www.unswpress.com.au

First published 2002

National Library of Australia
Cataloguing-in-Publication entry:

Nanotechnology: basic science and emerging technologies.

Bibliography.
Includes index.
ISBN 0 86840 437 3.

1. Nanotechnology. I. Wilson, Michael A.

620.5

Printer BPA, Melbourne

CONTENTS

AUTHORS

PROFESSOR MICHAEL WILSON is a former Chief Research Scientist at CSIRO and Director of the Centre for Materials Technology at the University of Technology, Sydney. He has served on the Prime Minister's working party for crime and science and is an acknowledged international expert in nanocarbons. Professor Wilson has published over 300 scientific papers in international journals. His expertise lies across numerous fields, including the development of methods of preparing nanotubes — the new forms of carbon after graphite and diamond, which are prepared from industrially cheap materials. He is also an expert in alumina production, soil chemistry, liquid fuels, petroleum exploration, nuclear magnetic resonance and some aspects of forensic science.

Wilson is on the editorial board of three international journals and is a referee for 25 different scientific journals. He has been a plenary (keynote) speaker at the American Chemical Society, Goldschmidt, Dahlem, International Humic Society, American Agronomy Society, Royal Australian Solid State Division Chemical Institute, and Australian Organic Geochemistry meetings. He also has awards from the Kyoto National Research Institute, Japan and the Australian Institute of Energy. He was awarded a DSc for research at age 41.

Wilson has extensive managerial experience as a former Chief Research Scientist and has a long record of turning his new science into practical money-making commercialised projects of benefit to Australia.

DR KAMALI KANNANGARA received her Bachelors Degree in Chemistry at the University of Colombo, Sri Lanka in 1985 and undertook graduate studies at the University of Hawaii, USA, under the supervision of Professor Marcus Tius. She was an East-West Centre scholar and received her PhD in 1994. In the research for her PhD thesis, she developed novel synthetic procedures that addressed new ways of achieving regio- and stereo-selectivity in tetrahydrocannabinol (THC) metabolites and a new methodology for benzoannelation of ketones. Dr Kannangara migrated to Australia in 1995 and joined the University of Technology, Sydney (UTS) as a Project Scientist, where she worked on synthesis of gas sensing conducting polymers. In 1997 she was appointed as a Research Scientist in the Department of Chemistry, Materials and Forensic Science, UTS, and is now Director of the Nuclear Magnetic Resonance (NMR) facility. The broad knowledge on organic chemistry that she acquired over the years is being applied to the fields of forensic chemistry, organic geochemistry and in the development of new sol-gel methods for monophasic nanocrystalline hydroxyapatite coatings.

PROFESSOR GEOFF SMITH is currently the Director of the Centre for Materials Technology and Professor of Applied Physics at the University of Technology, Sydney and is a world leader in materials for solar energy, daylighting, and windows. His multidisciplinary work ranges from fundamental new nanocomposite, nanohole, thin film and optical physics to new product development and commercialisation. His achievements include optimising electrodeposition of black chrome, new approaches to optical properties of nanocomposites, theory and realisation of angular selective thin films, luminescent daylight collection and light piping systems. Recent work includes the use of polymer nano- and micro- particles in lighting and a low cost clear solar control window using nanoparticles. He set the optical and material characteristics for the roof of Stadium Australia at the 2000 Olympics. Professor Smith has published over 140 journal and proceedings articles and is inventor on several patents. He has received awards from the Australian Research Council and Minister for Education and the World Renewable Energy Network, and was nominated for the Australia Prize (Energy) in 1998. Corporations from Germany, USA, Japan and Australia are currently linked into his group's research programs.

ASSOCIATE PROFESSOR MICHELLE SIMMONS is currently a Queen Elizabeth II Research Fellow and Director of the Atomic Fabrication Facility, part of the Special Research Centre for Quantum Computer Technology at the University of New South Wales, Sydney. She has over ten years experience in all aspects of semiconductor crystal growth, device fabrication and electrical and optical characterisation of quantum electronic devices. She has published over 140 papers in international journals and has been invited to present her work at nine international conferences and workshops. Dr Simmons completed her PhD in the design and fabrication of high efficiency solar cells in the Department of Applied Physics at the University of Durham, UK. After completing her PhD, Dr Simmons worked in the Semiconductor Physics Group at the University of Cambridge, UK, where she was in charge of the design, fabrication and characterisation of ultra high quality quantum electronic devices. This strong background in electronic device design led to several important discoveries, such as a possible 'metallic' state in semiconductors and the importance of spin polarisation in quantum electronic devices. In 1999 Dr Simmons came to Australia to take-up a prestigious QEII Fellowship in the fabrication of novel quantum electronic devices. Since then she has become the Director of the Atomic Fabrication Facility and a Program Manager in the Special Research Centre for Quantum Computer Technology at the University of New South Wales. In 2000 she was named the Australian Institute of Physics 'Women in Physics' lecturer and was awarded the GJ Russell Prize from the Australian Academy of Science. Her current research interests are to understand how quantum electronic devices work as they become purer and smaller and to use this knowledge to build the next generation of devices using quantum principles — in particular a silicon-based quantum computer.

DR BURKHARD RAGUSE received his PhD in physical organic chemistry from the University of Sydney in 1986 on host-guest complexation chemistry under the supervision of Associate Professor Ridley. After post-doctoral work with Professor Reetz at the Phillips University in Marburg, Germany, studying the mechanism and kinetics of organotitanium reactions by nuclear magnetic resonance, he joined CSIRO in 1988 to work on the Australian Membrane & Biotechnology Research Institute 'Ion Channel Switch' (ICS) biosensor project. The research involved the design and synthesis of the molecular biosensor components as well as subsequent physical characterisation of the function of the ICS biosensor. From 1993–98 Dr Raguse was a Project Manager within the CRC for Molecular Engineering, with concurrent projects in the areas of chemistry production, biophysics of membrane function and biosensor stability. He is a co-inventor of the ICS technology with over 16 core biosensor patents, which have resulted in over 50 granted patents in various countries. The work resulted in the formation of an Australian company (AMBRI Pty Ltd) to further develop and commercialise the technology. Since mid-1998 he has been working within the Biomimetric Engineering project at CSIRO Telecommunications & Industrial Physics as a Principal Research Scientist and is co-inventor on an additional four patents on the applications of nanotechnology. His current research interests include the development of nanoscale molecular assemblies and their physical characterisation, biomimetic information processing, and the biophysics of supramolecular aggregates.

PREFACE

With genetic engineering, nanotechnology looks like being the other big growth area this century. However, the educated public has little understanding of what it actually is, and textbooks for students are scarce. This book hopes to bridge the gap between highly detailed research publications and the generalities of the glamour books about the future.

Nanotechnology is as big as genetic engineering — or it soon will be. The first books on this subject were treated by many of us as science fiction. However, many of their predictions are now science reality. Using nanosystems, better and cheaper products can be made than using conventional materials. In the future we can expect computer controlled molecular machines much smaller than a speck of dust that have the accuracy and precision of today's target drug molecules. In nanomedicine, these will perform controlled surgery at the cellular and molecular level.

Nanotechnology will have an impact on war, crime, terrorism and the massive industries that go with them, namely security and law enforcement. The military has a great interest in nanotechnology, especially optical systems, but also in nanorobotics, nanomachines, smart weapons, nanoelectronics, virtual reality, massive memory, new ultra hard materials for armour, new energy-absorbing nanobased materials for stopping bullets dead, and bio-nanodevices to detect and destroy chemical and biological agents. Much of this is about protection against attack and minimising risk to military personnel. In the future, nanorobots may be able to repair defective airframes or the hulls of ships before any damage develops. Such a scenario is not yet possible, but it is feasible.

As for crime, the techniques of nanoscience will have a lot to offer forensic investigations, both for biological analysis, and materials and chemical studies. Portable instruments with sophisticated nanosensors will be able to perform accurate high level analysis at the crime site. These instruments should greatly improve conviction rates and the ability to locate real clues. Nanotechnology will stop money laundering by imprinting every computer digit.

Nanoscience may open up new ways of making computer systems and message transfer secure using special hardware keys that are immune to any form of hacking. Very few current computer firewalls are able to keep out really determined hackers. Nanoimprinting, which already exists, could be used to make 'keys', or even special nano-based biosensors coded with a dynamic DNA sequence. Nanoimprinting is already used to make banknotes virtually impossible to forge by creating special holograms in the clear plastic. Only if the master stamps were actually stolen would forgery be possible, and then a new hologram could be made.

A lot of these things are new and futuristic, but some well known products, such as paints, and processes that use chemical catalysts, such as petroleum and other chemical processing, rely on the properties of nanoparticles. They have been around for a long time. The difference is that now we are beginning to understand them at the fundamental level and so we are learning how to make them work a lot better. Previously progress was achieved very much by trial and error. Within 12 years the new chemical industry, based on particles with internal open nanostructures, will lead to a world-wide market of around A$200 billion per year. Totally new types of paints are starting to emerge as we learn to control the optical and thermal properties of the pigments using better nanoscience.

The first large growth areas are occurring right now, where existing large scale industrial production equipment can be adapted to use nanotechnology. Examples are nanoparticle-doped plastics, sunscreens, bathroom and window cleaners, car duco protection products, extremely hard new types of metals containing separate nanoparticles, and special industrial polishing powders. Of course there is also the microelectronics industry, which will become the nanoelectronics industry, with the next generation of chips having device features near and below 100 nanometres. In 20 years or so all devices and computer chips will be nanodevices! Nanoscale magnetic or optical memories are also coming soon. They will be extremely dense and cheap: many full stereo movies could be stored on one small disk.

Other applications of nanotechnology will include new home and health products, new industrial and manufacturing processes, and new ways to keep our environment clean and sustainable. This leads us into the world of nano-optics and nano-photonics where much current research is happening.

The problems of making the most of solar energy and natural light, and of cleaning up and in future controlling the quality and availability of our land, water and air have something in common. They all involve very large scale activities in terms of areas and volumes. What solutions can nanotechnology offer on such a large scale? The answer is the solutions we have been seeking for many years. The grand scale technologies not only have to be clever but they must also be very cheap to have any real impact. Until now we have struggled to develop solar technologies for widespread use because the options have been too expensive to make on the vast scale needed to achieve an economically significant uptake. Nanotechnology may overcome this hurdle and enable us to make the most of our great environmental resources at much lower cost and with minimal environmental effects. The costs of one kilowatt hour of solar electricity; of replacing 15 Wm^{-2} of lamp lighting in a building with visually comfortable daylight; and of a megalitre of pure water obtained from salty ground, river or sea water, are what ultimately counts, especially for the developing world. Thus, nanotechnology may even turn out to be a force for greater economic equality. When technologies cannot compete economically the only option is government regulation or subsidy. The difficulties that have been encountered in recent attempts to reduce our Greenhouse impact through the implementation of the Kyoto Agreement show just how hard it is to convince governments to adopt technologies that are not cost effective within political and economic parameters.

The public needs to be informed about many scientific issues, such as information technology, genetic engineering and genetic code. This book is about informing, but it aims to give a factual rather than a purely entertaining account. It is aimed at young scientists who have not completed a degree, but we also hope that other professionals will read it.

Several universities are developing courses in the embryonic area of nanotechnology. Over 20 courses are already being offered outside Australia. Flinders University was the first in Australia to offer a course in nanotechnology, and there were over 170 applicants for 25 places. Courses will follow in Australia at the University of Queensland, the University of Technology Sydney, Curtin University, the University of New South Wales and the University of Western Sydney. There is a massive scientific literature in this field already, which is unintelligible to most educated people or to junior level students. This book attempts to bridge the gap with hard scientific facts, but it does not throw out the fiction. Fiction is what dreams are made of and creates the scientific challenge.

All the authors are serious scientific researchers, but this has not made this book easier to write. It is tempting to scientifically clarify and reference all facts as in a research work, but this would make the manuscript unreadable to the intended audience. One problem we have faced is with references and figures. Each chapter could have 300 references, but this would put the book in the realms of a research text. However, it is important to introduce readers to the literature. We have therefore restricted ourselves to key references in each chapter from which other information can be gleaned, plus a few that will introduce the reader to literature searching. Similarly, we have avoided setting problems but have introduced some fun exercises, which are a different way to learn. These concentrate on finding out about terms that we have by necessity glossed over, and on stimulating the imagination — a key ingredient in nanotechnology.

It would not be incorrect to say that nanotechnology should be taught as a postgraduate course. As such, the task of writing an introductory text for undergraduates is impossible. However we have written this book to encourage people with imagination to take up science. In short, it is a marketing tool! However, it is not possible to write a text without assuming some chemistry and physics so we have taken the approach that the reader is also learning more chemistry and physics as they go along. At the start we assume very little, just an acquaintance with basic algebra, knowledge of light, heat and force, and molecular structures and molecular formulae. In Chapter 2 we discuss molecular nanotechnology and the techniques used to see atoms. This can be appreciated without too much additional science, but by Chapter 3 we expect the reader to have a good knowledge of first year chemistry. This chapter describes the processes of crystal assembly and re-assembly. By Chapter 4 we assume at least first year organic chemistry and we look at new engineered carbon systems called nanotubes. By Chapter 5 we assume this organic chemistry has advanced. We discuss the new form of molecular switches called rotaxanes and other ways that chemistry has been used to form organised nanosystems. In Chapter 6 we address nature's nanosystems and how to use them or make related new ones. We assume now that the reader has a basic knowledge of biochemistry. In Chapter 7 we look at optics from a first year physics level, but in Chapter 8, where we discuss electronics, lithography and molecular computing in more detail, we assume this has advanced. In Chapter 9 we look at likely new weird and wonderful applications. In the last chapter we let our hair down with some future prophesies. These two chapters should be readable by everyone.

One thing that we have left out is thin films. Although thin films are studied using all the tools of nanotechnology and some more, their inclusion in this book would expand the subject to the study of surfaces and their depositional properties. In this book we restrict our definition of nanotechnology to one where the bulk material under study is at least nanoscale in one dimension. This rules out thin films, except those on a nanoparticle and those that form part of an engineered nanostructure.

ACKNOWLEDGMENT

Special acknowledgment is also necessary to those who gave some of their teaching material. These include Matthew Phillips and Rick Wuhrer. We particularly thank Rick for allowing us to reproduce some material from his PhD thesis. Many thanks also for help from Professor Max Lu, Dr Wing Yeung, and Dr Costa Conn. Permission to reproduce a number of figures from the scientific literature is gratefully acknowledged. Much of the information we have found on the Internet. In these days when one patents before publishing it is pleasantly surprising to find people for whom public good is a major concern. We salute you.

1

BACKGROUND TO NANOTECHNOLOGY

This chapter describes the coming age of nanotechnology and reviews some simple chemistry and physics as a background to later chapters. You will find out about nanomachines and how atoms compose the individual building blocks of nanostructures. It is important to understand the way atoms join together and the energy that binds them. Unlike conventional materials, most of the atoms are on the surface in a nanomaterial. Hence it is also necessary to learn a little about surfaces.

1.1 SCIENTIFIC REVOLUTIONS

Let's talk some anthropology. Those of you who have seen the film '2001 a Space Odyssey' or its re-run will recall a few million years ago we worked out that rocks could be used to break things that were impossible to break with bare hands. The rock was our first tool. Once certain types of rocks became specialised, they were manufactured by craftsmen and different types of rock became useful for different purposes. The use of these tools increased the food supply and allowed the new specialists not to spend their time hunting. They were paid for their work by the hunters, first in food, and then by means of exchange units called money. Among the well fed, art also proliferated. The discovery of tools was probably the first step in humankind's evolution to control the planet and its own destiny. Other developments sprang from this, as shown in Table 1.1.

Table 1.1
History of useful science

Discovery type	Name	Age	Start date
Industrial	Tools	Stone	2 200 000 BC
Industrial	Metallurgy	Bronze	3500 BC
Industrial	Steam power	Industrial	1764
Automation	Mass production	Consumer	1906
Automation	Computing	Information	1946
Health	Genetic Engineering	Genetic	1953
Industrial	Nanotechnology	Nano age?	1991
Automation	Molecular assemblers	Assembler age?	2020?
All three	Life assemblers	Life age?	2050?

One valuable tool was fire. Around 5000 to 6000 years ago, someone put a rock containing copper ore on a campfire. The copper melted out and was collected. Now people had access to new metallurgical materials and the boundaries of technology expanded. Humankind could make new substances that were not visibly present in nature. Soon alloys — in which two types of metal are smelted together, such as bronze — were discovered. Then iron was made from iron ore, and from iron, steel. As food became abundant, more and more people were able to leave the land and work in trades.

Millennia passed, and all of industry until the 18th century was powered by the energy of human or animal muscles, or by natural energy sources like river water and the wind. However the discovery of steam power may well have been as big as the discovery of copper. Now there was an alternative source of energy. Trains and railways followed and then petroleum, cars, jet planes and spacecraft were produced.

Let us ponder for a moment. What led to the discovery of steam power? Were more scientists researching better horse-drawn vehicles or were they simply interested in the potential of steam as an energy source? Those who provided the money would probably have been more interested in the first proposition. We need to remember this when determining where we should spend our research dollars. The theme of steam power will be used again in the last chapter of this book.

Two different types of discovery followed. These were mass production and computers. Mass production was a very important step for humankind because it made the rate of supply faster than the rate an individual human could deliver. It was the product of the discovery and use of electricity. Computers, made useful by microtransistors, delivered mass production in mathematical transformation. Before this, any computation or information processing required the human brain and could only proceed at the speed a human could work.

DNA is short for deoxyribonucleic acid. It carries genetic information in living organisms called genes. RNA, or ribonucleic acid, puts this information to work in cells in various ways. A type of RNA called messenger RNA is very important. RNA and DNA are dealt with in Chapter 6. Genetic engineering, the ability to elucidate and modify the nature of genes in DNA, is a major revolution that is happening now. It is the first time we have had control over our own evolution and that of animals and plants rather than having to rely on breeding. It will increase longevity in a different way. Rather than reducing the chances of accident or disease it has the power to change us as a species. However, if the masses of long-lived people are unproductive, some would argue that we might be better rid of them. If better methods are developed for transferring and storing knowledge electronically between brains, increased longevity could even be a disadvantage.

Let us ponder again for a moment. During revolutions, wealth is made by pioneers. Knowledge of the old science is not important. Think of the sales opportunities for the people who had the first arrowheads, think of horses and carts, candles and electric lights, transistors and valves, and pharmaceutical medicines. Think of Ford, Gates, Nobel and those two rich cave people, Mrs Fire and Mr Wheel. A lot of people are already making plenty of money out of genetic engineering and gene therapy. So what else is coming?

The answer is nanotechnology. What is nanotechnology, and why will it make a lot of money? Why is it more important than all previous scientific advances? Nanotechnology is an anticipated manufacturing technology that allows thorough, inexpensive control of the structure of matter by working with atoms. It will allow many things to be manufactured at low cost and with no pollution. It will lead to the production of nanomachines, which are sometimes also called nanodevices. It is therefore an advance as important as the discovery of the first tool. However, rather than shape what nature offers, we can do it ourselves. Unlike metallurgy, natural substances are not used as the starting materials, but atoms — the ingredients of the universe.

Has this all started? Indeed it has. The concept of nanotechnology is attributed to Nobel laureate Richard Feynman in a lecture that he gave in 1959 and which was published in 1960 [1]: 'The principles of physics, as far as I can see, do not speak against the possibility of manoeuvring things atom by atom.'

If Feynman was the philosopher, then Drexler was the prophet. Feynman's definition was expanded by Drexler in a most stimulating and lateral thinking way in his book, *Engines of Creation, the Coming Age of Nanotechnology* [2]. This is essential reading. To quote Drexler (1990):

> 'nanotechnology is the principle of atom manipulation atom by atom, through control of the structure of matter at the molecular level. It entails the ability to build molecular systems with atom-by-atom precision, yielding a variety of nanomachines.'

Binnig and Rohrer expanded on Drexler's theories in a practical way. In 1981 they were the first to 'see' atoms and hence make nanotechnology possible. Scientists were soon able to pick up and move atoms to build structures. Originally the term nanotechnology was restricted to these original experiments, which had no immediate practical use. However, as soon as the significance of the discovery was appreciated, interest increased, and the term has been more broadly applied to refer to any technique that works and can be understood at a nanometre level. The origin of the term 'nano' comes from the Greek word for dwarf, but in scientific jargon, nano means 10^{-9}, so a nanometre is 10^{-9} metres, (that is, the size of ten or so atoms). Technology means the building of useful things from scientific principles. Thus nanotechnology means building useful things at the 10^{-9} level. So how is this different from chemistry?

Nanotechnology is not synonymous with chemistry, since it is more specific and concerned with observing atoms and molecules and manipulating them through visual observation at the nanoscale level. However, it may eventually encompass all of chemistry and a large part of physics and molecular biology. Feynman's and Drexler's definitions now define the field of molecular nanotechnology, which is sometimes also called molecular manufacturing. The latter is a poor description, since it is synonymous with synthetic chemistry. Other terms, such as molecular engineering, have also been used, but these terms have also broadened to include manipulating larger than atomic entities and the ability to design and manufacture materials and devices that are tens or hundreds of atoms across. We will try to cover all these areas in this book, albeit briefly. If we can keep the term 'molecular nanotechnology' to the manipulation of atoms or molecules one by one, then we will have achieved consistent terminology.

The discovery of nanotechnology in the broadest sense has immediate implications, since we can design a whole new range of machines from nanoscale objects, but not necessarily by breaking up matter into individual atoms. Rather, it may be done using bits of crystal or bits of biological material. The development and use of molecular nanotechnology — the building up from atoms — will be slower because it will take time to find the exact point where changing only a few atoms in a structure will make a difference. The single electron transistor, (Chapters 8 and 9) may be a case where molecular nanotechnology research will be commercialised faster. The scientific rewards for building nanomachines atom-by-atom should be greater than the shaping from top down approach, as researchers will achieve an ultimate level of control in assembling matter one atom at a time.

The greatest problem with molecular nanotechnology, however, is that producing one or two molecular nanostructures does not produce many goods. There are not enough of them. We need a machine that will produce vast numbers of nanostructures and also produce other machines for doing this. Machines of this type have been called 'assemblers'. When it is possible to make assemblers, molecular nanotechnology may cause a surge bigger than experienced by all the other industrial revolutions put together.

Drexler has described the assembler as a device having a nanorobot under computer control. An assembler is a nanomachine, but a very special one that can both build nanomachines and reproduce itself in the same process. It will be capable of holding and positioning reactive atoms and molecules in order to control the precise location at which chemical reactions take place. This general approach should allow the construction of large atomically precise objects by a sequence of precisely controlled chemical reactions, building objects molecule by molecule. If designed to do so, assemblers will be able to build copies of themselves — that is, to replicate.

By working in large teams, assemblers will be able to build objects cheaply. By ensuring that each atom is properly placed, they will manufacture products of high quality and reliability. Left over molecules would be disassembled and then re-used, thereby making the manufacturing process extremely clean. In an analogous fashion to a production line, an assembler will build an arbitrary molecular structure following a sequence of instructions. It will provide three-dimensional positional and full orientational control over the molecular component being added to a growing complex molecular structure. In addition, the assembler will be able to form any one of several different kinds of chemical bonds.

Is this science fiction? Not really. These structures already exist in biology and are called ribosomes. Ribosome nanomachines manufacture all the proteins used in all living things on this planet. A typical ribosome is relatively small (a few thousand cubic nanometres) and is capable of building almost any protein by binding together amino acids (the building blocks of proteins) in a precise linear sequence. To do this, the ribosome has a means of holding a type of RNA, which in turn is chemically bonded by a specific enzyme to a specific amino acid. It is also capable of holding a growing polypeptide, and then of causing the specific amino acid to react with, and be added to, the end of the polypeptide.

The instructions that the ribosome follows in building a protein are provided by messenger RNA. This is a polymer formed from the four chemical units adenine, cytosine, guanine, and uracil. A sequence of several hundred to a few thousand such units codes for a specific protein. The ribosome 'reads' this 'control tape' sequentially, and acts on the directions it provides.

The assembler requires a detailed sequence of control signals, just as the ribosome requires messenger RNA to control its actions. These may not have to be RNA or DNA. Many other sequenced 'tapes' of information can be built to make things other than proteins. In non-biological nanomachines, such detailed control signals can be provided by a computer. However, it must be a molecular computer or something similar to messenger RNA. Molecular computing is already a field that requires some thought because we cannot reduce electronic size much further without changing the way we count. Molecular and related quantum computing is dealt with in Chapters 5 and 8.

We can turn to nature to see what a nanomachine of the future will look like. There is a clue. All the information necessary for these duties is already contained in the smallest living forms. A replicating assembler with biocomputer control is a bacterium. Hence, an efficiently human-built nanomachine will be no bigger than a bacterium and probably as small as a virus. It could even be a virus.

Should we be concerned about runaway assemblers? Concerns about a 'grey goo' of assemblers consuming everything in the world have been raised by the alarmists. Without stock to work with or fuel, a production line stops, so this will probably be the fate of any early assemblers that run amok. It seems impossible that rogue assemblers could pose any risk. Problems are more likely to arise through the deliberate misuse of nanotechnology, such as the release of machines that consume flesh, water or the atmosphere; or a deadly virus-like nanomachine that kills. Thus, nanoterrorists could make the tragedy of the New York World Trade Center look small. Fortunately, a nanomachine can be disabled because we know its structure — unless it develops a virus-like ability to change its structure.

Are viruses nanomachines? Yes, but not man-made, and that really leads to a significant point. What will happen when biotechnology and nanotechnology combine? What will happen when we combine assemblers and genetic engineering? The answer is clear: the combination of assemblers, molecular computers and genetic engineering means new life forms. Not just the sorts of life forms created by modifying genes, such as genetically engineered crops, but life forms based on things other than DNA with their own intelligence. Philosophers may argue that structures that form these life assemblers may believe they are a kind of god. Perhaps this is the fate of *Homo sapiens*: just as we evolved from lower life forms and contributed to their demise, so we will produce life forms which make a choice about us, and indeed about everything. However, some would say that it would be naïve to believe someone else had not got there first somewhere else in the universe.

1.2 TYPES OF NANOTECHNOLOGY AND NANOMACHINES

What then, is a nanomachine? As discussed above, forty thousand years ago cave men and women used pieces of flint as tools to fashion materials into new objects such as pots and pans, which could be used as tools for boiling water. The definition of tool and machine then becomes blurred. Individual tools were used together for new functions. For example, a knife and a stick can be bound together with rope to form a spear. The two tools form a killing machine, which in turn is a tool. Such technology has been refined and further refined so that now we can break down objects and join bits of them together to form extremely complex machines. A silicon chip and some copper wire form a microcircuit that can be used with other components to form a computer. These are micromachines.

Life consists of a whole collection of machines. For example, apart from ribosomes, in each human there are huge numbers of copies of machines that convert carbohydrates to carbon dioxide and use the energy generated to perform life functions. This is not a lot different from a coal-fired power station, which does the same with coal rather than carbohydrates, except it is a lot bigger! Although it is smaller, the living machine is much more sophisticated because it has the capacity to reproduce itself. This is a mechanism of maintenance. A power station cannot do that.

There is no reason why we cannot use parts of living organisms as machines in much the same way as we use engineered machines such as power stations. Any substance in a living organism can be extracted and its chemical reactions can be used as a machine. Reactions that generate heat can be used as a source of energy.

Bits of life forms, such as membranes, which involve the combination of a number of chemical materials, can be used as machines to detect very small amounts of substances. However, there are many macroscopic examples that use collections of thousands of different nanomachines. For example, when light shines on a human eye, brain waves are generated. Thus, a living organism converts light energy to electrical energy as a brain response. A vat of eye tissue connected to the national electrical grid is not out of the question! Biological tissue can be grown in culture. These machines could be self-reproducing and therefore almost maintenance free.

Humankind has not been constrained by nature, and has built machines that nature never bothered to evolve at the macroscale, such as the wheel and axle, although increasingly we are finding at the nanolevel that these objects are used by nature. Humans have built these machines from macroscopic materials; first from wood and then from materials extracted from rocks, such as copper, then steel and then plastics.

Table 1.2
Prefixes that describe the sizes of things

Prefix	Symbol	Power	Name
Exa	E	10^{18}	quintrillion
Peta	P	10^{15}	quartrillion
Tera	T	10^{12}	trillion
Giga	G	10^{9}	billion
Mega	M	10^{6}	million
Kilo	k	10^{3}	thousand
Unity		10^{0}	one
Centi	c	10^{-2}	hundredth
Milli	m	10^{-3}	thousandth
Micro	m	10^{-6}	millionth
Nano	n	10^{-9}	billionth
Pico	p	10^{-12}	trillionth
Fento	f	10^{-15}	quartrillionth
Atto	a	10^{-18}	quintrillionth

As technology has developed, many machines and objects have been miniaturised. Machines that were a metre in size have been reduced to a hundredth the size (a centimetre) and even smaller. Table 1.2 lists the prefixes used to describe the size of things. Many machines in use today, such as those that use microelectronics, operate at the micro level. In fact, blood capillaries are a similar size to an integrated circuit component. There is no reason why we can't build machines one size smaller — at the nano level. Because this is the size of atoms, atoms or groups of atoms — molecules — must be used. Indeed, wheel-shaped molecules exist and so do molecules that are spherical, geared and sprocket shaped. These shapes are the components for humankind's mechanical machines, but in nanotechnology they are used at the atomic rather than at the engineering workshop scale. It is also possible to build complicated molecular machines involving lots of molecules, such as railway lines of nanometre dimensions with stations and trains that can be stopped and started. Molecules that can be used as electronic devices such as switches and transistors are most important.

Chemists have worked with molecules (which are nanostructures) for one hundred and fifty years. In the past, chemists, unlike nanotechnologists, have used conjecture to understand what is going on, since they have not been able to see atoms. The atomic theory, which says that matter is made up of 109 different types of atoms, is just that — a theory — but one that works. Now we know it is not a theory, because we can actually see atoms through the transmission electron microscope (TEM).

1.3 THE PERIODIC TABLE

The periodic table is an arrangement of the atoms (called elements) that groups them according to similar properties. They are listed from 1 through to 109 in a special way (Figure 1.1). From a molecular nanotechnologist's point of view, atoms are a collection of different basic building units from which we can make things. It seems strange that there is such a seemingly random number of atoms. Why only 109? Why not more or less? It's a bit like the number 42 in *The Hitch Hiker's Guide to the Galaxy* — seemingly meaningless. Most of the numbers in the universe are quite unique, such as π, the imaginary number i ($\sqrt{-1}$) and so on, but why this number? It is not really very mysterious. It appears that the stability of the nucleus decreases as the atomic weight increases after about element number 26, namely iron, and that this fades out infinitesimally. Thus in principle we can make element number 2050, it's just not very stable. That is much better, much more like we would expect the universe to behave. One thing we know is common in the Universe is that things 'tend to'. We can approach absolute zero but not actually get there; elements tend to become unstable, but instability is really a measure of the relative lengths of time for which they can exist. We can probably get close to all knowledge but not actually get there.

How can we expect to recall the millions of different ways these elements combine and hence how those combinations may be used in nanotechnological applications? Fortunately this is exactly what Figure 1.1, the periodic table, allows us to do. When atoms are arranged in terms of their atomic number in the way shown, they depict families of elements that have regular trends and properties.

We cannot go into too much detail here because this is not a chemistry book. However, each element has a symbol, H for hydrogen, Ca for calcium etc. Some are not obvious, such as Fe for iron. This is because the symbols do not always derive from the English name. Because an element's location in the table is a guide to its properties we should note the position of each element we meet for the first time. The vertical columns of the periodic table are called groups. These groups identify the principle families of elements. The horizontal rows are called periods. The members of each group show a gradual variation in their properties. The properties of sodium in group 1, for example, are a clue to the properties of the other elements in group 1, namely lithium, li; potassium, K; rubidium, Rb; caesium, cs; and Francium, Fr. These elements are called the alkali metals. They are all soft, silvery metals that melt at low temperatures. The important property of these atoms in nanotechnology is that a subatomic particle called an electron can easily be removed (or 'stripped') from them. Their size is also important. The group of elements next to the group 1 elements are the alkaline earth metals. These can easily be stripped of two electrons.

Figure 1.1

Periodic Tal

key

Atomic number → 1
Symbol → H
Atomic weight → 1.008
Name → hydrogen

1	2	3	4	5	6	7	8
1 H 1.008 hydrogen							
3 Li 6.941 lithium	4 Be 9.012 beryllium						
11 Na 22.99 sodium	12 Mg 24.31 magnesium						
19 K 39.10 potassium	20 Ca 40.08 calcium	21 Sc 44.96 scandium	22 Ti 47.87 titanium	23 V 50.94 vanadium	24 Cr 52.00 chromium	25 Mn 54.94 manganese	26 Fe 55.85 iron
37 Rb 85.47 rubidium	38 Sr 87.62 strontium	39 Y 88.91 yttrium	40 Zr 91.22 zirconium	41 Nb 92.91 niobium	42 Mo 95.94 molybdenum	43 Tc (98.91) technetium	44 Ru 101.1 ruthenium
55 Cs 132.9 caesium	56 Ba 137.3 barium	57-71 # *lanthanides*	72 Hf 178.5 hafnium	73 Ta 180.9 tantalum	74 W 183.8 tungsen	75 Re 186.2 rhenium	76 Os 190.2 osmium
87 Fr (223.0) francium	88 Ra (226.0) radium	89-103 * *actinides*	104 Rf (261.1) rutherfordium	105 Db (262.1) dubnium	106 Sg (263.1) seaborgium	107 Bh (264.1) bohrium	108 Hs (265.1) hassium

# lanthanides	57 La 138.9 lanthanum	58 Ce 140.1 cerium	59 Pr 140.9 praeseodymium	60 Nd 144.2 neodymium	61 Pm (146.9) promethium
* actinides	89 Ac (227.0) actinium	90 Th 232.0 thorium	91 Pa 231.0 protactinium	92 U 238.0 uranium	93 Np (237.0) neptunium

the Elements

11	12	13	14	15	16	17	18
							2 He 4.003 helium
		5 B 10.81 boron	6 C 12.01 carbon	7 N 14.01 nitrogen	8 O 16.00 oxygen	9 F 19.00 fluorine	10 Ne 20.18 neon
		13 Al 26.98 aluminium	14 Si 28.09 silicon	15 P 30.97 phosphorus	16 S 32.07 sulfur	17 Cl 35.45 chlorine	18 Ar 39.95 argon
29 Cu 63.55 copper	30 Zn 65.39 zinc	31 Ga 69.72 gallium	32 Ge 72.61 germanium	33 As 74.92 arsenic	34 Se 78.96 selenium	35 Br 79.90 bromine	36 Kr 83.80 krypton
47 Ag 107.9 silver	48 Cd 112.4 cadmium	49 In 114.8 indium	50 Sn 118.7 tin	51 Sb 121.8 antimony	52 Te 127.6 tellurium	53 I 126.9 iodine	54 Xe 131.3 xenon
79 Au 197.0 gold	80 Hg 200.6 mercury	81 Tl 204.4 thallium	82 Pb 207.2 lead	83 Bi 209.0 bismuth	84 Po (210.0) polonium	85 At (210.0) astatine	86 Rn (222.0) radon
111 (272)	112 (277)						

64 Gd 157.3 gadolinium	65 Tb 158.9 terbium	66 Dy 162.5 dysprosium	67 Ho 164.9 holmium	68 Er 167.3 erbium	69 Tm 168.9 thulium	70 Yb 173.0 ytterbium	71 Lu 175.0 lutetium
96 Cm (244.1) curium	97 Bk (249.1) berkelium	98 Cf (251.1) californium	99 Es (252.1) einsteinium	100 Fm (257.1) fermium	101 Md (258.1) mendelevium	102 No (259.1) nobelium	103 Lr (262.1) lawrencium

On the far right hand side of the periodic table are the noble gases. These combine with very few elements. They are therefore very unreactive and make useful materials for developing nanotechnological tools, especially the largest atoms such as radon, Rn and xenon, Xe. Next to the noble gases are the halogens. These are reactive gases or solids. We should also note that reactivity changes down a group in the periodic table.

Groups 3 to 12 contain the transition metals. They include iron, Fe; copper, Cu; silver, Ag; and gold, Au. Because they have high conductivity and useful electrical properties they are important candidates for nanotechnology. We should also mention group 14, which contains carbon and silicon. They are important in building nanotubes and nanoelectrical devices. Hydrogen, the simplest element, heads the periodic table.

The properties of an element can be predicted by knowing which family it belongs to, and which elements are its neighbours. This is the field of chemistry and will not be further discussed.

1.4 ATOMIC STRUCTURE

The seeds of the atomic theory go back at least to the time of the ancient Greeks. They developed the concept of elements as basic substances from which all forms of matter can be built. It was believed that when four major elements; earth, fire, air and water, were combined in correct proportions, they could produce all of the other substances. The idea that the elements are the building blocks of matter is consistent with our present day knowledge, but the difference is that we know there are different elements that make up all the matter on Earth. This includes some that do not exist naturally but which have been made in the laboratory.

What are elements? An element consists of only one type of atom. The ancient Greeks believed that matter could be divided endlessly, but this hypothesis was not based on the results of scientific experiments. It was only in 1807, when an English schoolteacher, John Dalton (1766–1844) revived the concept of atoms and proposed an atomic theory on facts and experimental evidence. The essence of the theory is expressed by the following postulates:

1 Elements are composed of indivisible particles called atoms.

2 All atoms of a given element are identical (now known to be incorrect); the atoms of different elements are different in some fundamental way(s).

3 Chemical compounds are formed when atoms combine with each other. A given compound always has the same relative number and types of atoms.

4 Chemical reactions involve reshuffling of atoms from one set of combinations to another. The atoms themselves are not changed in a chemical reaction.

Dalton's atomic theory was a milestone in the development of chemistry. The major premises of his theory are still valid today. However, some of the statements were modified to accommodate new observations that have been made since his time. Dalton considered the atom to be indivisible, but this statement must be clarified. 'Indivisible' means that atoms cannot be broken down further without changing the chemical nature of the element. For example, when a carbon atom is broken down into smaller particles (known as subatomic particles), it loses its chemical properties. We have all heard of atoms in connection with atomic bombs, and in terms such as 'splitting the atom'. The atomic bomb and the nuclear reactor both depend on self-sustaining nuclear fission chain reactions that release a tremendous amount of energy. In nuclear fission, a heavy nucleus splits into several much lighter particles, and many of these particles in turn may also be unstable. The energy released can be used for constructive or destructive purposes. German scientists Otto Hahn and Fritz Strassmann reported the first instance of a nuclear fission reaction in 1939. Subsequently, the first atomic bomb was used in warfare in 1945 at two Japanese cities, Hiroshima and Nagasaki.

There are three major fundamental subatomic particles: electrons, protons and neutrons. The protons and neutrons are collectively known as nucleons. In addition to these, there are a host of other subatomic particles — among them the positron, the meson, and the neutrino — but a discussion of these is beyond the scope of this book.

Electrons are tiny, very light particles that have a negative charge (-). Protons are much larger and heavier than electrons and have a positive charge (+). Neutrons are large and heavy like protons but have no charge. Since the total charge on atoms is neutral, they must have equal numbers of electrons and protons. Contrary to Dalton's original hypothesis, not all atoms of the same element have the same mass. These different forms are known as isotopes. Any mass differences that exist between the isotopes are due to the different number of neutrons in each atom. For example, carbon-12 and carbon-13 both have six protons (atomic number = 6) but the number of neutrons in each is 6 and 7 respectively.

To distinguish between different types of nuclei we write the number of protons (atomic number) as a pre-subscript and the total number of protons plus neutrons (mass number) as a pre-superscript. Thus:

$$^{1}_{1}H, ^{2}_{1}H, ^{3}_{1}H$$

are hydrogen (one proton) with 0, 1 and 2 neutrons respectively.

Likewise,

$$^{4}_{2}He, ^{6}_{3}Li, ^{12}_{6}C$$

are different elements with different numbers of protons in their nuclei.

Table 1.3
Some properties of subatomic particles

Particle	Symbol	Electronic charge units	Relative mass (amu)	Actual mass (g)
Proton	p	+1	1	1.673×10^{-24}
Electron	e^-	-1	5.45×10^{-4}	9.110×10^{-28}
Neutron	n	0	1	1.675×10^{-24}

The properties of the subatomic particles are summarised in Table 1.3.

The structure of the nucleus can mean that an atom can behave like a small magnet, and in a magnetic field it can be aligned either with or against the magnetic field. This magnetic property of the nucleus is called its magnetic moment. Only some nuclei have magnetic moments and this depends on the ratio of protons to neutrons. The hydrogen nucleus (1 proton, zero neutrons) and the ^{13}C nucleus (6 protons, 7 neutrons) have magnetic moments, but in ^{12}C (6 protons, 6 neutrons), there is an equal number of protons and neutrons and the nucleus does not have a magnetic moment. This can be important in detecting these nuclei using their magnetic fields. The phenomenon is known as nuclear magnetic resonance (NMR).

Some types of isotopes are known as radioisotopes, which undergo spontaneous nuclear changes that cause them to be transformed into other elements. It was Henry Becquerel who discovered that atoms of some elements are not stable and disintegrate naturally. Uranium, element number 92, Figure 1.1, was the first element to be found to be unstable. In natural radiation, atoms spontaneously emit radiation of various types, known as alpha (α), beta (β) and gamma (γ) particles. This phenomenon was later called radioactivity by Marie Curie, and was also found to occur in the elements polonium and radium. The existence of radioisotopes has had practical applications, such as in carbon dating and in chemotherapy for thyroid cancer. The characteristics of nuclear radiation are shown in Table 1.4. Some of these particles are the same as structures we have discussed already. For example the β particle is an electron. There are differences in how fast they move, which makes it convenient to give them different names.

Table 1.4
Characteristics of nuclear radiation

Radiation	Symbol	Mass (amu)	Electrical charge	Composition
alpha	α	4	+2	Identical to He^{2+}
beta	β	1/1837	-1	Identical to an electron
gamma	γ	0	0	electromagnetic waves of energy

The concept of radioactivity is crucial evidence that supports the notion that atoms are not indivisible particles and that under special circumstances they can be decomposed. But how are these subatomic particles distributed inside the atom? That is, what is the atomic structure? In 1911, Ernest Rutherford, a New Zealander, provided one of the most significant developments in the understanding of the structure of an atom. Prior to this time, it was thought that the atom had a uniform density. That is, each space contained an electron, neutron and proton. In Rutherford's experiment to test this model, he directed α particles, which carry a positive charge, at a thin sheet of metal foil. If the previous model was accurate the α particles should either pass through the foil virtually undisturbed or all be deflected back at 360 degrees. He was astonished by the experimental results. Although most of the α particles passed straight through, many of the particles were deflected at large angles. It was evident therefore, that the uniform model could not be correct. The large deflections of the positively charged α particles could only be caused by a very dense, positively charged nucleus which repulsed the α particles, surrounded by electrons. Neutrons and protons, collectively known as nucleons, lie within the nucleus. In Rutherford's experiment, most of the particles passed directly through the foil because the atom is mostly — but not totally — open space. The electrons occupy this space and move around a distance that is large relative to the nuclear radius.

Atoms are extremely small and the diameter of a single atom can vary from 0.1 to 0.5 nanometres depending on the type of element. One carbon atom, for example, is approximately 0.15 nanometre in diameter. To give you an idea of the size, consider a millimetre (1×10^6 nanometre) long line and it would take as many as 6.7 million carbon atoms to form a line of atoms across this one millimetre. As mentioned before, the electrons move around the nucleus and the electron density in an atom does not end abruptly at some particular distance from the nucleus. So what exactly is the diameter of the atom? One attempt to solve this problem would be to consider the radius of an atom as half the distance between neighbouring atoms when the element is present in its most compacted form. Even in this definition, there are some problems, because when atoms are attached to each other by chemical bonds, the distance between them is less compared to non-bonded atoms such as noble gases. However, the atomic radii of the elements can be compared if they are measured under circumstances that lead to similar kinds of bonds between their atoms. The general trend is that the atomic size increases down a group and a gradual decrease is observed when proceeding from left to right across the period in the periodic table. This will be discussed in a little more detail later in section 1.7.

1.5 MOLECULES AND PHASES

Phases are states that we define by their properties, such as liquids, gases and solids. Molecules are collections of atoms bound to each other that exist in these phases. Thus the oxygen that we breathe is a molecule made up of two oxygen atoms combined, and written as O_2. Likewise water, H_2O, is the combination of two hydrogen atoms and one oxygen atom. Molecules are formed by atoms in one of two ways. One way is by sharing electrons. This type of combination is called a covalent bond. Secondly, molecules can be formed by transferring electrons between each other. Since the atom losing the electron is positive and the atom gaining the electron is negative, they are represented by superscripts + and -. This is called an ionic bond. When an electron is transferred between two atoms, they are said to be ionised. An example is sodium chloride, NaCl, in which the sodium atom is positive, Na^+; and the chlorine atom is negative, Cl^-. A positive ion is called a cation and a negative ion is called an anion. When there is a covalent bond, as in chlorine, C_{12}, the bond is written as a straight line Cl–Cl. The line represents the two shared electrons. In covalent solids, the molecules are discrete and can be differentiated from each other. In ionic solids such as sodium chloride the atoms are arranged in a regular pattern in space called a lattice, such as in Figure 1.2, and there is not a single discrete molecule. There are a large number of different types of lattices depending on how the atoms pack together. This depends not only on charge but also on atomic size. Once an atom is charged its size differs considerably from that in a free atom.

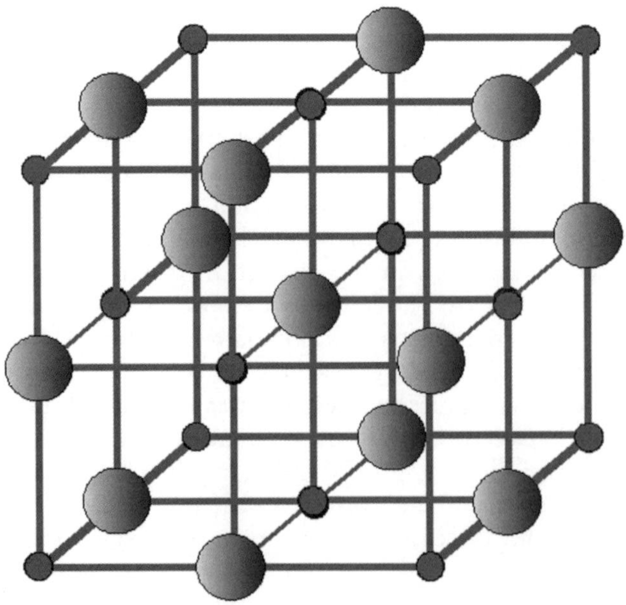

Figure 1.2

The sodium chloride lattice. The large spheres are chloride ions, and the small spheres are sodium ions. The lines between the atoms show the bonds that hold the structure together. They do not represent electrons as in covalent bonding.

A different type of bonding exists in metals. This is depicted in Figure 1.3. In metals all the electrons are shared by all the atoms at one time. Thus metals easily conduct electricity because an extra electron can be added or removed without removing it from a single discrete atom. Metallic bonding will not be discussed further here, however the ability of different materials to conduct electricity or not is important since it affects how we see atoms. This is discussed in Chapter 2.

Figure 1.3

Metallic bonding

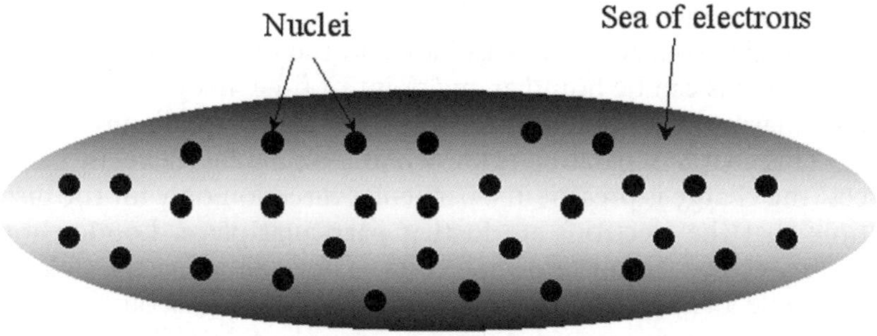

Figure 1.4

a) Origin of London forces between molecules or atoms.

b) Dipole-dipole interaction between two molecules in which electrons are not equally shared in a bond.

a

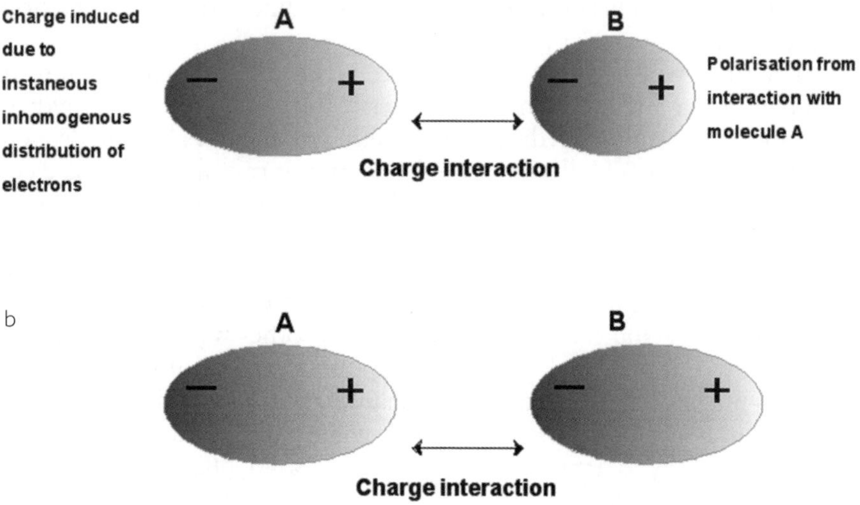

b

There are some other types of bonding which are important in nanotechnology. Molecules and atoms can also be drawn together by relatively weak forces collectively called van der Waals forces or interactions. These types of force are named after Johannes van der Waals who studied them in gases. There are three types of these forces: London forces (Figure 1.4a), which are sometimes called dispersive forces; dipole-dipole forces (Figure 1.4b); and hydrogen bonding. We will deal with each of these briefly.

The electrons move about a nucleus a bit like a wave on the sea. In the sea there is more water in the bulk of the wave than at the apex at any point of time in space, but this changes as the wave moves. In an atom, the electrons tend to 'wash' around the nucleus like a wave, and more electrons can be found at any point of time in a given space. This creates charge differences in space (depicted as an oval, Figure 1.4a, left side). This charge separation is known as a dipole moment. In London forces, the charge is present in a particular direction only for the instant in time that the electrons are in that particular place. London forces exist between all atoms and molecules. They can affect the charge of other molecules or atoms not otherwise instantaneously charged (depicted as a rounder oval, Figure 1.4a, right side). The electrons in the second atom are attracted to the part of the atom that is more positively charged. In nanotechnology, London forces can be used to pick up and move atoms.

Some molecules do not share their electrons equally in bonds. One nucleus can be greedier than the other. When the electrons are not shared equally there is a small charge difference across the bond. This is also a dipole but a more permanent one. When two molecules with dipoles approach each other they may interact, because the positive charge on one molecule may interact with the negative charge on the other molecule. This interaction is called a dipole-dipole interaction (Figure 1.4b).

The third type of van der Waals force is called hydrogen bonding. It occurs in water and in a number of other compounds containing hydrogen, such as selenium hydride, H_2Se, ethanol and DNA. A hydrogen bond forms when a hydrogen atom lies between two atoms whose electrons are not involved in bonding, such as oxygen, fluorine or nitrogen. The hydrogen is attracted to these electrons and the original bond is weakened. The hydrogen in water is pulled from one water molecule and partly bonds with another. In turn, the hydrogen atoms in this second water molecule are hydrogen bonded to other water molecules.

As noted above, molecules as compounds and atoms as elements come in phases. There are three simple phases; gases, solids and liquids, but mixtures of phases also exist. They will be discussed in Chapter 3. Nanotechnology is largely concerned with solid phases because the molecules or atoms do not move around so much and are hence easier to see. In solid phases, the molecules can be arranged in a regular order. They are said to be crystalline. The molecules can also be arranged in a disordered manner. In this case they are said to be amorphous and the solids are described as non-crystalline. The types of bonds mentioned above are important in these phases. When the bonds between molecules are strong, as in ionic bonding, solids are formed even at relatively high temperatures. When only London forces are present, solids can be formed only at very low temperatures. When the temperature rises, these intermolecular forces are broken. The weakest bonds are broken first. Without hydrogen bonding water would be a gas at room temperature.

Other variations in structure can exist. In a crystalline solid there is a three dimensional lattice and the molecules are all oriented in the same way. However, the same kind of orientation can also exist in some liquids. These are called liquid crystals. The molecules are organised to form a lattice but there is still enough motion that the structure is fluid. Other liquids can exist in which the molecules are oriented but they move so much in one direction that there is no lattice.

Electrons behave differently in different states. In the gaseous state, atoms and molecules are isolated and hence each molecule or atom has individual electronic properties. The electrons can move around on a molecule or atom in different energy states but they do not transfer between atoms unless they are activated in some way. When atoms come together in a solid, however, the situation is different. It turns out that only two electrons may occupy a given energy state at one time, so there are lots of energy states corresponding to each pair of electrons. Each of the energy states is so close that they overlap. This means that electrons occupy energy bands rather than discrete atomic energy states.

1.6 ENERGY

What is temperature? It is a property of a form of energy known as thermal energy. However there are other forms of energy.

One very important component of the universe is the stuff that protons and neutrons are made of. This stuff is known as matter, and it has a different property, which is recognised as mass. However matter is also just energy. To convert all forms of energy to the same unit we multiply by a conversion factor. For matter m this is the speed of light squared. Thus we use Einstein's equation $E = mc^2$, where c is the speed of light and m is mass.

Molecules form because they can save energy by doing so. Thus the energy of an ionic bond is typically about 42×10^{-20} Joules per bond. This amount of energy is required to break the bond. The energy of hydrogen bonds is about 3.3×10^{-20} Joules, and London and dipole forces are about one tenth of this.

Figure 1.5

The electromagnetic spectrum

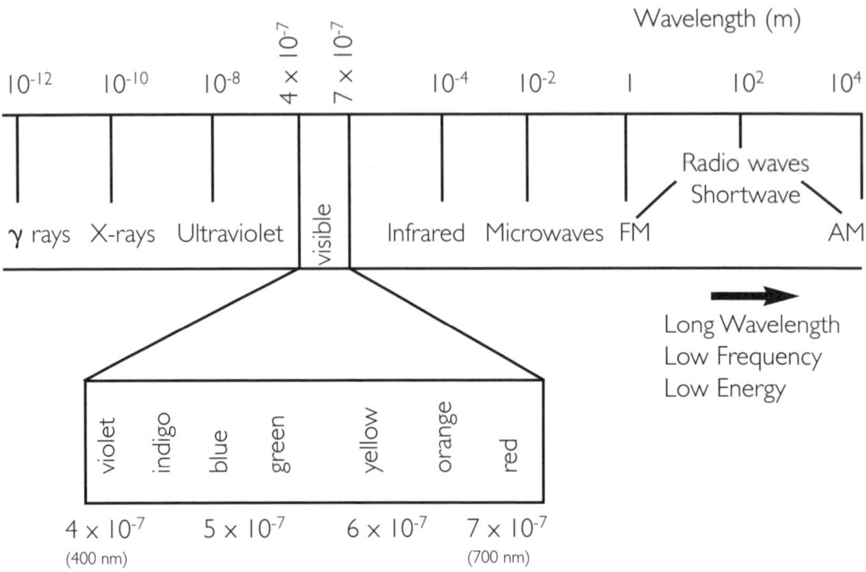

Light is also a form of energy that we call electromagnetic radiation. It can be described either as a wave, or as consisting of matter called photons. However, light is often described better as waves than as matter. This requires some further discussion. The wave properties of matter form the foundation for a theory called wave mechanics. The term 'quantum mechanics' is also used because wave mechanics predicts only certain energies, called quantum states. On a lake or an ocean the wind produces waves whose crests and troughs move across the water. The water moves up and down while the crests travel horizontally in the direction of the wind. The distance between the tops of the waves is the wavelength (λ). The waves also have amplitude and frequency. These are examples of travelling waves.

Another kind of wave is the standing wave. This is what happens on a guitar string. By confining certain points along the string, the string can vibrate in units of length determined by the position in which it is touched. In this type of wave, the crests, or points of maximum amplitude of the wave are called nodes and their positions are fixed. One, two, three, four or more nodes can be generated. A standing wave then, is a wave in which the crests and nodes are fixed.

When the air is vibrated by a guitar string, we hear notes that reflect the wavelength of vibration. Many notes can be played on a guitar string by shortening its effective length with the finger placed at frets, but even without shortening the string a variety of notes can be played. For example if the string is touched momentarily at its midpoint at the same time it is plucked, a range of wavelengths and notes can be formed. However there are restrictions on the wavelengths that can exist. Not just any wavelength is possible because the nodes at either end of the string are in fixed positions; the string must be plucked only in discrete positions to get a pleasant sound. These positions occur at certain points on the string where multiple units of the wavelength are possible within the length of the string. In effect, the wavelengths are quantised.

Quantisation is not a difficult thing to understand because it exists in lots of everyday life examples. While it is possible to have a price of an object of 83.568 cents, say for a given volume of petrol, this is not reality if you pay in coins. Money is not quantised, but in coinage, money is quantised, and the minimum quantum is the coin of smallest denomination.

All electromagnetic radiation flows through empty space at 3×10^8 metres per second, about 670 million miles per hour. If we were to monitor the electric field of light as it passed us as a wave, then the number of waves that hit us depends on the distance between the waves (wavelength λ) and it pushes in one direction with a set frequency υ. This frequency is what we see as colour. However our eyes detect only certain frequencies, which we call the visible or sometimes the optical spectrum. There are many other frequencies of electromagnetic radiation, such as radio waves, which have very long frequencies; and cosmic and γ rays which have very short frequencies. Figure 1.5 shows the wavelengths of different electromagnetic radiation. Clearly, as the wavelength shortens the frequency also increases. When the frequency is extremely high, it becomes a continuum. Then we no longer have a wave hitting us: the matter is continually there and the energy is better described as mass. Indeed, some forms of energy have properties of both matter and waves. The duality of energy as matter or electromagnetic radiation is important in nanotechnology, since at the atomic level these materials may interact, exhibiting any of the properties of the two states. The de Broglie equation expresses the relationship between wavelength λ, and mass as:

$$\lambda = h \;/\; \text{mass} \times \text{velocity}$$

where h is a constant called Planck's constant. According to this equation, a heavy particle travelling slowly could have the same wavelength as a light particle travelling quickly.

It is also possible to consider all other forms of energy in the electromagnetic spectrum as matter as well as waves. In the sea, water hits you and as the wave hits, there is a transfer of a packet of energy equivalent to matter energy. For electromagnetic radiation these packets of energy are called photons. A yellow light hits you with 10^{20} photons per second.

The electron is a particularly interesting form of energy which has both particle and wave properties. When atoms have been studied by exciting the electrons to a higher energy state and then observed, it is clear that they lose energy in only discrete amounts. This is true when energy is absorbed as well as when released. This phenomenon is also quantisation. The energy is quantised and the electron is also said to be quantised. Thus the electron, when described as a wave, can only have certain wavelengths.

Another very important group of energy properties should be reviewed here. These are the thermodynamic properties of free energy: enthalpy and entropy. These properties can be established without referring to atomic structure. Thus, in the mid-nineteenth century it was discovered that chemical reactions either gave up or took in energy. This energy, as in a motor car, can be used to do work. Not all reactions proceed in this way. Some reactions are spontaneous, and others need energy to make them go. Spontaneous changes do not have to be fast. They are just the natural way reactions go, like water flows downhill, not up hill, or that a pack of cards thrown in the air falls in a disorderly way, not back to a perfect pile. It is possible to go the other way, such as make water go uphill, but energy is needed to push it. Similarly, by putting in energy it is possible to put a scattered pack of cards back together.

The relationship between the amount of work that can be done by a chemical reaction is referred to as the change in free energy, ΔG. The value of ΔG has a negative sign because the reaction loses the work energy. The change in free energy, ΔG, is therefore dependent on the difference between the heat energy (called enthalpy) that can be obtained from a system (ΔH) and the energy that is consumed in driving the reaction against any forces which make it non-spontaneous. It turns out, not surprisingly, that spontaneity is temperature dependent. At low temperatures, motion is slower and there is more order. At higher temperatures, thermal energy causes rotation and vibration and there is a state much closer to natural chaos.

Thus:

$$\Delta G = \Delta H - T\,\Delta S \quad \text{eq 1.1}$$

where ΔS relates to the change in the spontaneity factor S and T is temperature. The Austrian physicist Ludwig Boltzmann established the link between S and the behaviour of atoms. His work was actively discredited and he took his own life. However, he left us with a very famous constant bearing his name, the Boltzmann constant.

Boltzmann proposed that the reason reactions are spontaneous is simply that disordered states are more numerous and thus more probable. It is possible that a pack of cards thrown in the air will come down with all the cards face up, but it is improbable. He proposed a property called entropy, S, which is a measure of the probability, W, of a system. The probability of the system and entropy are related logarithmically. Thus:

$$S = k \ln W \quad \text{eq 1.2}$$

where k is the Boltzmann constant. Entropy is also the spontaneity factor S in equation 1.1, so that when there is a change in entropy ΔS, there is a change in capacity to do work.

Entropy is easier to understand in molecular terms than in laboratory reactions. Imagine we are dealing with molecules of two different atoms on a surface and there are twenty of them. Since the two ends are different we could describe them as having heads and tails. If they were in perfect order all the ends of the atoms described as heads would be pointed in the same direction, that is, in one arrangement. In mathematics we say there is one arrangement, that is $W = 1$. Equation 1.2 shows that for $W = 1$, $S = 0$. If the twenty molecules were given energy by heating they could orient themselves in all possible directions $W = (2 \times 2 \times 2 \ldots)$ twenty factors $= 2^{20}$. Since $k = 1.38 \times 10^{-23}$ J K^{-1}, then $S = 1.38 \times 10^{-23}$ J K^{-1} $\times 20 \ln 2 = 1.9 \times 10^{-22}$ J K^{-1}. Creating disorder, for example through thermal energy, changes S from zero to this figure. Entropy has increased by a small amount per twenty atoms. For a large number of atoms, such as the number you could see as a small spill of powder in the laboratory (about 10^{21}), this is a very large number indeed. We shall see later that counting the orientation of molecules is useful in quantum computing (Chapter 8) and the formation of nanocomputers, and that the key to the process is the Boltzmann constant.

1.7 MOLECULAR AND ATOMIC SIZE

We have discussed at some length that nanotechnology involves working with materials around 10^{-9} metres in size, about the size of a few atoms. However the size of an atom is not a simple issue, since atoms are mainly space. The nucleus has been described as a football in the middle of a football stadium, where the stadium represents the space in which the electron moves. An atom has no rigid spherical boundary, but it may be thought of as a tiny, dense positive nucleus surrounded by a diffuse negative cloud of electrons. Moreover, electrons behave often either more like a wave or more like a particle, so it is not possible to say that an atom is defined by the space in which an electron orbits because it does not orbit. In short, there is the issue of the duality of matter.

That is not all. The size of an atom, whatever that really is, changes considerably when it loses an electron (ionises). Not only is the region of space that the lost electron occupied not part of the atom but the atom shrinks further because the remaining electrons are held closer to the nucleus. The nucleus now has a greater pull on the surrounding electrons because of the charge imbalance.

Despite all these factors we need to have an idea of how big things are if we are to undertake molecular nanotechnology. The relative sizes of various atoms are shown in Figure 1.6. However, as noted above, the sizes of atoms differ considerably in different environments and in different bonding arrangements. Some definitions are needed. The distances between atoms and ions have been determined very accurately, so it is useful to use these as a measure. Atomic radius is defined as half the distance between the nuclei of identical neighbouring atoms. The value of atomic radii depends on the type of chemical bond in which the atoms are involved (metallic, ionic, or covalent). When the neighbouring atoms are not alike, as in sodium chloride, part of the observed distance between atoms is assigned to one type of atom and the rest to the other type.

Figure 1.6

Relative sizes of different atoms

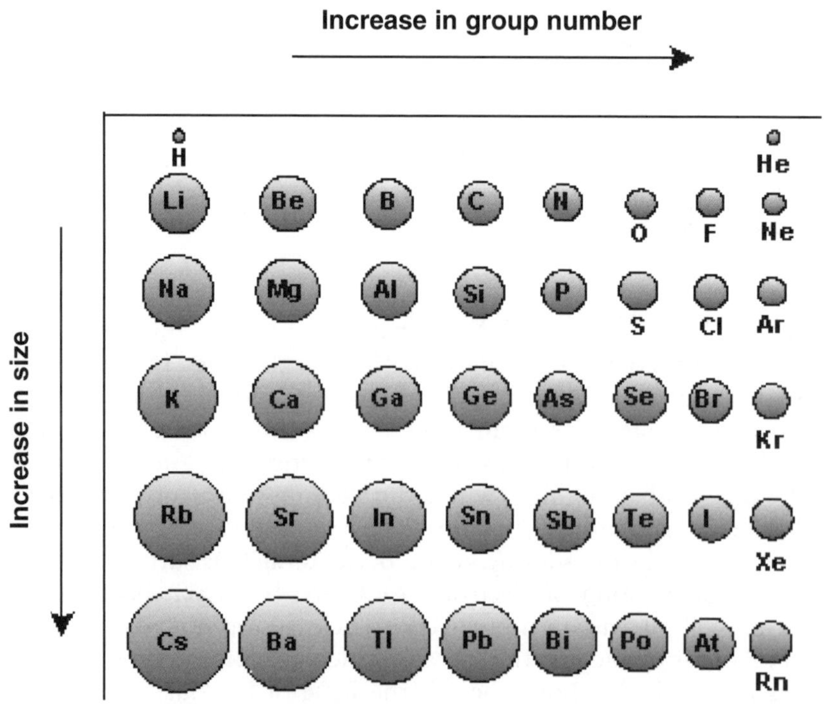

The radius of the sodium atoms bonded together in sodium metal is larger than the radius of sodium in the compound sodium chloride. In sodium chloride, each sodium atom has lost an electron to become a sodium ion (charged atom) of unit positive charge. On the other hand, each chlorine atom has gained one electron to become a chloride ion of unit negative charge. The ionic radius of chloride is nearly twice as great as the radius of a neutral chlorine atom. This is how the metallic radius and the ionic radius of atoms are described. The bonds between the pair of chlorine atoms in a chlorine molecule and between the carbon atoms in diamond are examples of covalent bonds. In these and similar cases, the atomic radius is designated as a covalent radius. Similarly we can have the van der Waals radius, which is the radius in which van der Waals forces are operative. Van der Waals radii are important in nanotechnology.

1.8 SURFACES AND DIMENSIONAL SPACE

Another very important element of the universe is space. If the universe had no dimensions it would not exist. If it were one dimensional it would be a line or a dot of infinitely small thinness. If two dimensional, all objects would be flat like a square and also of infinitely small thinness. In three dimensions there are objects that can be observed; a sphere, a box, a human. Of course it is also possible to have four dimensions. The best way to imagine this is by understanding how three-dimensional objects are represented in two dimensions on paper. We draw them in perspective as a cube. Thus a box is easily represented by additional lines to depict the third dimension. In a molecule of methane, one of the bonds appears to be in front of the plane of the paper and the other one appears to be behind it. To draw a four-dimensional object in three dimensions we need to construct a three-dimensional structure but include perspective. This is possible but it produces some strange shapes. It may one day become a new art form!

Imagine if you were a two dimensional object living in a two dimensional world but in reality there were three dimensions. It would be possible to disappear without breaking the laws of physics by moving into the third dimension. These concepts are important in nanotechnology because as objects get smaller, their dimensions become smaller and smaller. What are the dimensions of a row of atoms 1000 km long and 0.05 nm wide and high? Is that object one-dimensional? It is not, but it certainly approaches one dimension: that is, it tends to one-dimensional. It is possible that in nanotechnology we may discover objects that are better described as one- or two-dimensional, or even four-dimensional, in much the same way as matter may on some occasions be better described as a wave. Indeed, researchers looking for additional dimensions have suggested that to account for changes in the speed of light in space, there must be other dimensions and that they will be discovered on looking closer at atoms.

Nanotechnology is often concerned with single layers of atoms on surfaces. The surface is the edge of a three-dimensional object but in itself it is two-dimensional. If a row of atoms is arranged only in one coordinate, one side of the row of atoms makes contact with the surface but the other side is not in contact with anything. Such structures must have completely different properties from atoms buried in the bulk of the substance, and they behave so. There are in fact two surfaces: one at the edge of the contact material and one at the external edge of atoms on a surface.

In conventional materials most of the atoms are not at a surface. They form part of the bulk of the material, sometimes well crystallised and well laid out, and sometimes full of defects and impurities that destroy the strength of the material. In nanomaterials this bulk does not exist; indeed the principle difference in dealing with nanotechnological materials is the surface-to-bulk ratio. That is, the number of atoms bordering a surface divided by the total number of atoms. Materials with this property are unique and more than anything this is the reason why nanomaterials are different. This will be discussed in Chapter 3.

1.9 TOP DOWN AND BOTTOM UP

Essentially there are two different approaches to creating very small machines or devices. Firstly there is the increasingly precise 'top-down' approach of taking a block of material and whittling it away to the object that is wanted. It is no different from a bronze age scientist working with wood or stone; indeed the solidification of molten metal into shapes is more advanced. The size limit of the smallest features that can be created depends on the tools.

The second approach is called the 'bottom-up' approach, where individual atoms and molecules are placed or are self-assembled precisely where they are needed. Here molecular or atomic building blocks are designed that fit together to produce bigger objects. Scientists have already produced nanoelectronic components, such as molecular switches made of a few molecules, and molecular wires, in order to realise the next challenge of fabricating a molecular transistor. In parallel, biologists have been rapidly learning how certain mechanisms in nature work, such as understanding how a bacterium is propelled through water using a molecular motor, or understanding how electronic signals are generated across biological membranes (bio-sensors). It is the combination of the understanding gained from biological self-assembly, the chemical development of new molecular structures and the physical development of new tools of nanofabrication. All these subjects are dealt with in later chapters. Read on.

WHAT YOU SHOULD KNOW NOW

1 Some of the history of how humankind has used different types of scientific discovery for its benefit. Some discoveries have created new activities that produce better products to enrich our lives. These compose part of the industrial revolution. Others have allowed us to speed things up and are called automation. A third group have improved our health. You should now understand where nanotechnology fits into this perspective in human advancement.

2 Some knowledge of the periodic table of the elements. If you have not studied chemistry it is very important that you come to grips with it now. You could even study it as a separate subject.

3 Chemistry is concerned with the bonding and reactivity of atoms. You should now understand the differences in types of bonding, and in particular those weak forces called van der Waals forces that operate between atoms. You should also understand that atoms come in different sizes and have different properties.

4 You should now understand a little about physics, particularly the different forms of energy and what is meant by quanta. Quantum mechanics is important in the study of nanotechnology. You should now be able to appreciate the duality of particles and waves, and understand what we mean by particles such as electrons, photons, and α, β, and γ particles, and how they may also behave as waves.

1.10 EXERCISES

1 *Web search the following words: nanotechnology, Foresight Institute, atomic volume, molecular assembler, Richard Feynman, quantum mechanics.*

2 *Use plasticine balls and toothpicks to construct a lattice of sodium chloride and a molecule of methane (CH_4). Use the toothpicks as bonds and the plasticine balls as atoms. For methane assume all hydrogen atoms are as far apart as possible. Using the same principles construct a molecule of ethane (CH_3-CH_3).*

3 *Try to find a guitar and demonstrate to yourself that only certain wavelengths are allowed by the length of the string.*

4 *Take a lamp and shine it on a table. On the table place an almost two-dimensional object, such as a coin, and two different three-dimensional objects such as an orange and a round food tin. Note the shape of these objects in a two dimensional world (a circle and a dot). Notice that the light in the third dimension casts a shadow of the three-dimensional objects in the two-dimensional world. Clearly if you were a two-dimensional object an experiment to test for the existence of the third*

dimension would be to look for a shadow. How would you therefore look for evidence of the fourth dimension in a three dimensional world?

1.11 REFERENCES

1 Feynman, RP (1960) A lecture in engineering science. In *California Institute of Technology* February edn. Also usually available at various moving sites on the Internet.

2 Drexler, K Eric (1990) *Engines of Creation*. Fourth Estate, London, 296 pp.

2

MOLECULAR NANOTECHNOLOGY

This chapter describes how we 'see' atoms using atomic microscopy. You will find out how we can pick up atoms and move them around, or spray them onto surfaces to form nanomaterials.

2.1 ATOMS BY INFERENCE

The atomic theory predicts that atoms form substances by combining together in ratios of whole numbers: that is, one atom combines with one or two of another, not 1.325 or some other complicated number. We have already described the different chemical bonds in Chapter 1, and how they are useful in understanding how atoms join together. The chemical bond is associated with the concept of valency. Valency is the number of atoms a specific atom might join up with in forming a substance. A good working model we learned as children was the proposition that atoms had 'hooks' that they could use to hook up with other atoms; and valency was simply the number of hooks. Different atoms had different numbers of 'hooks'. It is not quite as simple as this, however, because atoms can sometimes have variable valency. Nevertheless, it is fairly easy to predict the ratios of atoms that would combine in a given reaction if we know how they have combined in reactions that have already been studied. This information is used to predict the common properties of atoms as exhibited by their group number in the periodic table. We all know now, for example, that two atoms of hydrogen combine with one atom of oxygen to form water. Thus, knowing that the valency of hydrogen is one and that selenium forms H_2Se, we would predict that Na_2Se exists because Na_2O exists.

If the valency is known, then experimenting makes it easy to work out the relative weights of atoms by finding out what weight of an element combines with a known weight of another element. Provided the ratio of atoms in the product is certain (called the stoichiometry) then the weight ratio is known. For example, fluorine and hydrogen combine together in the weight ratio of 19 to 1: that is, 1 gram of hydrogen combines with 19 grams of fluorine. Since the product is HF, which contains one fluorine atom and one hydrogen atom, it follows that the ^{19}F fluorine atom is 19 times heavier than hydrogen.

It is a little harder to find out how many actual atoms are in a gram of an element. For hydrogen, this number is believed to be about 6.02×10^{23} and is called Avogadro's number, in honour of a famous scientist who did much of the pioneering work on the ratios in which atoms combine. Avogadro's number of atoms is called a mole. Thus two moles contain 12.04×10^{23} atoms. This number is big. If you went into a shop and ordered a mole of bread rolls (Avogadro's number 6.02×10^{23}) instead of a dozen, they would cover the whole of the Earth to a height of 37 km. Now atoms do not weigh much so we need to count out a lot to weigh them. Rather than count atoms in tens we count them in units of 6.02×10^{23}, or moles, because this allows us to work with smaller and simpler numbers, particularly when weighing things out. One atom of ^{1}H hydrogen weighs 0.166×10^{-23} grams. Therefore, one mole weighs $0.166 \times 10^{-23} \times 6.02 \times 10^{23} = 1$gram. If we want to weigh out atoms in quantities that can be measured in the laboratory, we can weigh them out in grams and express the quantity of atoms in moles. Thus 1 gram of hydrogen atoms is weighed out to get 1 mole or 6.02×10^{23} atoms, and 19 grams of fluorine atoms are weighed out to get the same number. This allows us to make sure we have nearly the right number of atoms of each reactant when undertaking a chemical reaction to make a product. In fact, both of these substances are gases and are hard to weigh out — and they normally don't hang around as atoms anyway — but the principle applies to all chemical reactions.

Recently tools have been developed that allow us to see atoms. It is therefore now possible in theory to actually measure Avogadro's number. This means that for the first time we can count actual atoms and in principle we could count out a mole, but we would be counting for a long time — about 10 000 billion years if we counted at our normal rate of a few units per second, 1, 2, 3 etc. As noted in Chapter 1.2, the first tool that allowed us to see and count atoms was the transmission electron microscope (TEM).

2.2 ELECTRON MICROSCOPES

Electron microscopes are instruments that use a beam of highly energetic electrons to examine very small objects. This examination can yield information on morphology, topography and crystallography, which we now need to define. Morphology is the shape and size of the particles making up the object. In nanotechnology there are nanoparticles of different shape and size, of the order of up to 100 nm, although it is difficult to determine where microtechnology starts and nanotechnology finishes. Many particles may have nanomorphology in one or two dimensions but micromorphology in others. Examples are the small platelets that make up phyllosilicate clay minerals.

Topography, on the other hand, is the surface features of an object or 'how it looks', including its texture, or hardness. Crystallography describes how the atoms are arranged in the object. They may be ordered in a regular lattice, thereby producing a crystal, or they may be randomly organised, in which case they are said to be amorphous. The way in which the atoms are ordered can affect the properties of a material, such as its conductivity, electrical properties and strength.

Electron microscopes were developed because light microscopes are limited by the physics of light to 500 x or 1000 x magnification and a resolution of 0.2 micrometres. In the early 1930s this theoretical limit had been reached. The transmission electron microscope (TEM) was the first type of electron microscope to be developed and — not surprisingly — the first designs were mimics of the light transmission microscope. Thus a focused beam of electrons was used instead of light to 'see through' the specimen and gain information on its structure and composition in just the same way that light was used.

The stepwise stages in obtaining an image in a transmission electron microscope are:

1 a stream of electrons is formed by an electron source and accelerated toward the specimen using a positive electrical potential;

2 the stream is confined and focused using metal apertures and magnetic lenses into a thin, focused, monochromatic beam;

3 the beam is focused onto the sample using a magnetic lens;

4 interactions occur inside the irradiated sample, affecting the electron beam;

5 these interactions and effects are detected and transformed into an image.

2.3 SCANNING ELECTRON MICROSCOPE (SEM) [1, 2]

The first scanning electron microscopes (SEMs) were developed in 1942 and the first commercial instruments were released around 1965. Its late development compared with TEM was due to the electronics involved in 'scanning' the beam of electrons across the sample. The scanning electron microscope has become more popular than the TEM. This is probably because the SEM can produce images of greater clarity and three-dimensional quality and requires less sample preparation. However, transmission electron microscopes can produce images of greater magnification. The SEM is especially useful because it has a rather large depth of field, that is, more of the image being magnified is in focus. In addition the SEM has an extremely wide range of magnification, producing images in the range of 10 to 100 000 times their normal size. The SEM produces a sharp, three-dimensional view of a specimen, and is very helpful in analysing its shape and structure.

Development of the SEM began in 1935 with the work of Max Knoll at the Technical University in Berlin. Knoll used an electron beam to scan specimens in a modified cathode-ray tube with conventional lenses, but the instrument could not see samples at high magnification rates. The problem was the width of the electron beam. Zworykin, Hiller, and Snyder used electrostatic lenses to produce a narrower electron beam of 50 nm. This was improved by Oatley and Smith, who successfully made a SEM with an electron beam of less than 20 nm by using electromagnetic, rather than electrostatic, lenses.

In a SEM (Figure 2.1), an electron gun emits a beam of electrons, which passes through a condenser lens and is refined into a thin stream. From there the objective lens focuses the electron beam onto the specimen. This objective lens contains a set of coils, which are

Figure 2.1

Highly simplified view of a scanning electron microscope

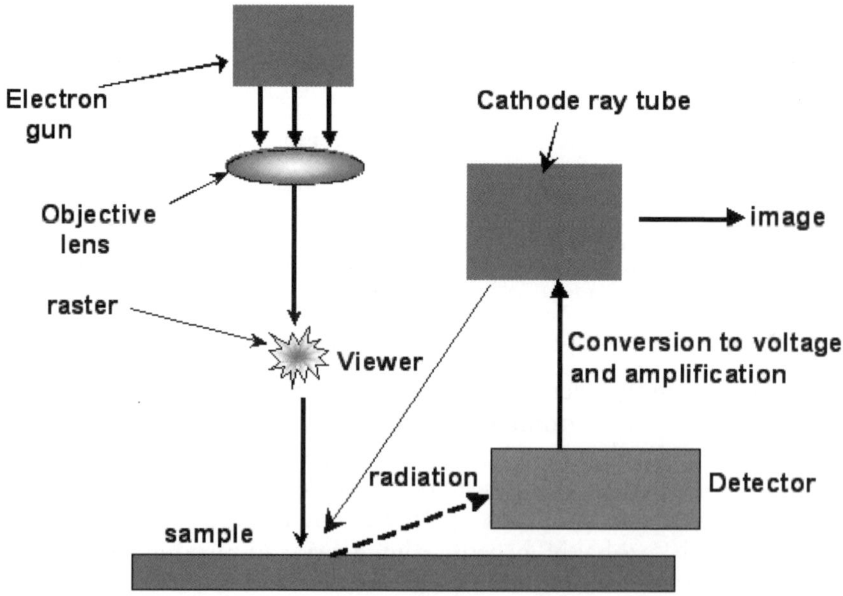

energised with varying voltages. The coils create an electromagnetic field that exerts a force upon the electrons in the electron beam, which in turn redirects the electrons to scan the specimen in a controlled pattern called a raster. The electromagnetic field of the coils also causes a spot of light on a cathode-ray tube to move along at the same rate as the scanning electron beam. When the electrons from the beam hit the specimen, a series of interactions deflect secondary particles to a detector, which then converts the signal to voltage and amplifies it. This voltage is then applied to a cathode-ray tube and converted to an image. The intensity of the image (brightness) is determined by the number of secondary particles that hit the cathode-ray tube, which is dependent upon the angle the electrons bounce off the specimen. Thus the image of the specimen depends on its topography.

When an electron from the beam encounters a nucleus in the specimen, the resultant attraction produces a deflection in the electron's path, known as Rutherford elastic scattering. A few of these electrons will be completely backscattered, re-emerging from the surface of the sample. Since the scattering angle is strongly dependent on the atomic number of the nucleus involved, the primary electrons arriving at a given detector position can be used to yield images containing information on both topology and atomic composition. For mineral samples, this latter information is often good enough to identify them. Some of the beam's electrons can also interact with the electrons in the sample. The amount of energy given to these secondary electrons as a result of the interactions is small, and so they have a very limited range in the sample of a few nanometres and often do not escape. The electrons that are a very short distance from the surface are able to escape and are observed by the detector. The images from these secondary particles contain a lot of detailed information. This means that using secondary particles as imaging data boosts higher-resolution topographical images, making this the most widely used detection method of the SEM imaging options.

One drawback of SEM is that it cannot be used to analyse specimens that give off any type of vapour. This vapour would interact with the electrons. This disadvantage can be overcome by using cryogenic SEM, where the specimen is frozen and coated with gold so that the vacuum tube can remain relatively free of vapour. In recent years, the SEM has been made with a controlled vacuum chamber, where the top of the chamber is a vacuum, and the very bottom of the chamber near the sample is kept at low vacuum. This allows the natural expulsion of particles from wet samples, so the microscope can analyse crystallising and drying. A small area of low vacuum does not distort the SEM image significantly. This instrument is called an environmental SEM.

2.4 MODERN TRANSMISSION ELECTRON MICROSCOPE (TEM) [3, 4]

Although developed first, the TEM is now preferred to the SEM for most applications in nanotechnology. Magnifications of 400 000 times can be easily obtained for many materials and atoms can be imaged at magnifications greater than 15 million times. A TEM works much like a slide projector. A projector shines a beam of light through the slide, and as the light passes through it is subjected to changes by the structures and objects on the slide. These effects result in only certain parts of the light beam being transmitted through certain parts of the slide. This transmitted beam is then projected onto the viewing screen, forming an enlarged image of the slide. TEMs work the same way except that they shine a beam of electrons (like the light) through the specimen (like the slide). Whatever part is transmitted is projected onto a phosphor screen for the user to see.

Materials for TEM must be specially prepared to thicknesses that will allow electrons to be transmitted through the sample, much like light is transmitted through materials in conventional optical microscopy. The preparation of these samples is often the key to a good TEM picture, but it is one reason why TEM is not as popular as SEM. There is no quick procedure.

Figure 2.2

Highly simplified view of the structure of the transmission electron microscope

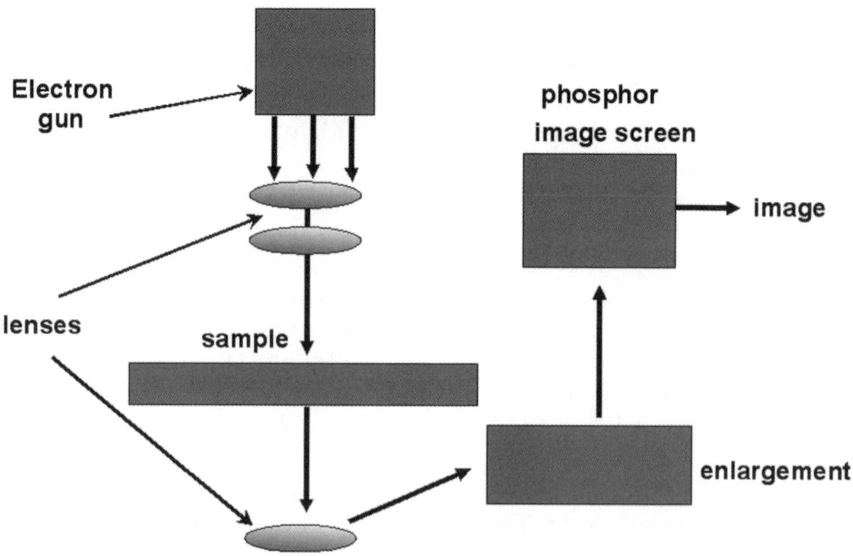

The components of a TEM are outlined in Figure 2.2.

1 There is an electron gun that produces a stream of electrons.

2 This stream is focused to a small, thin, coherent beam by the use of condenser lenses. The first lens largely determines the 'spot size', which is the general size range of the final spot that strikes the sample. The second lens actually changes the size of the spot on the sample, altering it from a wide dispersed spot to a pinpoint beam. The size of the beam is restricted by a condenser aperture.

3 The beam strikes the specimen and parts of it are transmitted.

4 This transmitted portion is focused by the objective lens into an image.

5 A selected area aperture enables the user to examine ordered arrangements of atoms in the sample. Because a focussing electron beam often results in a large amount of energy concentrated on a sample during analysis, special cooling stages are available with many instruments.

6 The image is passed through lenses and enlarged.

7 The image strikes the phosphor image screen and light is generated, allowing the user to see the image. The darker areas of the image represent the areas of the sample through which fewer electrons were transmitted (they are thicker or denser). The lighter areas of the image represent those areas of the sample through which more electrons were transmitted (they are thinner or less dense).

Images obtained from a TEM are two-dimensional sections of the material under study, but applications which require three-dimensional reconstructions can be accommodated by these techniques. Since the energy of the electrons in the TEM determine the relative degree of penetration of electrons in a specific sample, this can be exploited to produce three-dimensional images. It should be emphasised that the radiation can damage the sample. The electrons can cause other electrons to be excited and this can lead to chemical reactions.

Like SEM, when electrons interact with a specimen there are also emissions related to the spacing of energy levels in the atoms that in turn allow those atoms to be characterised and quantified. Thus the chemical composition of the material can also be measured. This can also be done with spatial resolution, so not only can the sample be seen in high magnification, but each of the observed substructures can be chemically characterised. In some instruments, such as a TEM equipped with an energy dispersive spectrometer (EDS), elemental analyses can be obtained from areas as small as a few nanometres in diameter. Because of low count rates, these analyses usually have a relative error between 5% and 10%. On the other hand, very precise accurate chemical analyses (relative error ~ 0.5%) can be obtained from larger areas of the solid (0.5–3.0 micrometres in diameter) using an electron microprobe with wavelength dispersive spectrometers (WDS). Analyses at a lower precision and accuracy (1–2% relative) may be obtained from SEMs equipped with EDS.

Both SEMs and TEMs are often equipped with an EDS or WDS to chemically characterise biological or non-biological materials. Another type of spectrometer, which has maximum sensitivity for light elements, can also be attached to a TEM in order to obtain site-specific information about an element. Thus, the type of element and the nature of its bonding in a crystalline or amorphous solid can be determined.

2.5 SCANNING PROBE MICROSCOPY (SPM) — ATOMIC FORCE MICROSCOPE (AFM) [5–8]

Scanning probe microscopy is a collective term that encompasses a number of newly developed microscopy technologies, the most popular of which are Atomic Force Microscopy and Scanning Tunnelling Microscopy. They have a common operation.

A diagrammatic representation of an atomic force microscope is shown in Figure 2.3a. The atomic force microscopy (AFM) generates a topological image by systematically moving a sharp tip about 2 μm long held at the apex of a cantilever, across a surface within air or liquid, with little sample preparation. The extension of a crystal (called a piezo in Figure 2.3a) is responsible for the movement of the tip across the surface. As the tip tracks the surface, the force between the tip and the surface causes the cantilever to bend. A device called an optical lever measures the deflection of the cantilever. The optical lever of most machines consists of a laser beam reflected from the gold coated back of the cantilever on to a positional sensitive diode. The positional sensitive diode can measure changes in position of the incident laser beam as small as 1 nm, thus giving sub-nanometre resolution.

Nano-sized tips can be made about 500 nm long and 100 nm wide, but this may even be reduced further. One approach (chemical force microscopy) has been to coat the top surface of the cantilever with various materials to make it specific for different chemicals. Thus, when the cantilever is coated with short sections of one half of the DNA helix, it will recognise other parts of the DNA helix by hydrogen bonding. If DNA structures are on the surface, they bind with the cantilever, thus bending it. As discussed above, the degree of bending can then be measured with a laser. This technique then has the potential to measure DNA sequences of interest as the degree of bending depends on the sequence. It offers the potential for very fast DNA analysis and also for trace measurements of forensic importance. You can imagine with development that one analysis might identify your whole genetic make-up, since it could rapidly recognise different sequencing or it could

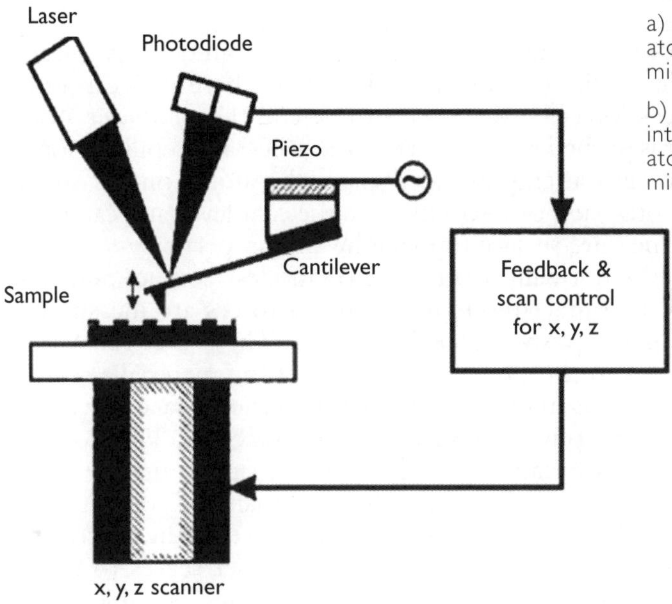

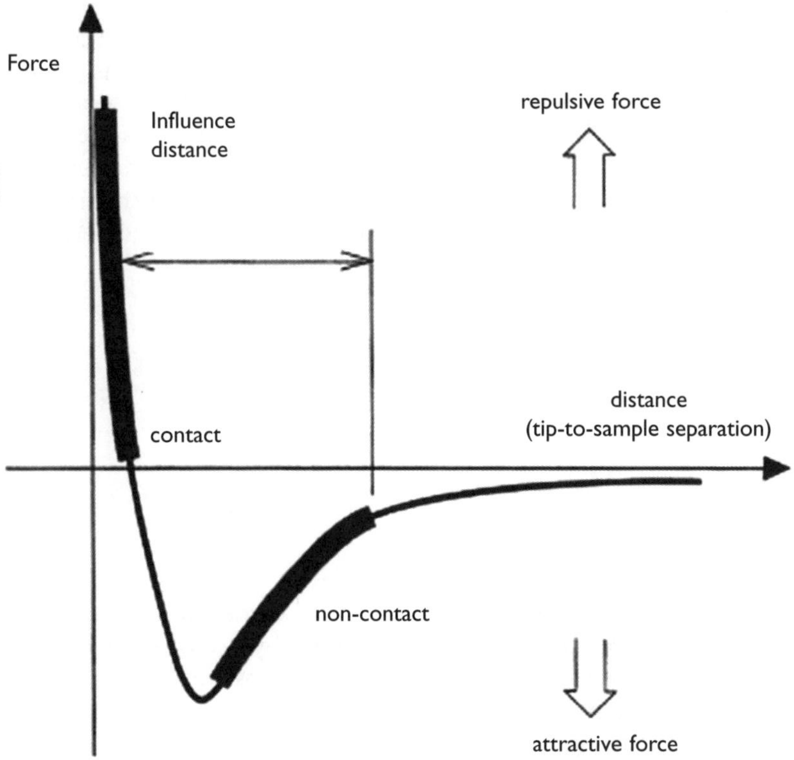

Figure 2.3

a) Diagram of atomic force microscope.

b) Surface interactions of atomic force microscope tip.

detect how many people have sat on a car seat during its lifetime using hair analysis.

The choice of how distant from the surface the cantilever is placed is important. As the cantilever approaches the surface there are weak attractive van der Waals forces between the surface and the cantilever atoms. As the cantilever is pushed into the surface these become repulsive forces, since the cantilever is attempting to displace the atoms on the surface (Figure 2.3b). Both types of forces displace the cantilever and can therefore be used to measure surface topography and properties.

Thus the AFM normally works in two modes: contact and non-contact modes. In contact mode the repulsive forces are measured as the sample cantilever pushes into the sample. These are interactions between the atoms in the tip and the atoms of the material's surface. Consequently the tip is in physical contact with the surface during the analysis and can cause physical damage to soft materials. This does not work on soft samples because surface atoms are ripped up like butter on a knife. However a way around this is a variant of contact mode called 'tapping mode'. The cantilever is vibrated so that the tip contacts the surface intermittently. This intermittent contact serves to reduce the lateral forces incident on the soft sample, reducing possible surface damage or tip contamination but maintaining resolution.

In non-contact mode the attractive van der Waals forces are measured by oscillating the cantilever with a small amplitude some 5 to 10 nm from the surface of the sample. Because non-contact mode measures the weaker attractive forces, the lateral resolution is often said to be less than that achieved with contact mode. This is not correct, since non-contact AFM is used to demonstrate true atomic resolution.

The tips are normally made from silicon and typical specifications for two types are shown in Table 2.1. However, carbon nanotubes (Chapter 4) are useful as tips. A typical ultra-high resolution AFM cantilever and conical tip are shown in Figure 2.4.

Table 2.1
Specifications of ultralever tips

Properties	Weak spring	Strong spring
Tip Radius (nm)	5	5
Force or Spring Constant (N/m)	1.1	1.6
Cantilever Thickness (μm)	0.8	0.8
Resonant Frequency (kHz)	140	170

Prior to AFM analysis of thin film surfaces, the scanner is calibrated in the x, y and z directions. This is done by use of known gratings which are structures deliberately made with steps and prior calibration. They are available in a number of sizes: 25.5 nm, 106 nm and 512 nm are typical and are normally structured so that one or two dimensions

Figure 2.4

A Silicon cantilever.

a) Cantilever with conical tip.

b) Close up view of tip, which is conical in shape.

c) TEM image of the tip.

a

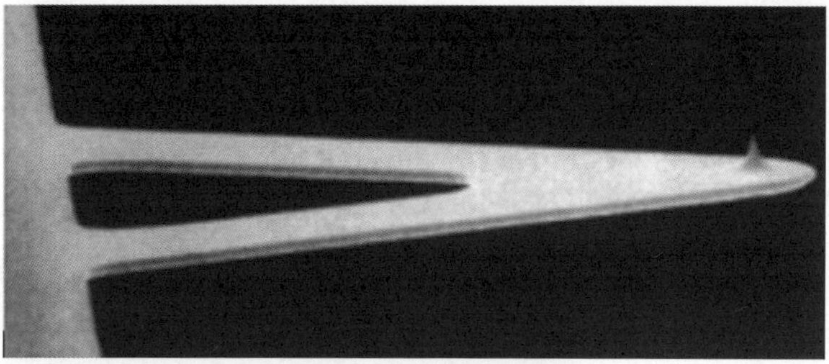

b

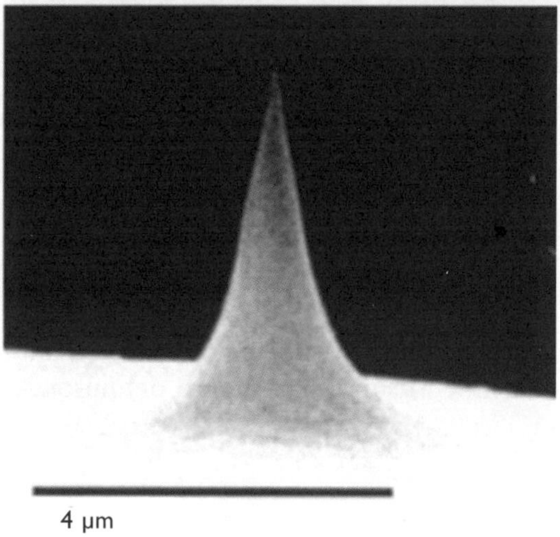

c

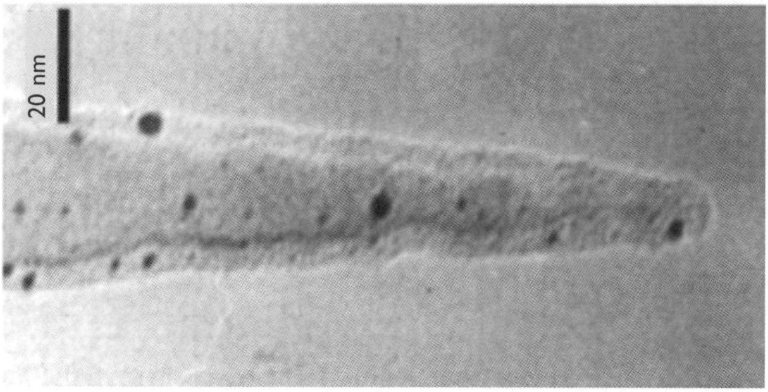

can be calibrated on one sample. Sensibly, the one chosen is closest to the size of the features measured. Mica and gold are some of the best samples for calibration.

Setting up the calibration is a little more difficult than described above. The surface chemistry of the tip, ambient environment, such as water and pH, and charging all affect calibration. Resolution is getting better and better but is normally about 10–50 nm. In principle, as the cantilever tip moves up an edge, the deviation should be proportional to the grating height. However in practice the plot is not necessarily a straight line because the forces operating under heavy contact or light contact or in the contact zone are different. Operation with light force is used for soft samples, because it minimises the total force between the tip and the sample and improves linearity. For cases where the cantilever force is much less than the sample's surface strength, the slope of the curve mostly reflects the spring constant of the cantilever, which can then be used to relate deviation to topography.

A range of useful information can then be obtained apart from just the simple topographical traces of cantilever deviation against distance moved. The average roughness is the average deviation of the data, referenced to the average of the data within the included areas. The average roughness is determined using the standard definition:

$$R_{average} = \sum_{n=1}^{N} \frac{Zn - \bar{Z}}{N} \qquad \text{eq 2.1}$$

where $\bar{Z}$ is the mean height and N is given as the number of data points within the selected area.

The root-mean-squared roughness (rms roughness) is the standard deviation of the data, determined using the standard definition:

$$R_{rms} = \sqrt{\frac{\Sigma (Zn - \bar{Z})^2}{N - 1}} \qquad \text{eq 2.2}$$

where $\bar{Z}$ is the mean height and N is the number of data points within the selected area.

The quantities $R_{average}$ and R_{rms} are drawn from elementary statistics, but they also have a basis in common experience. A surface that is rough has higher $R_{average}$ and R_{rms} values than a surface that is smooth.

The mean height is the average height within the included selected area:

$$\bar{Z} = \frac{1}{N} \sum_{n=1}^{N} Zn \qquad \text{eq 2.3}$$

where N is the number of data points within the included area and $\bar{Z}$ is the mean height.

The median height is the height value that divides the height histogram into two equal areas. At the median value, 50% of the data points have higher values and 50% have lower values.

It is possible to make quite different surface topographies that have similar roughness depending on grain boundaries, so values should not be used without topographical observation. For example, a flat terrace with a lot of crevices (grain boundaries) will appear to have the same average roughness as a rougher surface which is free of grain boundaries.

There are other AFM imaging modes that allow the mapping of surface properties. These include lateral force microscopy, magnetic force microscopy, scanning electrochemical microscopy and pulse force microscopy, which maps mechanical properties. The principles of these techniques are similar to those described above but they are designed to measure individual properties. Most instruments map these surface properties as they obtain images depending on requirements.

2.6 SCANNING TUNNELLING MICROSCOPE (STM) [9–12]

The scanning tunnelling microscope (STM) was invented in 1981 by Binnig and Rohrer at IBM Zurich. Five years later they were awarded the Nobel prize in physics for their invention, along with the discoverers of the electron microscope. The STM was the first instrument to generate real-space images of surfaces with atomic resolution.

STMs are similar to AFMs except they use a sharpened, conducting tip with a bias voltage applied between the tip and the sample. When the tip is brought within about 1 nm of the sample, electrons from the sample begin to pass through the 1 nm gap into the tip or vice versa, depending upon the sign of the bias voltage. Like a wave at sea can pass over rocks without moving the rocks, so the electrons can pass over the energy barrier to the sample. In other words, this process invokes the wave properties of an electron to move across an energy barrier at lower energy than if it were a particle. The process is called tunnelling. The resulting tunnelling current varies with tip-to-sample spacing, and it is the signal used to create an STM image. For tunnelling to take place, both the sample and the tip must be conductors or semiconductors. Unlike AFMs, STMs in principle cannot image insulating materials. However if a tunnelling current can be produced from a water layer, then the insulating material can be seen. This is particularly important for biological materials.

Much of the other operations are like AFMs. In constant-height mode, the tip travels in a horizontal plane above the sample and the tunnelling current varies depending on topography and the local surface electronic properties of the sample. Similarly, the tunnelling current measured at each location on the sample surface constitutes the topographic image. In constant-current mode, STMs use feedback to keep the tunnelling current constant by adjusting the height of the scanner at each measurement point. For example, when the system detects an increase in tunnelling current, it adjusts the voltage applied to the scanner to increase the distance between the tip and the sample. In constant-current mode, the motion of the scanner constitutes the data set. If the system keeps the tunnelling current constant to within a few percent, the tip-to-sample distance will be constant within a few tenths of a nanometre. The change in voltage necessary to keep this constant distance is of course also a measure of topography, but it can mean other things.

Each mode has advantages and disadvantages. Constant-height mode is faster because the system doesn't have to move the scanner up and down, but it provides useful information only for relatively smooth surfaces. Constant-current mode can measure irregular surfaces with high precision, but the measurement takes more time.

More accurately, the tunnelling current corresponds to the electronic density of states at the surface. STMs actually sense the number of filled or unfilled electron states near the surface, within an energy range determined by the bias voltage. Rather than really measuring physical topography, it measures a surface of constant tunnelling probability.

The sensitivity of STMs to local electronic structure can cause misinterpretation. For example, if an area of the sample has oxidised, that is, has lost electrons, the tunnelling current will drop precipitously when the tip encounters that area. In constant-current mode, the STM will instruct the tip to move closer to maintain the set tunnelling current. The result may be that the tip digs a hole in the surface. Nevertheless, the sensitivity of STMs to electronic structure can be a tremendous advantage in some studies since it can tell us about electron deficient sites or electron-rich mines.

Both AFM and STM can be used in systems that have adsorbed molecules, such as water. The response is often different to that without a water layer because the water layer exerts a capillary force that is very strong and attractive. As the scanner pulls away from the surface, the water holds the tip in contact with the surface, bending the cantilever strongly towards the surface. At some point, depending upon the thickness of the water layer, the scanner retracts enough so that the tip springs free. This is known as the snap-back point. Snap-back point varies with adsorbent; and the positions and amplitudes of the snap-back points depend upon the viscosity and thickness of the layers present on the surface.

Liquid cells for the STM allow operation with the tip and the sample fully submerged in liquid, providing the capability for imaging hydrated samples. When imaging in solution a further control is afforded by the choice of solution, since in some solutions there is a double layer of ions on the surface that can be studied or moved. A liquid environment is useful for a variety of STM applications, including studies of biology, geologic systems, corrosion, or any surface study where a solid-liquid interface is involved. Electrochemical cells (EC) also provide a controlled environment for STM operation. Applications of EC-STM include real-space imaging of electronic and structural properties of electrodes, including changes induced by chemical and electrochemical processes, phase formation, adsorption, and corrosion, as well as deposition of organic and biological molecules in electrolytic solution.

2.7 NANOMANIPULATOR [13]

As described above, the scanning tunnelling electron microscope has a computer-controlled probe that skates across the surface, and the mechanical deflections are transferred to electrical energy like a vinyl record needle. The probe can also be programmed to push against the surface like a finger. When pushing, the charge separating the flexing tip and its fixed mount creates an electrical current that is proportional to the pressure exerted on the tip. By transmitting this current to the proper computer interface a human can actually do the touching. This instrument is called a nanomanipulator. In its current form it combines a scanning tunnelling electron microscope with 3D graphics rendering, a program, and a probe that fits over one finger like a finger bandage. When the scientist gets to a bump on the surface he or she feels it. In this way scientists can feel surfaces. They have even felt viruses. A multipurpose nanomanipulator has been built to perform nanomanipulation studies under vacuum inside a scanning electron microscope as well. It can also be used to build simple objects and may eventually allow the operator to 'feel' when they pick up and move atoms. If this was virtually converted to fingertips, we could indeed build structures as if we build things in our everyday environment. In other words, we could use atoms as if they were really small objects for building, like Lego bricks. Specific devices for picking up atoms have already been invented and are called nanotweezers. Using nanotweezers is a bit like picking up Lego bricks with chopsticks.

2.8 NANOTWEEZERS [14–16]

Two researchers at Harvard University have created a nanoscale grasping device ideal for measuring and manipulating molecular structures. The device is used with the scanning-probe microscope and consists of carbon nanotube tips (see Chapter 4) to form tweezer-like structures. The tweezer structure can be closed with an applied electrical field like a pair of chopsticks to produce a device that grasps and moves molecules or atoms.

In Chapter 4 we discuss carbon nanotubes in great detail. Nanotubes are specific carbon structures that grow to create hollow tubes only one or two nanometres in diameter. The tubes are mechanically robust and are also good conductors of electricity, making them ideal for scanning-probe techniques. In nanotweezers, nanotubes are connected to a gold electrode. An even finer version of the device is being researched, where the narrowest possible tubes would be directly grown on the gold electrodes using vapour deposition, called chemical vapour deposition or CVD. This is also discussed in Chapter 4.

However, even with the finest carbon nanotubes the tips are quite crude. The design of tips for the SPM that incorporate individual molecules specifically synthesised for the purpose of lifting atoms is a likely next step, and one that seems essential if we are to make progress in using SPMs to guide chemical reactions in a selective way. Nevertheless, quite a few structures have already been built using the current primitive technology. These include corrals to hold electrons and structures with missing atoms.

2.9 ATOM MANIPULATION [17,18]

Single atom manipulation can be achieved with appropriate selection of the charge (polarity), the magnitude and duration of a voltage pulse applied between the STM tip and the sample surface, as well as the tip-to-sample separation. By placing a tungsten tip above silicon atoms and applying voltages of -5.5 V to the surface for 30 ms, silicon atoms can be lifted from the surface. This is shown in Figure 2.5a.

The method can also be selective. Hydrogen can be removed from hydrogen silicon bonds by scanning a tip over the surface while applying a continuous potential difference of several volts, or by applying rapid pulses. The mechanism of extraction is subtle and depends on whether the sample is positively or negatively charged. When the sample is positively charged the electrons are involved in breaking the bond, since they are transferred. When the sample is negatively charged a strong electric field is created between the sample and the surface. This field is great enough to pull the bond apart.

Atoms can also be deposited once they have been lifted. Figure 2.5b shows a very famous experiment in which atoms were deposited to spell out the smallest advertisement in the world for IBM using xenon atoms. The smallest counting device, an abacus, has been made out of xenon atoms. These atoms are called adatoms, but molecules can be used as well. Carbon monoxide molecules have been used to make a pictorial man. Nevertheless, if the moving of atoms or molecules could be sped up, it could be used as a mechanism for storing numbers. Atoms have also been moved into a straight line, but probably the most astounding picture is the assembly of iron atoms in a circle so that the wave pattern of the electrons between them can be observed (Figure 2.6; see also colour plate 1). The corral shows the wave nature of the electrons on the metal surface, rather like the concentric rings you obtain in a cup of tea when you shake it.

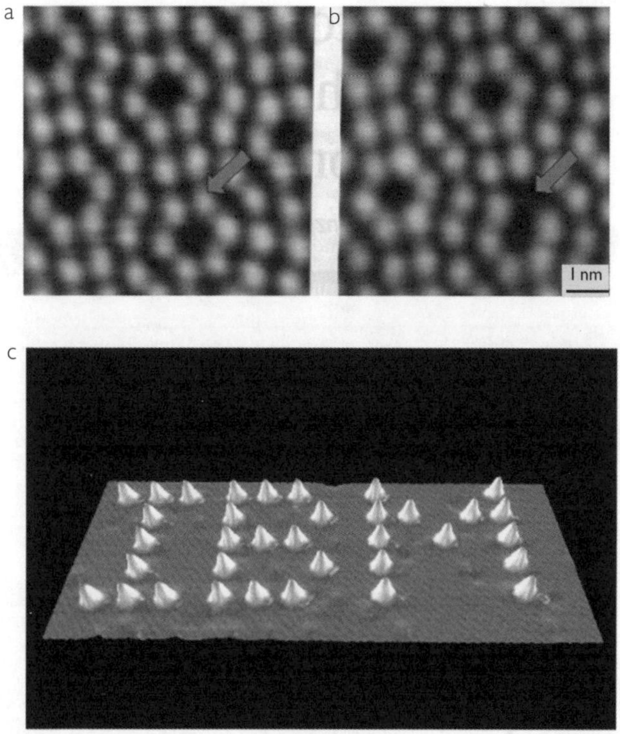

Figure 2.5

a) Manipulation of silicon atoms on a surface before removal, and

b) after removal of an atom. Huang D (2001) Artificial Atomic Surface structures created with STM single-atom manipulation. *Microscopy and Analysis* 19: 5–7.

c) IBM advertisement made of xenon atoms. Reproduced courtesy of International Business Machines Corporation. Unauthorised use not permitted.

The wave behaviour of electrons can be explored in other ways. Scientists have been able to create an atomic mirage. An ellipsoid bowl has two radii at each end due to the two parts of circles that complete its structure. When water is placed in an ellipsoid bowl and a drip allowed to fall at one of the two centres of radius, that is, the first focal point, then waves move out to the edge, rebound and refocus at the second radius — that is, at the second focal point. Thus the drip can be seen to appear at the second radius focal point as a small crest. The phenomenon is not unique. It is similar to the way in which light is focussed by optical lenses or mirrors. What is new is that this is also true for electrons. If an ellipse of atoms is formed and if a further atom is placed at the radius of one ellipse, electrons are added to the pool. They appear at the second radius focal point, like in the water experiment! In effect there is a mirage of the atom [18]. Figure 2.7a shows how waves are refocussed at the opposite focal point (ae) of an ellipse when disturbed at point -ae. Figure 2.7b shows the point of an image on placing a cobalt atom at a focal point. See also colour plate 2 and the discussion below.

The size and shape of the elliptical corral determines the energy and spatial distribution of the confined electrons. These effects are observed because the corral is of a size that is similar to the de Broglie wavelength of the electron (Chapter 1). While the water analogy above is good, the

Figure 2.6

a) A corral of 48 iron atoms around a sea of electrons, showing wave patterns. Crommie MF, Lutz CP, Eigler DM (1993) Confinement of electrons to quantum corrals on a metal surface. *Science* 262: 218–20.

b) Another view of a corral of 48 iron atoms showing waves of electrons focussing at the two radii. The imagery is reproduced courtesy of International Business Machines Corporation. Unauthorised use not permitted.

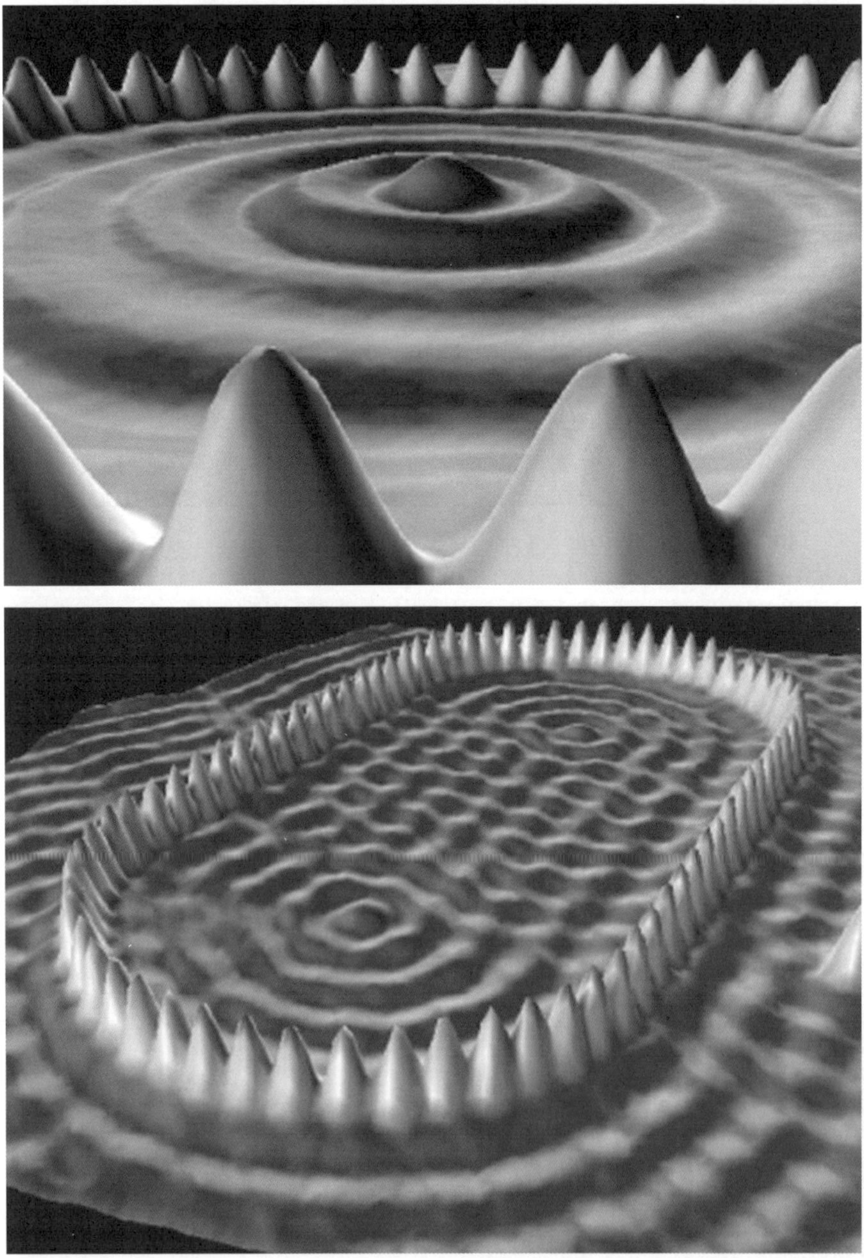

phenomenon is a quantum one, and hence drawing an analogy with light (electromagnetic radiation) is a better way of understanding it. The properties of light are transferred when it is refocussed. A property of atoms which relates to their magnetic moment (Chapter 1) is a phenomenon called 'Kondo resonance'. By positioning a cobalt (Co) atom, at one radius of the ellipse, a strong Kondo signature is detected at the *second* radius focus. If we regard the atom as matter rather than electromagnetic radiation then this begs the question: is it an image or really a 'reformed' atom? Atoms are, after all, primarily electron waves and nuclei. Whatever the answer, the same electronic states in the surface electrons at the mirage are those surrounding the introduced atom. This is a fundamentally new way of transferring information. It transfers electrons on the atomic scale but uses the wave nature of electrons instead of conventional wiring and hence has electronic implications. It will be discussed again briefly in Chapter 8.

Figure 2.7

Refocussing of wave behaviour of electrons by placing an additional cobalt atom within a corral of cobalt atoms on a plane of copper.

a) Geometry.

b) Actual image before cobalt introduction. HC Manoharan, Lutz CP and Eigler DM (2000) Quantum mirages formed by coherent projection of electronic structure. *Nature* 403: 512–13.

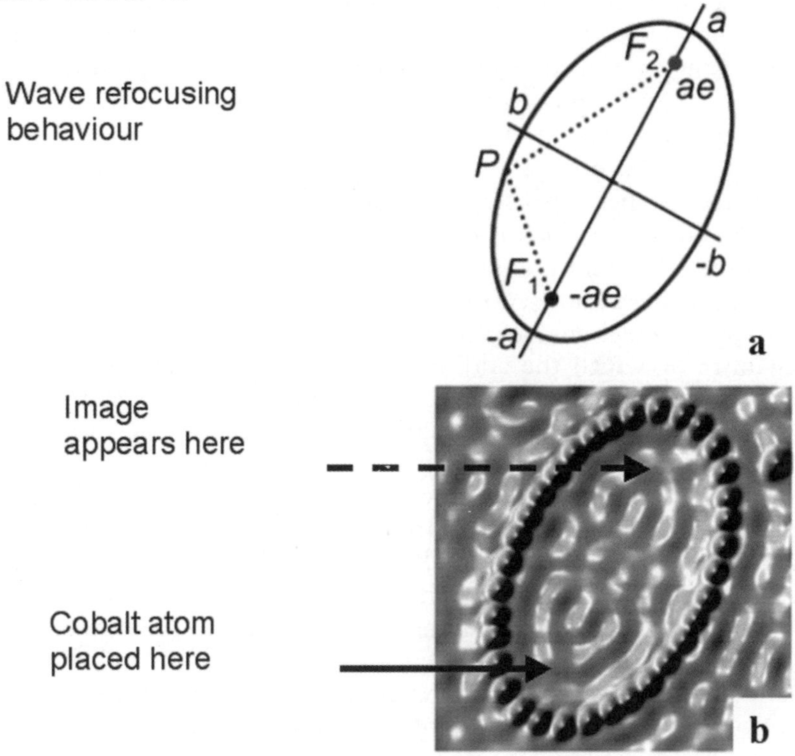

2.10 NANODOTS [19]

Nanodots and quantum dots are small lifted pieces of raised surface up to several atoms thick that are created by attaching atoms to the SPM tip so that a strip of atoms are lifted and then deposited again. This is a little like pulling a bit of plasticine from a surface to make a bump. Using a platinum tip, nanodots of about 1–2 nm in diameter have been created. Nanodots are useful in molecular computing, which will be discussed in Chapter 8.

2.11 SELF ASSEMBLY

Rather than having to shift atoms one by one it would be better if we could get them to line up by themselves. Self assembly is the process whereby molecules line up where you want them to go naturally. This is easier said than done since they do not always go in the places you want them to. This is because molecules are controlled by energetics and have a driving force for negative free energy, ΔG. They will go to a site of lowest free energy, which is determined by bond strength and entropy factors. Recall Chapter 1, equation 1.1: $\Delta G = \Delta H - T\Delta S$, and if ΔS (products – reactants) is negative, then $- T\Delta S$ is positive, and ΔG becomes less negative. Molecules will line up on a surface if the surface is smooth and they will also align on a surface even though this involves a decrease in entropy if there are sufficient energy savings involved. They are more likely to form monolayers than multilayers because the energy saving, ΔH, involved in forming the second layer, is less.

One of the most powerful ways of attracting molecules so that their free energy is decreased is charge. The electrical attraction between positively charged organic ions and negatively charged metal ions (or the reverse) is therefore a useful assembly mechanism. The other is hydrogen bonding, which is the way DNA keeps its double helix and how nature self-assembles the double helix.

Chemists have known about self assembly for a long time because they have known about adsorption and about catalysis. Catalysis is the mechanism by which the addition of another material (the catalyst) speeds up the existing reaction. In adsorption, molecules align themselves on surfaces. When interactions between molecules are small, such as with helium atoms or hydrogen atoms, the mechanism of adsorption is easily described by various adsorption equations, such as the Langmuir isotherm [24]. In nanotechnological terms, this is because the surface consists of long stretches of even surfaces that allow molecules to fit next to each other easily and without interaction with each other, but with a common interaction with the surface. This is a little like a lot of people sitting on a half empty bus. Where these equations do not work are where there are multiple surfaces, lots of intermolecular interactions and lots of competition for available space.

This is equivalent to people crushing into a rush hour train. There are interactions between people and lots of new 'surfaces' when room must be found for more people at the next station. However, complex models can be built to describe these situations [25].

Likewise, catalysis has been known for a long time. In principle the catalyst should also be recovered in its original form, since it is not a reactant. In reality, catalysts become tired since they are often slightly modified in the reaction. Normally they can be used over and over again before they become tired and can be regenerated with a simple chemical treatment. For useful purposes many chemical reactions are too slow to be bothered with if a catalyst is not present.

There are many catalysts that act in solution or in the gas phase by unzipping a molecule at a vulnerable site, therefore allowing a reagent to react when it otherwise could not. These are called homogeneous catalysts. They are part of the field of chemistry but not part of nanotechnology until we can actually take a snapshot of the atoms as they react. However, solids on which molecules can come and react are called heterogeneous catalysts, and nanotechnology is needed in order to understand and optimise molecular reactions on the surface.

Catalysis by solids has often been described as a black art, but this is because we do not understand surfaces. One role of a catalyst is to act as a template where two molecules are held in close proximity so they can react. Another role is to allow dissociation. We know that on the surface of a solid it can be energetically more favourable for a molecule to dissociate. For example, H_2 can dissociate on platinum and remain quite stable so that each hydrogen atom can react independently. This is because the molecule can get involved with catalyst lattice interactions or in some cases enter the lattice. Surface area is also important because the larger the surface area the larger the number of interactions that are possible.

In the future we will undertake catalysis by building trenches for the reacting molecules to fit or structures for the catalysing material to sit on where it can be exposed readily (Figure 2.8a). This will also require knowledge of how molecules sit on surfaces. Both static and rotating molecules on surfaces have already been observed. Figure 2.8b shows high vacuum STM images of a copper surface covered with hexa-tertbutyldecacyclene molecules. In Figure 2.8b, left the molecule appears as a hexagon shaped object, but when moving in Figure 2.8b, right it appears as a doughnut [26]. Now we can readily look into the private lives of molecules on surfaces and make them react.

Figure 2.9 shows a typical surface. The atoms can be seen as small spheres on the surface. The surfaces consist of crevices, valleys, mountains and holes. What a great landscape to explore! It is like being on another planet. Like any building site the surface will need to be flattened to start work. First we need to bring in a nanobulldozer to make the surface flat. For other reasons, we may need to drill a series of

Figure 2.8

a) Support structures for nanodevices built using dip pen nanography. Hong S, Zhu J and Mirkin JCA (1999) Multiple ink nanolithography: towards a multiple-pen nanoplotter. *Science* 286: 523–25. Hong S, Mirkin CA (2000) A nanoplotter for soft lithography with both parallel and serial writing capabilities. *Science* 288: 1808–11. b) Left: static adsorbed molecules of hexagonal shape on a surface; right: one molecule has begun rotating. Gimzewski JK, Joachim C, Schlittler RR, Langlais V, Tang H, Johaannsen I (1998) Rotation of a single molecule within a supramolecular bearing. *Science* 281: 531–33.

a

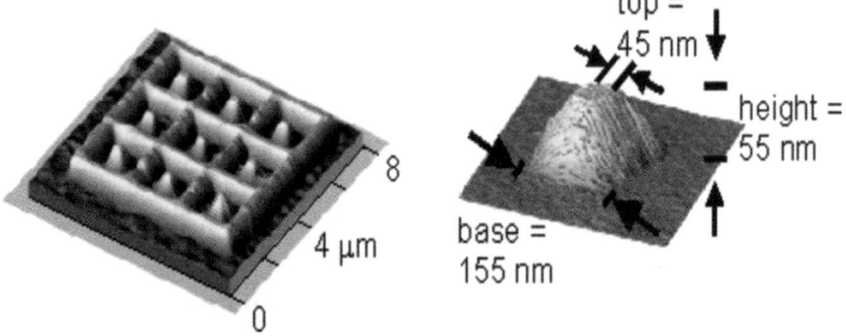

b

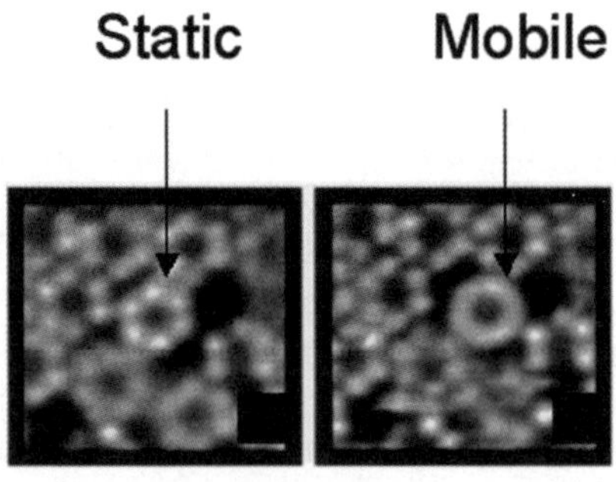

identical holes where molecules could fit down like rods in a socket. These might be used to build superstructures or building frames for other molecules that form the components of nanofactories. However they could also be used as traps for unwary molecules where catalysis can occur in their decomposition. If this is achieved, there is no doubt that we will be able to speed up our current catalysed industrial processes dramatically. If a series of nano-designed catalyst sites could be aligned in a row, and molecules forced from one row to the next row of nanocatalysts, we would have the assembly production lines envisaged by Drexler.

Figure 2.9

The lonely desert of a copper surface. The imagery is reproduced courtesy of International Business Machines Corporation. Unauthorised use not permitted. Crommie MF, Lutz CP, Eigler DM (1993) Imaging standing waves in a two-dimensional electron gas. *Nature* 363: 524–27.

2.12 DIP PEN NANOLITHOGRAPHY (DPN) [20–23]

Dip pen nanolithography (DPN) is a direct-write technique that is used to create nanostructures on a substrate of interest by delivering collections of molecules via a capillary from an AFM tip to a surface, as in Figure 2.10a. Indeed the work to date makes the writing of IBM look small time. A whole scientific paper has been written in molecules!

Alkylthiols (also called alkanethiols) are compounds that possess a hydrocarbon tail group that is 1–4 nm in length and a headgroup of a thiol group (a sulfur-containing functional group –SH). In the first work, alkylthiols have been used as delivery molecules and a gold substrate has been used to collect them because it is known that thiols form a monolayer in which the thiol head groups form relatively strong bonds to the gold, and the alkane chains extend roughly perpendicular to the surface. The thiol lattice formed is identical to that of a monolayer obtained via solution deposition of alkanethiols on gold. This experiment is important because it allows binding structures to be placed at will on a surface so that other units can be bound on the surface. If we were building circuits, for example, or nanomotors, we would need to be able to hold them in a predictable way so that they could be joined up. DPN could be used as a glue pen that puts the glue at points where we want to fix things, such as in Figure 2.10b.

Figure 2.10

a) Diagram representing dip pen nanolithography technique.

b) Pattern formed using dip pen nanography. Hong S, Zhu J, Mirkin JCA (1999) Multiple ink nanolithography: towards a multiple-pen nanoplotter. *Science* 286: 523–25. Hong S, Mirkin CA (2000) A nanoplotter for soft lithography with both parallel and serial writing capabilities. *Science* 288: 1808–11.

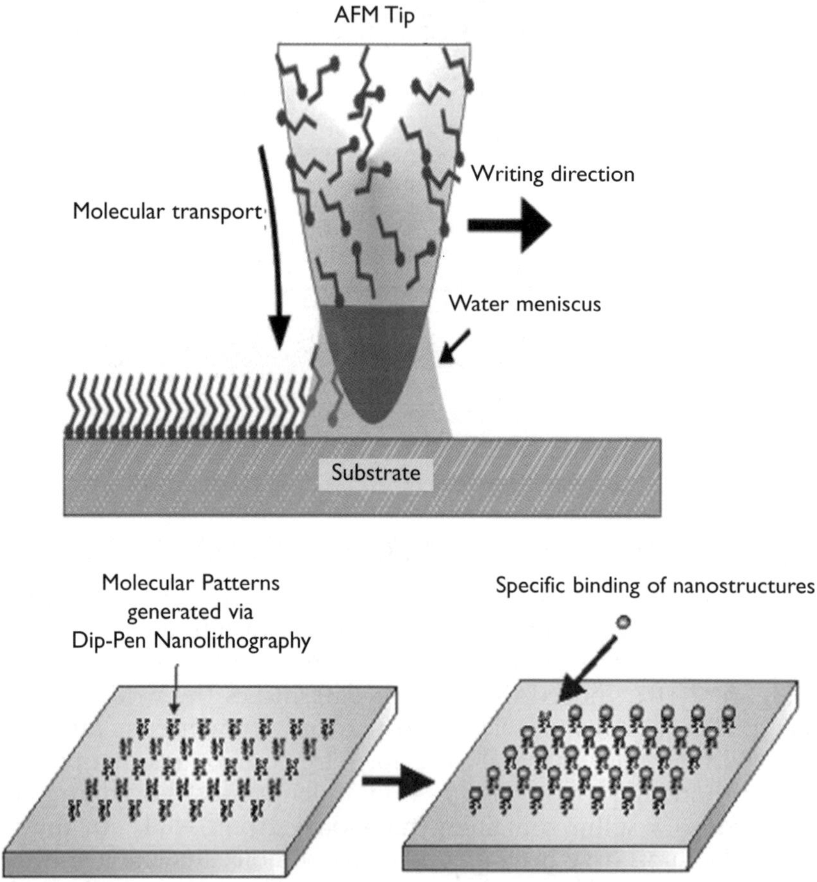

Thus one of the most important attributes of DPN is that the same device is used to image and write a pattern. With the aid of software a DPN nanoplotter is able to write any type of complicated pattern. Indeed, multiple pen DPN is being developed to spray different molecules.

DPN has also been used to form solid, three-dimensional nanostructure patterning areas that do not etch away under treatment, called nanoresists. Resists are discussed in more detail in chapter 8. A variety of different feature sizes and shapes can be prepared. Thiols protect the gold from acid etch solution and then by etching away the unprotected gold surface shapes can be built, as in Figure 2.8a. In particular DPN is perfect for building customised structures in arrays consisting of several thousand components that can be tested for a certain process, for instance catalysis.

WHAT YOU SHOULD KNOW NOW

1 The different types of microscopes for seeing atoms, how they operate, and the things they measure such as topography and roughness. In a scanning electron microscope an electron gun emits a beam of electrons, which passes through a condenser lens and is refined into a thin stream. Scanning tunnelling electron microscopes use a sharpened, conducting tip with a bias voltage applied between the tip and the sample. This generates the image as the tip is brought within about 1 nm of the sample and electrons from the sample begin to pass through the 1 nm gap into the tip or vice versa. Atomic force microscopy (AFM) generates a topological image by systematically moving a sharp tip about 2 mm long held at the apex of a cantilever, across a surface within air or liquid with little sample preparation.

2 How atoms are manipulated one by one and moved to form interesting new structures called corrals. How in these structures the wave properties of the electrons can be pictorially represented and used to generate mirage atoms like a refocussed splash of water.

3 How we can use self-assembly to generate ordered surfaces and build the support blocks for millions of nanomachines.

4 How the black art of catalysis may become much more sophisticated in the future by carefully building new surfaces to undertake specific functions.

2.13 EXERCISES

1 *Web search the following words: Nanomanipulator, quantum dot, nanodot, nanotweezers, atomic force microscopy, scanning tunnelling electron microscopy.*

2 *Think of as many ways you could write or make a mark on a conventional surface. Now think how you might do this on a nanosized surface. Make up an appropriate word, for example nanoink. Web search the word. Think of ways you might alter a cantilever on an AFM or STM to undertake this work. Write this up as a proposal.*

3 *Make a list of some properties of things that we would like on a macroscale. Now list the effect of doing this on a nanoscale. For example, non-stick frying pans are useful, but it would be nice to have non-sticking blood in people with blood clotting conditions. We like to have doors that open but if all the parts of the door were hinged on a nanoscale, we would be able to walk through it. That is, we could walk through walls.*
4 *From the references given in the text pick one area of interest, find the reference in a research library and read it in detail.*
5 *Write an essay on adsorption. Understand what the Langmuir isotherm means and find out its limitations. Detail multilayer adsorption. Try to understand reference 25.*

2.14 REFERENCES

1 Goldstein JI, Newbury DE, Echlin P, Joy DC, Romig AD, Lyman CE, Fiori C & Lifshin E (1992) *Scanning Electron Microscopy and X-ray microanalysis*, 2nd edn. Plenum Press, New York, 673 pp.
2 Newbury DE, Joy DC, Echlin P, Fiori CE & Goldstein JI (1986) *Advanced Scanning Electron Microscopy and Microanalysis.* Plenum Press, New York, 454 pp.
3 Joy DC, Romig Jr AD & Goldstein JI (1986) *Principles of Analytical Electron Microscopy.* Plenum Press, New York, 464 pp.
4 Robards AW & Wilson AJ (1999) *Procedures in Electron Microscopy*, Current Core Edition and Updates. Plenum Press, New York, 700 pp.
5 Blackman GS (1990) Atomic Force Microscopy studies of lubricants on solid surfaces. *Vacuum* 41: 1283–86.
6 Ducker WA, Cook RF & Clarke DR (1990), Force measurements using an AC Atomic Force Microscope. *J. Appl. Phys.* 67: 4045–52.
7 Weisenhorn AL, Hansma PK, Albrecht TR & Quate CF (1989) Forces in Atomic Force Microscopy in Air and Water. *Appl Phys. Lett.* 54: 2651–53.
8 Howard R & Benatar L (1996) *A practical guide to scanning probe microscopy.* Park Scientific Instruments, Sunnyvale CA, USA, 74 pp.
9 Avouris Ph, Walkup RE, Rossi AR, Akpati HC, Nordlander P, Shen P-C, Abeln GG & Lyding JW (1996) Breaking Individual Chemical Bonds via STM-Induced Excitations. *Surface Science* 363: 368–77.
10 Bai C (1999) *Scanning Tunneling Microscopy and its Application*, 2nd edn. Springer-Verlag, Telos, 265 pp.
11 Binnig G & Rohrer H (1987) Scanning Tunneling Microscopy — from Birth to Adolescence. *Reviews of Modern Physics* 59: 615–25.
12 Golovchenko J (1986) The Tunneling Microscope: A New Look at the Atomic World. *Science* 232: 48.
13 Sincell M (2000) Nanomanipulator lets chemists go mano with molecules. *Science* 290: 1530.
14 Falvo MR, Taylor RM, Helser A, Chi V, Brooks Jr FP, Washburn S & Superfine R (1999) Nanometre-scale rolling and sliding of carbon nanotubes. *Nature* 397: 236-38.
15 Guthold M, Falvo MR, Matthews WG, Paulson S, Washburn S, Erie DA, Superfine R, Brooks Jr FP & Taylor, RM (2000) Controlled Manipulation

of Molecular Samples with the nanomanipulator. *IEEE/ASME Transactions on Mechatronics* 5: 189–98.

16 Dai H, Hafner JH, Rinzler AG, Colbert DT & Smalley RE (1996) Nanotubes as Nanoprobes in Scanning Probe Microscopy. *Nature* 384: 147–51.

17 Huang D (2001) Artificial Atomic Surfaces structures created with STM single-atom manipulation. *Microscopy and Analysis* 19: 5–7.

18 Manoharan HC, Lutz CP & Eigler DM (2000) Quantum mirages formed by coherent projection of electronic structure. *Nature* 403: 512–15.

19 Patanè A, Polimeni A, Henini M, Eaves L, Main PC & Hill G (1999) $In_{0.5}Ga_{0.5}As$ quantum dot lasers grown on (100) and (311)B GaAs substrates. *J. Cryst. Growth* 201/202: 1139–42.

20 Piner D, Zhu J, Xu F, Hong S & Mirkin CA (1999) Dip-Pen Nanolithography. *Science* 283: 661–63.

21 Hong S, Zhu J & Mirkin CA (1999) A New Tool for Studying the In-Situ Growth Processes for Self-Assembled Monolayers Under Ambient Conditions. *Langmuir* 15: 7897–900.

22 Hong S, Zhu J & Mirkin CA (1999) Multiple Ink Nanolithography: Towards a Multiple-Pen Nanoplotter. *Science* 286: 523–25.

23 Hong S & Mirkin CA (2000) A Nanoplotter with Both Parallel and Serial Writing Capabilities. *Science* 288: 1808–11.

24 Any good physical chemistry text will describe the Langmuir and other adsorption equations. For example, Moore WJ (1983) *Basic Physical Chemistry*. Prentice Hall, London, 711 pp.

25 Zhu HY, Ni LA & Lu GQ (1999) A pore size dependent equation of state for multilayer adsorption in cylindrical mesopores. *Langmuir* 15: 3632–41.

26 Gimzewski JK, Joachim C, Schlittler RR, Langlias V, Tang H & Johaannsen I (1998) Rotation of a single molecule within a supramolecular bearing. *Science* 281: 531–33.

3

NANOPOWDERS AND NANOMATERIALS

There are six widely known methods to produce nanomaterials other than by direct atom manipulation. These are plasma arcing, chemical vapour deposition, electrodeposition, sol-gel synthesis, ball milling, and the use of natural nanoparticles. In this chapter you will find out about different types of nanomaterials and how they are formed, what methods are used for different materials and how the new materials can be used.

3.1 WHAT ARE NANOMATERIALS?

All materials are composed of grains, which in turn comprise many atoms. These grains can be visible or invisible to the naked eye, depending on their size. Conventional materials have grains varying in size anywhere from hundreds of microns to centimetres. Nanomaterials, sometimes called nanopowders when not compressed, have grain sizes in the order of 1–100 nm in at least one coordinate and normally in three. Typical nanomaterials (zinc oxide and cerium oxide) are shown in Figure 3.1 under transmission electron microscopy. Under normal conditions these oxides form crystals that can be as large as a cubic millimetre in size, but these photographs show that as nanoparticles they are less than 100 nm in length. They are not amorphous, because the atoms are still arranged in discrete crystals.

Nanomaterials are not new. However the understanding that certain preparations of oxides, metals, ceramics (a ceramic is defined as an inorganic substance that can be heated into a useful hard structure) and other substances are nanomaterials is relatively recent. Carbon black is a nanomaterial that is used in car tyres to increase the life of the tyre and provide the black colour. This material was first discovered in the early 1900s. Fumed silica, a component of silicon rubber, coatings, sealants and adhesives, is also a nanomaterial. It became commercially available in the 1940s. With the advent of the advanced microscopic analysis techniques described in the last chapter, new nanomaterials have been developed much more systematically and with a greater understanding.

Figure 3.1

Nanocrystals of

a) cerium oxide, CeO_2 and

b) zinc oxide, ZnO. By permission of Advanced Powder Technologies, Perth, Australia.

a

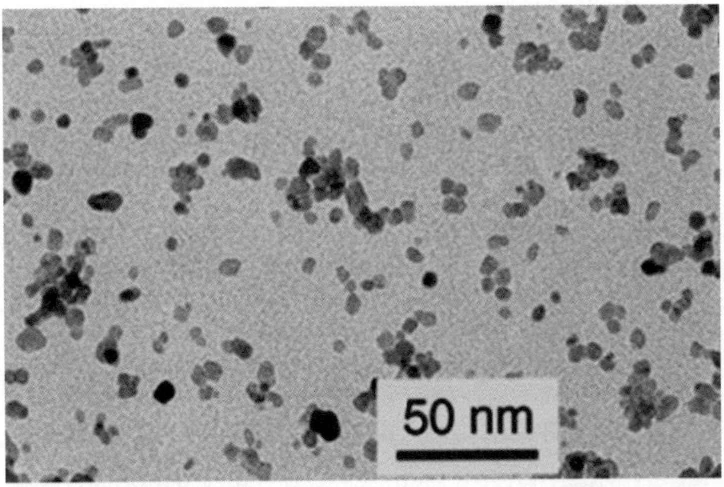

b

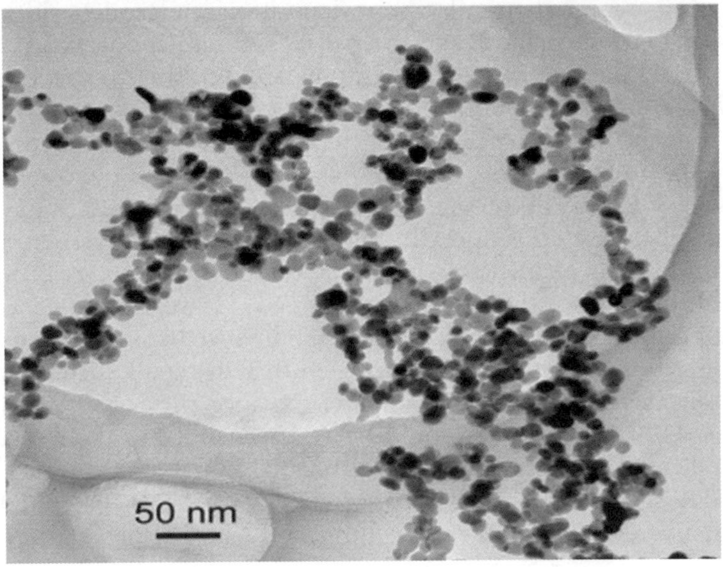

Interest in nanomaterials was already evident before Feynman's speech [1] or Drexler's book [2] defined the new age of nanotechnology. However, it was realised only at about this time that the particle size of carbon black and fumed silica is of nanometre dimensions. In typical nanomaterials, the majority of the atoms are located on the surface of the particles, whereas they are located in the bulk of conventional materials. Thus the intrinsic properties of nanomaterials are different from conventional materials, since the majority of atoms are in a different environment. Nanomaterials represent almost the ultimate in increasing surface area.

Substances with high surface areas have enhanced chemical, mechanical, optical and magnetic properties, and this can be exploited for a variety of structural and non-structural applications. In aerospace and automotive applications, for example, materials made from metal and oxides of silicon and germanium exhibit superplastic behaviour, undergoing elongations from 100 to 1000 per cent before failure. This is because the individual nanosized particles can expand relative to each other. Some nanomaterials are exceptionally strong, hard and ductile at high temperatures. However, they are chemically very active because the number of surface molecules or atoms is very large compared with the molecules or atoms in the bulk of the material. Sometimes, to retain the desired properties of the nanomaterial, a stabiliser must be used to prevent further reaction. This enables them to be wear-resistant, erosion-resistant and corrosion-resistant, but this resistance is usually imparted by some sort of protection mechanism.

3.2 PREPARATION

There are six widely known methods to produce nanomaterials. These are plasma arcing, chemical vapour deposition, electrodeposition, sol-gel synthesis, ball milling, and the use of natural nanoparticles. In the first two methods, molecules and atoms are separated by vaporisation and then allowed to deposit in a carefully controlled and orderly manner to form nanoparticles. The third method, electrodeposition, involves a similar process, since individual species are deposited from solution. The fourth process, sol-gel synthesis, involves some prior ordering before deposition. In ball milling, known macrocrystalline structures are broken down into nanocrystalline structures, but the original integrity of the material is retained. However, the nanoparticles can reform into new materials, which involve breaking the original crystallite bonds. It might be construed that materials such as layered phyllosilicate clays are not nanomaterials since they are natural substances. However nothing could be further from the truth. Layered phyllosilicates are natural nanostructures, although they are normally aligned in such a way that their nanostructure cannot be exploited. The surface area to bulk atom ratio is huge and when dispersed they form nanocrystallites. Thus chemical treatment can generate nanomaterials with different properties.

We will deal briefly with each of the methods of preparing nanomaterials in turn.

3.3 PLASMA ARCING [3]

Plasma is an ionised gas. A plasma is achieved by making a gas conduct electricity by providing a potential difference across the two electrodes so that the gas yields up its electrons and thus ionises. In a vacuum or in an inert gas the electrodes can be made volatile. Heat is produced and the electrodes or even other materials can be volatilised and ionised using this heat. Plasma arcing has been important in forming carbon nanotubes (Chapter 4). A typical plasma arcing device consists of two electrodes. An arc passes from one electrode to the other. The first electrode (anode) vaporises as electrons are taken from it by the potential difference. To make carbon nanotubes, carbon electrodes are used. Atomic carbon cations are produced. These positively charged ions pass to the other electrode, pick up electrons and are deposited to form nanotubes.

The electrodes can be made of other materials but they must be able to conduct electricity. An interesting variation is to make the electrodes from a mixture of conducting and non-conducting materials. During heating, the non-conducting material is vaporised and ionised so that it also becomes part of the plasma arc and is transported and deposited on the cathode.

Plasma arcing can be used to make deposits on surfaces rather than new structures. In this way it resembles chemical vapour deposition (see below) except that the species involved are ionised. As a surface deposit, the nanomaterial can be as little as a few atoms in depth. It is not a nanomaterial unless at least one dimension of the bulk particle of the surface deposit is of nanometre scale. If this is not true, it is a thin film and not a nanomaterial. Each particle must be nanosized and independent, other than interacting by hydrogen bonding or van der Waals forces.

A variation on plasma arcing is flame ionisation. If a material is sprayed into a flame ions are produced, and these can also be collected and deposited in nanocrystallite form.

3.4 CHEMICAL VAPOUR DEPOSITION [4]

This method involves depositing nanoparticulate material from the gas phase. Material is heated to form a gas and then allowed to deposit as a solid on a surface, usually under vacuum. There may be direct deposition or deposition by chemical reaction to form a new product which differs from the material volatilised. This process readily forms nanopowders of oxides and carbides of metals if vapours of carbon or oxygen are present with the metal.

Production of pure metal powders presents a greater challenge, but it has been achieved using microwaves. In this method, microwaves tuned to metal excitation frequencies are used to melt and vaporise the reactants to produce a plasma at temperatures up to 1500°C. Then the plasma enters a reaction column cooled by water, which facilitates the formation of nanometre size particles. The residence time of reactants in the plasma can be controlled to ensure complete conversion. Metal concentration in the gas phase (proportional to partial pressures — see chemistry texts), flow rate of metal vapour, and temperature, alter the particle grain sizes and their distribution. After cooling to 700°C or another appropriate temperature these nanoparticles are filtered from the exhaust gas flow at high temperature, where they fall into a container or bottle below. Depending on the material, slow introduction of a stabiliser helps reduce their reactivity by forming a thin protective coating.

Chemical vapour deposition can also be used to grow surfaces. An object to be coated is allowed to stand in the presence of the chemical vapour. The first layer of molecules or atoms deposited may or may not react with the surface. However, these first formed depositional species can act as a template on which material can grow. The structures of these materials are often aligned, because the way in which atoms and molecules are deposited is influenced by their neighbours. This works best if the host surface is extremely flat. During deposition, a site for crystallisation may form in the depositional axis, so that aligned structures grow vertically. This is diagrammatically shown in Figure 3.2a, with a real structure formed from carbon nanotubes in Figure 3.2b. This is therefore an example of self assembly (Chapter 2.12). It can be seen from Figures 3.2a and 3.2b that the properties of the surface will be different in the z axis than the x, y plane. This gives the surface unique characteristics.

Figure 3.2

Aligned cylindrical molecules formed by chemical vapour deposition.

a) Model.

b) Aligned carbon nanotubes. Reproduced with permission from NASA Ames, the Center for Nanotechnology.

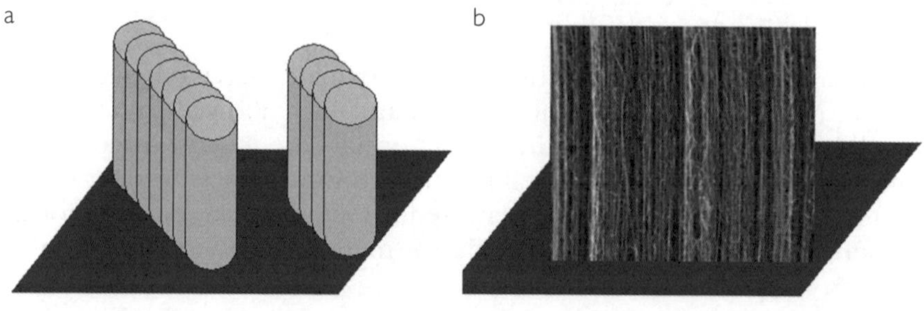

Chemical vapour deposition is particularly useful when nanosites have first been engineered on a surface and the material is subsequently placed on the sites. This is normally done by electron lithography (Chapter 8) or etching (Chapter 2.11). A material is coated on a surface and then parts of the coating are evaporated using a strong electron beam or chemicals. In particular variants, the electron beam can be used to polymerise certain compounds — such as compounds with triple bonds — thereby creating heterogeneous areas. Chemical vapour deposition can be used to build structures on these surfaces (Figure 3.3a), so there is structure in both the micro- and the nano- domain (Figure 3.3b). A variation on this is masking. In masking, a layer is placed on a surface over the structures that we do not want to remove. This protects the structure and the mask is removed after treatment.

Alternatively, dip pen nanolithography (see Chapter 2.12) can be used to write on the surface. The main advantage of chemical vapour deposition over dip pen nanolithography is that it can build nanostructures rapidly by self assembly. The disadvantage is that the molecules cannot be directed. In dip pen nanolithography, the molecules can be precisely placed.

Figure 3.3

a) Diagram to illustrate methodology for forming nanopillars by chemical vapour deposition.

b) Example with carbon forests. Reproduced with permission from NASA Ames, the Center for Nanotechnology.

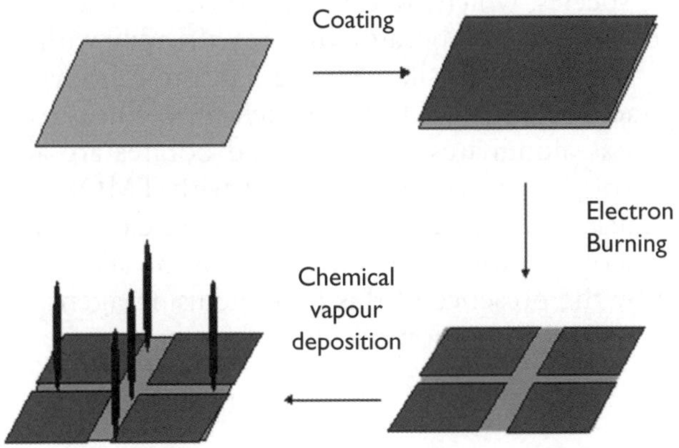

3.5 SOL-GELS [5]

Sol-gel is a useful self assembly process for nanomaterial formation. A characteristic of a solution is that it is clear. That is, you can see through it. Solutions are clear because molecules of nanometre size are dispersed and move around randomly. In colloids the molecules are much larger and range from 20 μm to 100 μm in diameter. Colloids are suspensions of these sized molecules in a solvent. Colloid particles are thus much larger than normal sized molecules or nanoparticles, but they still cannot be seen with a light microscope. However, when mixed with a liquid, colloids can often look cloudy or even milky, whereas nanosized molecules in solution always look clear. A colloid that is suspended in a liquid is called a sol. A suspension that keeps its shape is called a gel. Thus sol-gels are suspensions of colloids in liquids that keep their shape.

The sol-gel process, as the name implies, involves the evolution of networks through the formation of a colloidal suspension (sol) and gelation of the sol to form a network in a continuous liquid phase (gel). The precursors for synthesising these colloids normally consist of ions of a metal but sometimes other elements surrounded by various reactive species, which are called ligands. Metal alkoxides and alkoxysilanes are most popular because they react readily with water. The most widely used alkoxysilanes are tetramethoxysilane (TMOS) and tetraethoxysilane (TEOS), which form silica gels. However, alkoxides such as aluminates, titanates, and borates are also commonly used in the sol-gel process, often mixed with TMOS or TEOS. Additionally, because water and alkoxides are immiscible, a mutual solvent such as an alcohol is used. The miscibility of the alkoxide and water is facilitated by the presence of this homogenising agent.

Sol-gel formation occurs in four stages:

1 hydrolysis;

2 condensation and polymerisation of monomers to form particles;

3 growth of particles;

4 agglomeration of particles followed by the formation of networks that extend throughout the liquid medium resulting in thickening, which forms a gel.

SILICA GELS

All four of the above processes are affected by the initial reaction conditions. Thus the characteristics and properties of a particular sol-gel network are related to a number of factors that affect the rate of hydrolysis and condensation reactions, such as:

- pH
- temperature and time of reaction
- reagent concentrations
- nature and concentration of catalyst
- H_2O/Si molar ratio $[H_2O/Si]$
- aging temperature and time
- drying.

Of the factors listed above, pH, nature and concentration of catalyst, H_2O/Si molar ratio $[H_2O/Si]$, and temperature have been identified as most important. By controlling these factors, it is possible to vary the structure and properties of the sol-gel-derived inorganic network over wide ranges. As the number of siloxane bonds increases, the individual molecules are bridged and jointly aggregate in the sol. When the sol particles aggregate, or inter-knit to form a network, a gel is formed. Upon drying, trapped volatiles (such as water or alcohol) are driven off and the network shrinks as further condensation can occur. With hydrolysis under basic conditions (pH greater than 7) and $[H_2O/Si]$ values ranging from seven to twenty-five, dispersed spherical nanoparticles are produced.

HYDROLYSIS

In hydrolysis the reaction occurs through the addition of water, which results in the replacement of alkoxide groups (OR) with hydroxyl groups (OH). Subsequent condensation reactions involving the silanol groups (Si-OH) produce siloxane bonds (Si-O-Si) plus the by-products water or alcohol. Under most conditions, condensation commences before hydrolysis is complete. However, conditions such as pH, H_2O/Si molar ratio, and catalyst can force completion of hydrolysis before condensation begins. Nevertheless, regardless of pH, hydrolysis occurs by the attack of the oxygen contained in water on the silicon atom. This is evidenced by the reaction of isotopically labelled water (water with oxygen of atomic weight 18 instead of 16) with TEOS, which produces only unlabelled alcohol in both acid- and base-catalysed systems:

$$H_2{}^{18}O + ROSi(OR)_3 \rightarrow R\,OH + H^{18}OSi(OR)_3$$

where R = alkyl groups of various types.

Although hydrolysis can occur without the addition of external catalysts, it is most rapid and complete when they are employed. Mineral acids (HCl) and ammonia (NH_3) are most generally used, however, other catalysts are acetic acid (CH_3COOH), potassium hydroxide

(KOH), amines, and hydrofluoric acid (HF). It has been observed that the rate and extent of the hydrolysis reaction is most influenced by the strength and concentration of the acid or base catalyst.

Under acidic conditions, it is likely that an alkoxide group is protonated in a rapid first step. Electron density is withdrawn from the silicon atom, thus making it more susceptible to attack from water. The water displaces a protonated alkoxy group:

$$H_2O + ROH^{+}Si(OR)_3 \rightarrow ROH + H_2O{+}Si(OR)_3$$

It has been proposed that H_2O displaces ROH with simultaneous attack on the opposite side from which the ROH leaves.

Under basic conditions, it is likely that water dissociates to produce hydroxyl anions in the first step and the hydroxyl anion then attacks the silicon atom. It has been proposed that -OH displaces -OR by attack from the opposite side, similar to acid catalysis:

$$HO^{-} + ROSi(OR)_3 \rightarrow RO^{-} + HOSi(OR)_3$$

Hydrolysis occurs until all -OR (alkoxy) groups are replaced by -OH (hydroxyl). However, as noted, condensation to form siloxane bonds (Si-O-Si) often begins before all groups are exchanged and therefore the reaction is quite complex.

CONDENSATION AND POLYMERISATION OF MONOMERS TO FORM PARTICLES

Polymerisation to form siloxane bonds occurs by either an alcohol-producing or a water-producing condensation reaction. The monomer HOSi $(OR)_3$ reacts:

$$2\ HOSi\ (OR)_3 \rightarrow (OR)_3\ SiOSi(OR)_3 + H_2O$$

$$2HOSi(OR)_3 \rightarrow (OR)_2OH\ SiOSi(OR)_3 + HOR$$

Like hydrolysis, it is generally believed that the acid-catalysed condensation mechanism involves a protonated silanol species. Protonation of the silanol makes the silicon more susceptible to attack. Subsequently the trimer (three silicon) and tetramer (four silicon) species are formed. The reaction is not simple and there is a wide range of silicon species in solution, some of which produce cyclic structures. Cyclisation occurs because of the proximity of the chain ends and the substantial depletion of the monomer population. It should therefore be favoured by dilute solution. Rings, like linear species, may open and accept more monomers. A typical sequence of condensation products is monomer, dimer, linear trimer, cyclic trimer, cyclic tetramer, and higher order rings. This sequence of condensation requires both depolymerisation (ring opening) and the availability of monomers that are in solution equilibrium and/or are generated by depolymerisation. The most basic silanol species (silanols contained in monomers and at the end of chains) are the most likely to be protonated. Therefore, condensation reactions may occur preferentially between neutral species and protonated silanols situated on monomers, and the end groups of chains.

Silica solubility is also important when the pH is changed. In polymerisations below pH 2, the condensation rates are proportional to the $[H^+]$ concentration, but because the solubility of silica is quite low below pH 2, precipitation occurs. Formation of primary silica particles results and it is not possible for a network to grow. Thus, developing gel networks consist of extremely small primary particles, not exceeding 2 nm in diameter.

Between pH 2 and pH 6, condensation preferentially occurs between more highly condensed species and those less highly condensed. This suggests that the rate of dimerisation is low, however, once dimers form, they react preferentially with monomers to form trimers, which in turn react with monomers to form tetrameters. Further growth occurs by addition of lower molecular weight species to more highly condensed species and aggregation of the condensed species to form chains and networks. The solubility of silica in this pH range is again low and particle growth stops when the particles reach 2–4 nm in diameter.

Above pH 7, polymerisation occurs in the same way as in the pH 2 to pH 6 range. However, in this pH range, the condensed species are ionised and therefore they are mutually repulsive so reaction is slower. Due to the greater solubility of silica and the greater size dependence of solubility above pH 7, particles grow in size and decrease in number as highly soluble small particles dissolve and re-precipitate on larger, less soluble particles. Growth stops when the difference in solubility between the smallest and largest particles becomes indistinguishable. This process is referred to as Ostwald ripening. Particle size is mainly temperature dependent, because solubility of larger particles is higher at higher temperatures, so that at higher temperatures larger particles are formed before they drop out of solution.

Particle size can also be affected by adding surfactants (often organic quaternary ammonium salts). These have polar structures, which can interfere with particle growth by binding to small particle surfaces.

ZIRCONIA AND YTTRIUM GELS

Similar processes can be developed with zirconia. A slurry of the hydroxide is dissolved in acid. Yttrium nitrate is added to form a sol, which gels and forms powders. Surfactants are used to control the size of the spheres and particles:

$$Zr(OH)_4 + HNO_3 \rightarrow \text{sol} \rightarrow \text{Yttria sol} \rightarrow \text{spheres} \rightarrow \text{gel} \rightarrow \text{powder}$$

ALUMINOSILICATE GELS

Aluminosilicate gels are special because they form tubular structures. The overall pathway proposed is shown in Scheme 1. Tetraethoxysilane or other alkoxy species hydrolyse sequentially, as shown below, to give rise to species containing alkoxy groups, for example (CH_3CH_2 O):

$$H_2O + (CH_3CH_2O)_4Si \rightarrow (CH_3CH_2O)_3SiOH + CH_3CH_2OH$$
$$H_2O + (CH_3CH_2O)_3SiOH \rightarrow (CH_3CH_2O)_2Si(OH)_2 + CH_3CH_2OH$$

Normally aluminium is introduced as aluminium chloride. In this form aluminium has an octahedral structure since it is surrounded by six molecules of water (scheme 1). This will react rapidly with various ethoxy-substituted silica tetrahedra (Scheme 1, Step 1) but some will rapidly transform into tetrahedral aluminium (Scheme 1, Step 2). These species would rapidly lose OEt (ethoxy) groups and polymerise, possibly with silicon or alumina groups (the latter is shown in scheme 1, step 3). In this step we have written:

$$Si(OEt)_3OAl(OH)_3 \rightarrow Si(OEt)_3OAl(OH)_2OAl(OH)_3$$

It is probably more complex than this since a range of molecules may be present and in some of them, many of the ethoxy groups are substituted by silicon tetrahedra. Evidence suggests that any tetrahedral aluminium must then rehydrate (Scheme 1, step 4) and lose any remaining ethoxy groups (step 5). The final stages of the process involve the conversion of the structure shown on the right side of line 6, Scheme 1 into a polymer of that form, called protoimogolite. This structure naturally curves due to the binding of oxygen across two aluminium atoms, and can then form an aluminosilicate tube called imogolite (Figure 3.4). This last step appears to be a crystallisation process.

Scheme I

Possible reaction paths in imogolite formation. Et is an ethoxy group (CH_3CH_20).

1 $Si(OEt)_3OH + Al(H_2O)_6^{+3} \longrightarrow Si(OEt)_3OAl(H_2O)_5^{2+} + H_3O^+$

2 $Si(OEt)_3OAl(H_2O)_5^{2+} \longrightarrow Si(OEt)_3OAl(OH)_3^- + 2H_2O + 3H^+$

3 $Si(OEt)_3OAl(OH)_3^- + Al(H_2O)_6^{+3} \longrightarrow 2H^+ + 4H_2O + Si(OEt)_3OAl(OH)_2Al(OH)_3$

4 $Si(OEt)_3OAl(OH)_2Al(OH)_3 \longrightarrow Si(OEt)_3OAl(OH)_2(H_2O)_2Al(OH)_3(H_2O)_2$

5 $Si(OEt)_3OAl(OH)_2(H_2O)_2Al(OH)_3(H_2O)_2 \longrightarrow Si(OH)_3OAl(OH)_2(H_2O)_2Al(OH)_3(H_2O)_2$

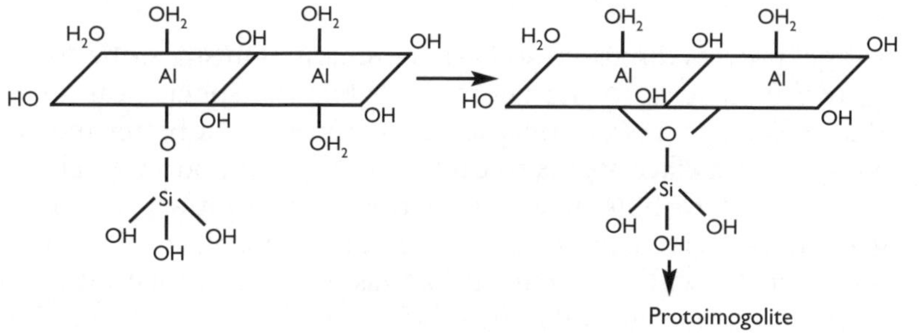

Figure 3.4

Cross section of an imogolite nanotube.

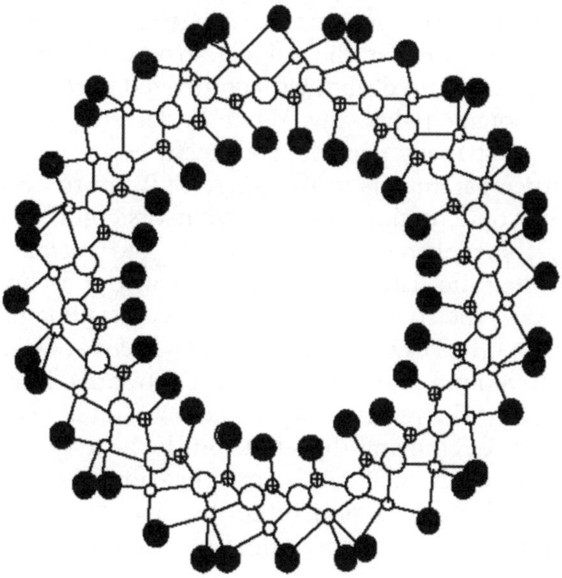

The external tube diameter of imogolite has been shown to be approximately 2.5 nm by TEM and the tubes are several micrometres long. Electron micrographs have shown that the tubes exist in a high degree of order as self-aligned bundles.

Adsorption has been used to measure the internal diameter of the intra-tube pore. Nitrogen studies of the pore volume indicate that the intra tube pore diameter is 1.5 nm.

Imogolite pore measurements have been calculated with the aim of exploiting charge and intra-tube pores in nanotechnology as useful voids for adsorbing and storing things. Imogolites have also been shown to be adsorbents for anions such as chloride, chlorate, sulfate and phosphates. Because the structure can be dissolved away with hydrofluoric acid, if the tube is filled with a few atoms and then dissolved away, rows of atoms can be formed. This is an alternative to dip pen nanotechnology or chemical vapour deposition (Chapter 2.12).

FORMING NANOSTRUCTURED SURFACES USING THE SOL-GEL PROCESS

The sol-gel process discussed so far is summarised in steps one to three of Figure 3.5, that is, to the gel stage. While the spheres can be collected as gelled spheres or collapsed gels (xerogels), a better means of exploiting their surface area is to capture the gel on a surface. This way a greater surface to bulk area ratio is obtained. Another possibility is aerogels. Aerogels are composed of three-dimensional, continuous networks of particles with air (or any other gas) trapped at their interstices. Aerogels are porous and extremely light, yet they can withstand 100 times their own weight.

Another very clever way of maximising the surface area is colloidal crystallisation on surfaces. In this process, water is very carefully removed so that the sol gel structure is not lost in the precipitate. Nanostructured silica with controlled pore size, shape and ordering can be obtained this way. When surfactants are mixed with water, long range spatially periodic architectures are created in a nanofoam, with lattice parameters in the range of 2 nm to 15 nm. The hydrolysis and polycondensation of silica precursors confined to the aqueous domains of these phases affords a simple and versatile method of producing monolithic nanostructured silica whose architecture is effectively the cast of the structure of the liquid crystalline phase used in the synthesis [6]. For example, hexagonal patterns of silica with cylindrical pores 2.0–3.0 nm in diameter (Figure 3.6) are obtained on a surface by conducting the reaction in a hexagonal liquid crystalline phase and subsequently removing the surfactant by heating (called calcination). In this material, the silica walls are approximately 1.2 nm thick. The hydrolysis and condensation reactions, which at room temperature proceed over a period of approximately six hours, are confirmed by analysis

Figure 3.5

Summary of sol-gel process

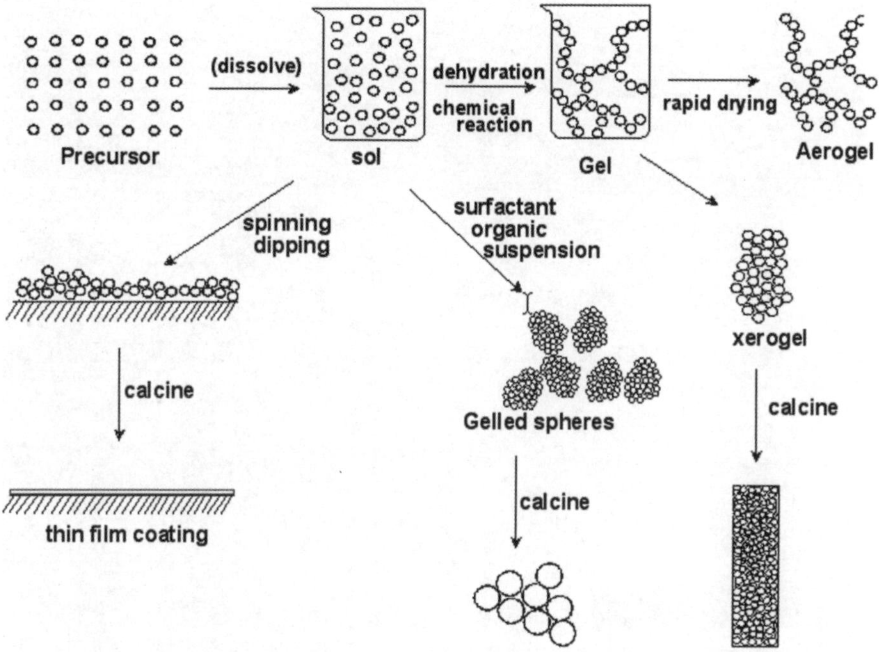

[7–10]. These studies demonstrate the continuity of the nanostructure from the fluid liquid crystalline reaction mixture, through the gelled system to the calcined material. Similarly, by conducting the reactions in a cubic liquid crystal phase, it is possible to obtain silicas with the corresponding nanostructures of this phase. These silicas have specific surface areas of around 1500 m^2g^{-1}. The combination of high surface area, large pores and uniform pore diameter makes these materials of considerable interest for applications in catalysis and chemical separations, especially if catalysts such as transition metals (such as V, Co etc) are incorporated as framework species and appropriate salts are included in the reaction mixtures.

Another significant advantage of the liquid crystal method is to produce nanostructured metals. This development involves dissolving the metals and again retaining the liquid crystal structure pattern [9]. This development is significant because it creates metallic catalysts that have surface areas greater than those of conventional or even colloidal metals. This gives them a greater potential as catalysts. Metals of particular interest are platinum, Pt, and palladium, Pd. There are also big cost benefits. Almost every platinum atom takes part in the reaction and not just the surface atoms as in a conventional solid.

Figure 3.6

Hexagonal patterns of silica with 2–3 nm pores prepared by sol-gel.

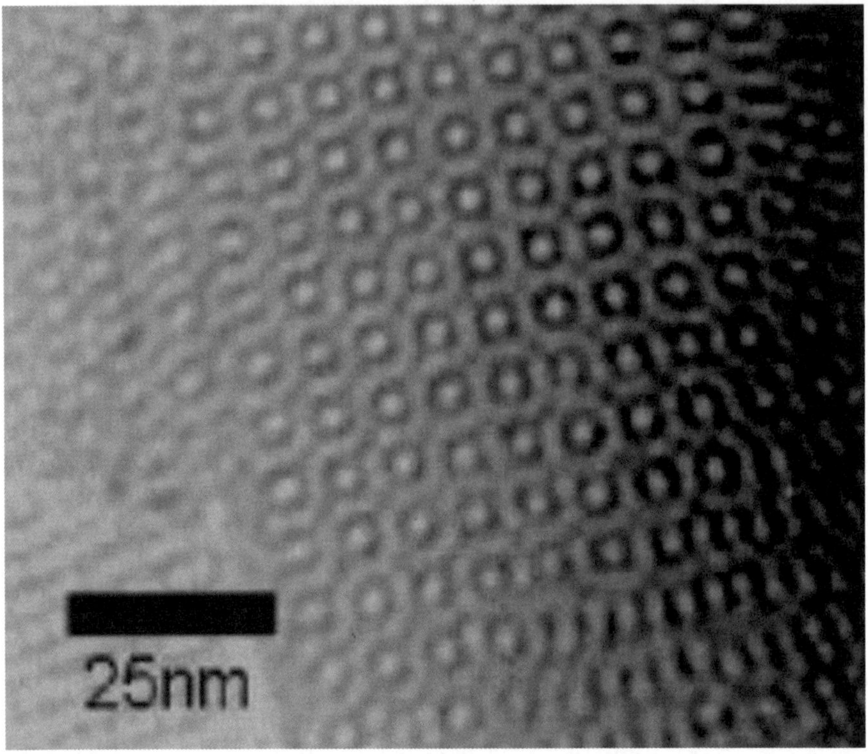

TRAPPING BY SOL-GELS

In another useful development, organic, inorganic and bio-organic molecules are doped (entrapped) in silica glass using sol-gel procedures. Most existing organic and bio-organic molecules cannot be doped in glass, because glass is prepared at elevated temperatures (about 1000°C). However, due to the relatively low temperature needed for the preparation of sol-gel matrices (in some cases room temperature), these molecules can now be entrapped in sol-gel glass.

Sol-gel technology offers the following advantages:

- The glasses are inert.
- The glasses are much more stable to heat than plastics.
- The glasses are much more stable to decomposition by sunlight.
- The stability of the encased molecules or enzymes is higher in the sol-gel glasses than in plastics. (This is due to the high rigidity of the glass cage within which the molecule is doped, in comparison to plastic cages).

- The sol-gel glasses are highly transparent, rendering this technology suitable for all optical requirements. This transparency includes the ultra-violet range, which is far beyond the transparency of most commercial plastics.
- Compared to using the surface of porous glasses for the attachment of molecules, this technology offers much better protection of organic molecules. Molecules that are adsorbed on porous glass surfaces are easily leached out; molecules that are anchored on the surface can be dislodged when the anchor decomposes; and there is no protective cage as there is with sol-gel glasses.

In addition to the unique advantages of the sol-gel technology, it also shares the common advantages of the plastics technology:

- It is a low temperature process.
- The product can be obtained in any form (plates, discs, powders, thin films, etc.).
- It can be attached to most other materials (plastics, paper, metal, etc.).
- It can be polished to optical quality.

Sol-gel technologies can also produce encapsulated active ingredients for sunscreens. With the increasing awareness that ultraviolet (UV) sunlight is the primary cause of skin aging, wrinkles, blotchy pigmentation and skin cancer, people are using sunscreens more often. These preparations contain concentrations of up to 40% active ingredients. A direct and undesired consequence of the increasing use of sunscreen is that an increased amount of the active ingredients penetrate through the epidermis into the body. Sunscreens function by absorbing, reflecting or scattering UV light, and while they are performing this task, photodegradation products and free radicals are formed. It is safer to isolate this photochemistry from body tissues by confining it within the glass particles. The active ingredients of many sunscreens are therefore encapsulated as nanoparticles in smooth, cosmetically acceptable microparticles (1000 nm). The particle loading (content of active ingredient) is around 80 per cent. The product is presented in the form of an aqueous suspension containing about 40 per cent microparticles. When added to a sunscreen lotion formulation the product has a pleasant feel and leaves no white residue.

3.6 ELECTRODEPOSITION [11]

Electodeposition has been used for a long time to make electroplated materials. By carefully controlling the number of electrons transferred, the weight of material transferred can be determined in accordance with Faraday's law of electrolysis. This states that the number of moles of product formed by an electric current is directly proportional to the number of moles of electrons supplied. Since the quantity of electricity passed (measured in coulombs) is current (amps) x time (sec) and Faraday's constant F (96 485 coulombs is currently the most accurate estimate) is the charge per mole of electrons (1 mole of electrons = 96 485 coulombs), then the number of moles of electrons is charge supplied /F. Thus if we wished to copper plate an object with 5 g of copper, we would need 5 / 63.546 (atomic weight of copper) = 0.07868 moles of copper or 2 x 0.07868 moles of electrons if we were reducing Cu^{2+} ions. This is 1.52 x 10^4 coulombs or 4.2 amps for one hour. If the surface of the object to be coated is 2 m^2, each m^2 would have 2.5 g of copper or 0.03934 x 6.022 x 10^{23} (Avogadro's number) atoms = 0.237 x 10^{23} atoms. Each nm^2 would therefore have 0.237x 10^{23}/ 10^{18} = 23 700 Cu atoms.

If the surface was perfectly flat, then knowing the diameter (Chapter 1, Figure 1.6) of a copper atom it should be possible to calculate the height and number of atoms in the copper layer. In practice the surface is rocky (Chapter 2, Figure 2.15) so a coverage factor needs to be included which is related to root-mean-squared roughness or average roughness (Chapter 2.5).

In nanotechnology the aim is to place only a single layer or more of coverage on a surface by electrodeposition in a very controlled way. Thus, hypothetically if an atom of 10 nm diameter, which packs cubically (that is, at the corners of a square), is to be deposited as a monolayer at 50% coverage of a 1 cm^2 surface, then 0.5 x 10^{12} atoms are needed. If it is prepared from a divalent cation this requires 1x 10^{12} electrons or 0.166 x 10^{-11} moles (=16 016 x 10^{-11} coulombs) or 160.16 milliamps per microsecond. The current and time must be carefully measured and any other factors involved in consuming current, such as impurities, must be known in great detail. Hence the necessity for super clean rooms.

Nevertheless, nanostructured films of platinum can be produced by electrodeposition from liquid crystalline mixtures. The films obtained are mechanically robust, remarkably flat, uniform, and shiny in appearance. They also have surface areas comparable with those of the platinum blacks deposited from conventional electroplating baths, and exhibit quite different and favourable electrical properties than conventional platinum deposits. The concept of electroplating from liquid crystalline mixtures can be used for other metals including Pd, Ni, and Au, organic polymers (for example, polyaniline), oxides and semiconductors. The unique nature of nanostructured films from liquid crystals makes them of considerable interest for a very wide range of applications. These include batteries, fuel cells, solar cells, windows that can disperse heat and change properties depending on the environment (electrochromic windows), sensors, field emitters, and photonic devices.

An electrochromic device consists of materials in which an optical absorption band can be introduced, or an existing band can be altered by passing a current through the materials, or by the application of an electric field. Nanocrystalline materials, such as tungstic oxide ($WO_3.xH_2O$) gel, are used in very large electrochromic display devices. The reaction governing electrochromism (a reversible colouration process under the influence of an electric field) is the double-injection of ions (or protons, H^+) and electrons, which combine with the nanocrystalline tungstic acid to form a tungsten bronze. These devices are mostly used in public billboards and ticker boards to convey information. Electrochromic devices are similar to the liquid-crystal displays (LCDs) commonly used in calculators and watches. However, electrochromic devices display information by changing colour when a voltage is applied. When the polarity is reversed, the colour is bleached. The resolution, brightness, and contrast of these devices greatly depend on the grain size of the tungstic acid gels. Hence, nanomaterials are being explored for this purpose. More optical uses of nanomaterials are described in Chapter 7.

Electrodeposition can be used to fill holes to make dispersed nanomaterials. Nanoholes have been strategically placed in membranes. These materials start with the bombardment of a polymer sheet by energetic heavy ions accelerated by a cyclotron. The heavy ions pass through the sheet and leave minuscule damage tracks. Chemical etching is used to turn these tracks into holes (nanoholes) with diameters in the 10 to 100 nanometre range. Filling nanosized holes in polymer membranes with various combinations of metals (Figure 3.7) produces nanocomposites, which have different uses. For example, if some holes are filled with a conducting metal like gold, then they can be charged and this can influence the nature of ions that will go through the unfilled holes. If there is a device at the other end that responds to charge then the device becomes a specific ion detector. Other nanocomposite materials can be produced with designer optical, electrical, magnetic or chemical properties. Such materials can be used as screening materials for heat, light or radiation, for example as thermo liners in ovens and utensils, and in mobile phones. However, if they are compound-specific they can be used as the active sensing material in so-called intelligent materials for 'artificial noses' or sensors of biological material. The most important development is the manufacture of multipurpose chips, which will be able to sense a host of substances at once and thus provide very specific and effective diagnoses.

Certain non-carbon types of nanotubes (Chapter 4) can be made by electrodeposition and also by sol-gel processes.

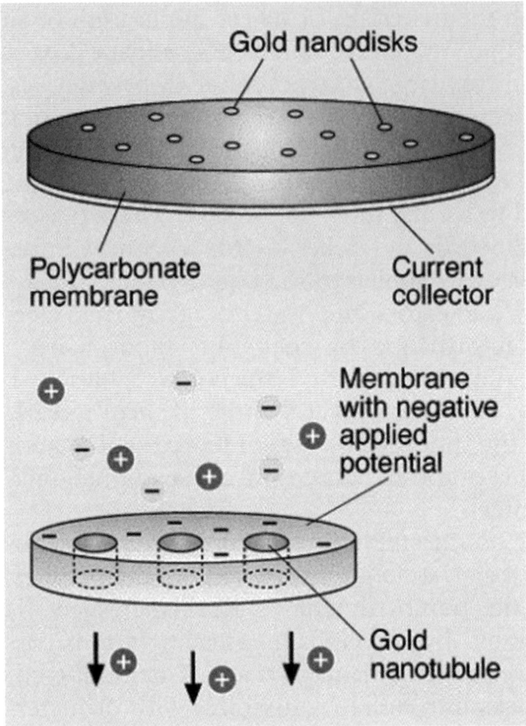

Figure 3.7

Metal plastic nanocomposites for selective ion separations

3.7 BALL MILLING [12, 13]

In the introduction to this chapter, we mentioned that we have known for one hundred years that the structure of very finely divided powders is quite different to their macrocrystalline counterparts, but it is only lately that we have realised why. These early nanomaterials were made by a simple method called ball milling, which is better described as mechanical crushing. In this process, small balls are allowed to rotate around the inside of a drum and drop with gravity force on to a solid enclosed in the drum. Ball milling breaks down the structure into nanocrystallites. The significant advantage of this method is that it can be readily implemented commercially. Ball milling can be used to make a variety of new carbon types, including carbon nanotubes [12, 13]. This is dealt with in Chapter 4. It is useful for preparing other types of nanotubes, such as boron nitride nanotubes [14] and a wide range of elemental and oxide powders. For example, iron with grain sizes of 13–30 nm can be formed. Other crystallites, such as iron nitrides, can be made using ammonia gas. A variety of intermetallic compounds based on nickel and aluminium can be formed with the empirical formulae Ni_xAl_{100-x}, where $47 < x < 61$.

Ball milling is the preferred method for preparing metal oxides (Table 3.1). Their uses range from pigments to capacitors to coatings

Table 3.1
Uses for nanometal oxides

Market	Particles required	Nanotechnology advantages
Polishing slurries	Aluminium Oxide	Faster rate of surface removal reduces operating costs
	Cerium Oxide	Less material required due to small size of particles
	Tin Oxide	Better finishing due to finer particles
Capacitors	Barium Titanate	Less material required for a given level of capacitance
	Tantalum	Higher capacitance due to reduction in layer thickness and increased surface area resulting from smaller particle size
	Alumina	Thinner layers possible, thus significant potential for device miniaturisation
Pigments	Iron Oxide	Lower material costs, as opacity is obtained with smaller particles
	Zirconium Silicate Titanium Dioxide	Better physical-optical properties due to enhanced control over particles
Dopants	Wide variety of materials required depending on application	Improved compositional uniformity
		Reduction in processing temperature reduces operating and capital costs
Structural ceramics	Aluminium Oxide	Improved mechanical properties
	Aluminium Titanate	Reduced production costs due to lower sintering temperatures
	Zirconium Oxide	
Catalysts	Titanium Dioxide	Increased activity due to smaller particle size
	Cerium Oxide	Increased wear resistance
	Alumina	
Hard coatings	Tungsten Carbide	Thin coatings reduce the amount of material required
	Alumina	
Conductive inks	Silver	Increased conductivity reduces consumption of valuable metals
	Tungsten	Lower processing temperatures
	Nickel	Allows electron lithography

to inks. All of these applications rely on the increased surface to bulk ratio, which alters the chemical properties of the metal oxide.

To successfully prepare metal oxides, it is important to keep the crystallites from reacting, and to have an understanding of the kinetic energy transferred during crushing. Much of the commercial know-how is in the nature of the additive. However, a by-product can sometimes be useful. In the production of nanocrystalline Zirconia (ZrO_2) zirconium chloride is treated with magnesium oxide during milling to form zirconia and magnesium chloride:

$$ZrCl_4 + 2MgO \rightarrow ZrO_2 + 2MgCl_2$$

The by-product, magnesium chloride, acts to prevent the individual nanocrystallites of zirconia agglomerating. It is washed out at the end of the process.

Ball milling techniques could be improved by applying a greater knowledge of the energetics involved in the process. The exact energy of delivery to each crystal needs to be determined and methods developed to ensure that each crystal receives the same energy, rather than relying on 'cook and look' processes.

3.8 USING NATURAL NANOPARTICLES [15]

As noted above, there has been considerable interest in producing pores in materials that moderately sized small molecules of 10 nm or so can fit into and react on the surface. Bombarding materials with energetic heavy ions accelerated by a cyclotron is an expensive way of doing things. Some of the most successful materials, called zeolites, have been synthesised by conventional chemistry, and operate at nanopore size. There is a wealth of data on these materials beyond the scope of this book. Phyllosilicates, which consist of layers of silicate tetrahedra and aluminium octahedra, have potential but consist of many individual stacked plates. It is possible to pillar these platelets using species such as the polycation $[Al_{13}O_4(OH)_{24}]^{7+}$, which is about 1 nm in size, and other materials. These bricks open the platelets and make large pores (Figure 3.8) and the pillars can be converted to oxide by thermal treatment. However they are not very thermally stable at high (500°C) temperatures often needed for catalytic reactions. Other complications can arise because the number of pillars per platelet can differ and it is possible to make the pillars longer or shorter [15].

Figure 3.8

The formation of nanostructured phyllosilicates. Courtesy of Professor Max Lu, University of Queensland, Australia

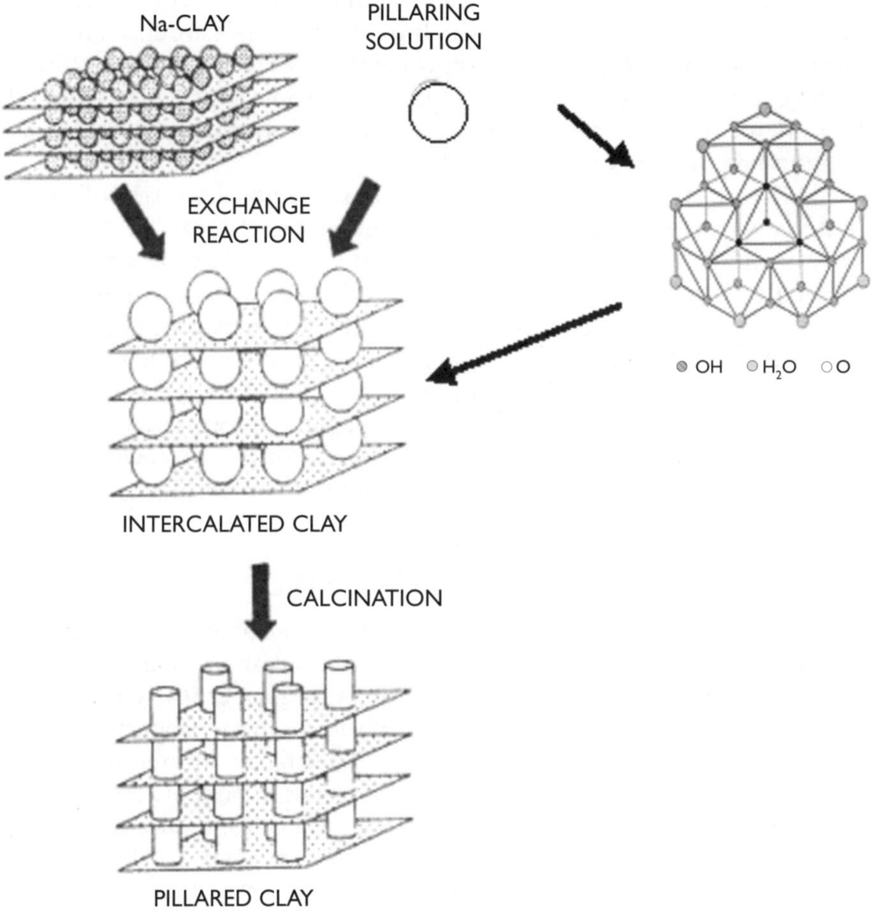

There have been some interesting new developments in altering pore spaces [16]. In a novel approach, quaternary ammonium salts (surfactants) which form micelles are used as centres around which silica species condense. These templates are removed by heating to create a continuous network of pores that mimic the size and shape of the template. The advantage of this approach is that the pore volume is controlled by the volume fraction of the template constituents, and the pore size is controlled by the size of the surfactant micelles. This approach has been used with pillared clays [16]. Interestingly, the porosity of the products depends on the quantity rather than the size of the surface of the surfactant molecules. These materials have been called composite clay nanostructures [16]. The quaternary ammonium salts form micelles in the interlayer regions and the pillaring agents form pillar materials beside them because of the affinity of surfactant molecules for the surface of the pillar precursors. The micelles act as templates, preventing the intercalated framework from collapsing during the dehydration process in which the framework hardens. The surfactants can be removed at temperatures between 150 and 250°C leaving a highly porous product. One interesting application is to fill these new pores with organic materials, thereby forming nanosilicocarboalumina composites.

Different surfactants can be used. For example, a series of compounds with $(CH_2CH_2O)_n$ with different values of n can be added to assist in understanding the mechanism. It appears that the size of the molecule is important, but in the reverse of what would naïvely be expected, it is found that larger surfactant molecules produce smaller pores. A surfactant with large n is expected to have a much stronger interaction with the precursor surface compared to a surfactant with small n. The strong interaction must influence the formation and configuration of the micelles. The mean diameter of the framework pores decreases with n because the stronger interaction results in micelles with a smaller diameter.

3.9 APPLICATIONS OF NANOMATERIALS

The uses of materials are determined by their mechanical and chemical properties. Young's modulus, which describes elasticity, hardness, ease of fracture, and conductivity, are all important properties. These studies are supported by an understanding of the chemical properties of the material, such as elemental analysis and molecular or atomic structure. Using the tools described in Chapter 2 it is now possible to describe these properties by actually looking at the molecules. Figure 3.9 shows TEM of a Si_3N_4 based ceramic material. Each atom can be seen in rows. The different material in the middle is the early stages of a phase transformation to a different arrangement of atoms, known as a nanoscaled nucleation. These structures and dislocations determine the strength of materials, their possible applications and their macroscopic properties. Corrosion, chemical modification and other influences can all be studied the same way.

It is fairly easy to understand why nanomaterials are more flexible. Each of the nanocrystallites, (particles) can move past each other so that stretching is easy. However, nanosize particles can also enhance hardness. Crystallites can pack together along boundaries in macrocrystallites and prevent the structure from unzipping. Super strength materials can be produced using nanocomposites with, say, a nanocrystalline particle embedded in a non-crystalline matrix. For example, a hardness in excess of 60 Gigapascals (GPa) has been reported for nanocrystals of titanium nitride embedded in thin films of silicon

Figure 3.9

TEM of a Si_3N_4 based ceramic material. The different material in the middle is the beginnings of a phase transformation — a nanoscaled nucleation. Courtesy of Dr John Drennan, University of Queensland, Australia.

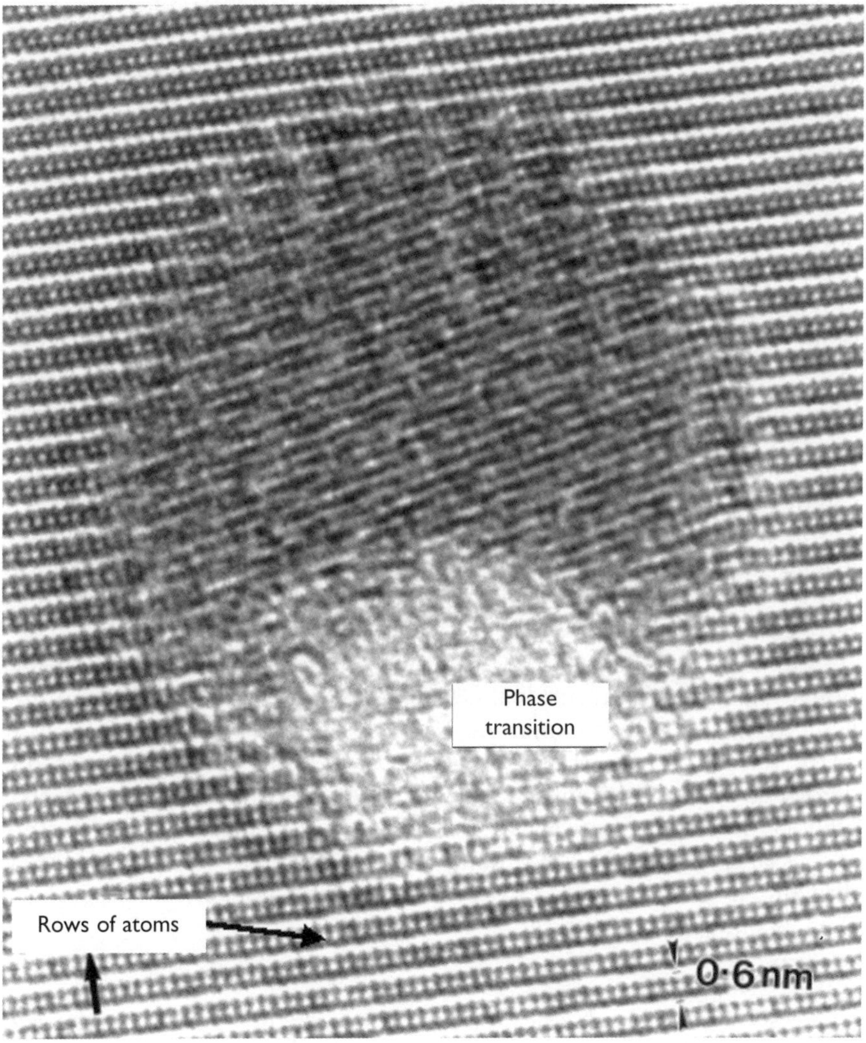

nitride (we measure hardness by the pressure required to indent the material). Superhardness is classed as > 40 GPa; and in some nanocomposite systems, measurements of > 80 GPa or ultra hardness, have been reported. These materials are therefore almost as hard as diamond (hardness > 100 GPa).

Future applications are discussed in more detail in Chapter 9 and some special optical uses are described in Chapter 7. Current applications include, but are not limited to, the following:

INSULATION MATERIALS

Nanocrystalline materials synthesised by the sol-gel technique result in aerogels (Chapter 3.5). Since they are porous and air is trapped at the interstices, aerogels are currently being used for insulation in offices, homes, and so on. By using aerogels for insulation, heating and cooling bills are drastically reduced, thereby saving power and reducing the attendant environmental pollution. They are also being used as materials for 'smart' windows (Chapter 7), which darken when the sun is too bright and lighten when the sun is not shining brightly.

MACHINE TOOLS

Some nanomaterials are harder than conventional materials. Cutting tools and drills made of nanocrystalline materials, such as tungsten carbide, tantalum carbide, and titanium carbide, are much harder, much more wear-resistant, erosion-resistant, and last longer than their conventional (large-grained) counterparts. They also enable the manufacturer to machine various materials much faster, thereby increasing productivity and significantly reducing manufacturing costs. Nanocrystalline silicon nitride (Si_3N_4) (Figure 3.9) and silicon carbide (SiC), have been used in automotive applications as high-strength springs, ball bearings, and valve lifters, because they are readily machineable and have excellent physical, chemical, and mechanical properties. They are also used as components in high temperature furnaces.

On the other hand, some nanomaterials are softer than conventional materials. Conventional ceramics are very hard, brittle, and difficult to machine. Zirconia, ZrO_2, a hard, brittle ceramic, has even been rendered superplastic by nanocrystalline grains. It can be deformed to great lengths (up to 300% of its original length). Nanocrystalline ceramics can be pressed and sintered into various shapes at significantly lower temperatures, whereas it would be very difficult, if not impossible, to press and sinter conventional ceramics even at high temperatures.

PHOSPHORS

The resolution of a television or a monitor depends greatly on the size of the pixel. These pixels are essentially made of materials called phosphors, which glow when struck by a stream of electrons inside the cathode ray tube (CRT). The resolution improves when the pixels, or the phosphors, are smaller. Nanocrystalline zinc selenide, zinc sulfide, cadmium sulfide, and lead telluride synthesised by the sol-gel technique improve the resolution of monitors. The use of nanophosphors would reduce the cost of making high resolution televisions. Carbon nanotubes (Chapter 4) are also candidates. Similar arguments apply to low cost flat panel displays.

BATTERIES

Nanocrystalline materials synthesised by sol-gel techniques are used as separator plates in new generation batteries because of their aerogel structure, which can hold considerably more energy than conventional plates. Nickel metal hydride (Ni-MH) batteries made of nanocrystalline nickel and metal hydrides require far less frequent recharging and last much longer.

HIGH POWER MAGNETS

The strength of a magnet increases with increased surface area per unit volume. It has been shown that magnets made of nanocrystalline yttrium-samarium-cobalt nanoparticles possess very unusual magnetic properties due to their extremely large surface area. Typical applications for these high power rare-earth magnets include quieter submarines, automobile alternators, land-based power generators, motors for ships, ultra-sensitive analytical instruments, and magnetic resonance imaging (MRI) in medical diagnostics.

MOTOR VEHICLES AND AIRCRAFT

Motor cars waste significant amounts when thermal energy generated by the engine is lost. This is especially true in the case of diesel engines. To prevent this waste, the engine cylinders (liners) are currently coated with nanocrystalline ceramics, such as zirconia and alumina, so that they retain heat much more efficiently and result in complete and efficient combustion of the fuel.

In aircraft, fatigue strength is critical. The fatigue strength increases with a reduction in the grain size of the material. Nanomaterials provide such a significant reduction in the grain size over conventional materials that the fatigue life is increased by as much as 300 per cent. Furthermore, components made of nanomaterials are stronger and can operate at higher temperatures. This means that aircraft can fly faster and more efficiently using the same amount of aviation fuel. In spacecraft, elevated-temperature strength is crucial for components such as rocket engines and thrusters, and particularly for surface material that will be in contact with the atmosphere on re-entry. Likewise, the amount of fuel that can be carried on board is crucial so the ratio of fuel to total weight is important. Nanomaterials, such as nanocrystalline tungsten-titanium diboride-copper composites, are potential candidates for increasing combustion efficiencies and hence reducing fuel use in these applications.

MEDICAL IMPLANTS

For an implant to effectively mimic a natural human bone, the surrounding tissue must penetrate the implants to give the implant the strength it needs. Since conventional materials are relatively impervious, human tissue does not penetrate the implants, and they are not as effective. Furthermore, these metal alloys wear out quickly, necessitating frequent, and often very expensive, surgery. However, nanocrystalline zirconia ceramic is hard, wear-resistant, corrosion-resistant (biological fluids are corrosive), and biocompatible. Nanoceramic aerogels can also withstand up to 100 times their weight. The products last longer and the patient needs less frequent implant replacements. This leads to a significant reduction in surgical expenses. Nanoceramics can also be made of apatite, a calcium phosphate material from which bone is derived, thereby mimicking nature's own process [17, 18]. The sol-gel prepared nanotitania (TiO_2) forms a chemical bond with the living bone in the body, although the bond is not very strong. However, impregnating the titania gel with hydroxyapatite is a promising way to increase its bioactivity. Nanocrystalline silicon carbide (SiC) is a candidate material for artificial heart valves primarily due to its low weight, high strength, extreme hardness, wear resistance, inertness (SiC does not react with biological fluids), and corrosion resistance.

OTHER MEDICAL USES

Gel technology has also been used in DNA separations. There has been a quest for an artificial gel that would replace the organic gels used to separate fragments of DNA for analysis. Traditionally this has been done by a process called gel electrophoresis. Enzymes are used to chop DNA strands into many short pieces of varying length. The sample is placed at one end of a column of organic gel and an electric field is applied to force the DNA to move through the gel. As they slowly pass through the tiny pores of the material, DNA fragments of different lengths move at different speeds and eventually collect in a series of bands as a ladder-like structure that can be photographed using fluorescent or radioactive tags. The resulting image is a list of the lengths of the fragments, from which genetic information can be determined. Silicon-based nanostructures with pores comparable to the size of a large DNA molecule can also perform these separations. These methods could also be applied to the study of other large organic molecules, including proteins, carbohydrates and lipids, for many of which electrophoresis is useless.

As noted, nanoparticles of iron can be used to form strong magnets. However, the size of the particle can be put to good use medically in other ways. Nanoparticles of the iron oxide Fe_3O_4 have diameters in the 5–100 nm range. These particles are magnetic and thus can be followed by a magnet in the body using nuclear magnetic resonance imaging (this is known as magnetic resonance imaging (MRI) so that the use of the word 'nuclear' does not scare the patient). In MRI, protons are excited with short pulses of radio-frequency radiation. The free induction decay as the protons relax is measured and deconvoluted by means of a mathematical process called a Fourier transform, which provides an image of the tissue that corresponds to proton density. Areas of high proton density, usually in the form of water or lipid molecules, have a strong signal and appear bright. Areas of bone or tendon, which have a low proton density because of the lack of water and lipids, have a weak signal and appear dark.

Traditionally, a major limitation of MRI has been its inability to distinguish differences in soft tissue types (such as healthy parts of the liver from diseased regions), as the relative proton densities can be very similar. Other regions, such as the bowel, are hard to image because air pockets and faecal matter make the proton density inconsistent. Various contrast agents have been developed to circumvent these imaging problems.

Contrast agents work by changing the strength of the MRI signal at a desired location. For example, iron oxide contrast agents change the rate at which protons decay from their excited state to the ground state, allowing more effective decay through energy transfer to a neighbouring nucleus. As a result, regions containing the superparamagnetic contrast agent appear darker in MRI than regions without the agent. For instance, when iron oxide nanoparticles are delivered to the liver, healthy liver cells can take up the particles, but diseased cells cannot. Consequently, the healthy regions are darkened, while the diseased regions remain bright.

These iron oxide particles have many advantages over other contrast agents. Unlike agents like perfluorochemicals, oils, and fats, they are miscible with aqueous systems, which means they can mix with material in the bowel and be used in small volumes. (Immiscible agents must be used in sufficient quantity to displace intestinal matter). This miscibility also allows them to be used intravenously. Compared with other magnetic contrast agents they are much more potent.

Nanoparticles are commonly used for laboratory studies as well as *in vivo*. Tagged to a biomaterial of interest they can be dispersed effectively, coated with a monoclonal antibody for a cell-surface antigen. The antibody-tagged particles are then incubated with a solution containing the cells of interest. The microparticles bind to the surfaces of the desired cells, and these cells can then be collected in a magnetic field. Methods of this type have been used to isolate or remove numerous cell types, including lymphocytes (cells that control immune response) and tumour cells.

Particles with truly nanoscopic dimensions have also been used in the 'First Response' home pregnancy test, which uses conventional micrometre-sized latex particles in conjunction with gold nanoparticles (less than 50 nm diameter). Gold nanoparticles have a characteristic visible absorption band called the plasmon resonance absorption, which makes them pink (plasmons are described in Chapter 7). The micro- and nanoparticles are derivates with antibodies to human chorionic gonadotrophin, a hormone released by pregnant women. When mixed with a urine sample containing this hormone, the micro- and nanoparticles coagulate and the resulting clumps are coloured pink.

Nanogold particles have been used extensively as specific staining agents in biological electron microscopy (Chapter 2). The small sizes of these particles allow them to be physically close to the structures that they stain, providing high resolution. Because these nanoparticles are gold, which has a high backscatter coefficient, they appear bright in a scanning electron microscope image. In contrast, the high density of gold makes them appear dark in a transmission electron microscope image. Site-specific staining is obtained by labelling the nanogold particles with antibodies directed against a protein in the region of interest. Antibody-labelled gold nanoparticles are commercially available, and unlabelled gold nanoparticles can also be purchased and then labelled with a specific antibody by the investigator.

YOU SHOULD NOW KNOW:

1 That a plasma is an ionised gas. A plasma is achieved by making a gas conduct electricity by providing a potential difference across the two electrodes so that the gas yields up its electrons and thus ionises. In a vacuum or in an inert gas the electrodes can be made volatile. Heat is produced and other material can be volatilised and ionised using this heat. Plasma arcing has been important in forming carbon nanotubes.

2 Chemical vapour deposition is a method of making nanoparticulate material from the gas phase. Material is heated to form gas and then allowed to deposit as a solid on a surface, usually under vacuum. There may be direct deposition or deposition by chemical reaction to form a new product that differs from the material volatilised. Chemical vapour deposition is particularly useful for making aligned materials on surfaces

3 The sol-gel process, as the name implies, involves the evolution of networks through the formation of a colloidal suspension (sol) and gelation of the sol to form a network in a continuous liquid phase (gel). The precursors for synthesising these colloids consist of ions of a metal or metalloid element surrounded by various reactive ligands. Sol-gel processes can preserve the voids in solution in the solid state thereby producing nanoholes.

4 In nanoelectrodeposition, the aim is to place only a single layer or more of coverage on a surface in a very controlled way. Nanostructured films of platinum can be produced by electrodeposition from liquid crystalline mixtures. Electrodeposition can be used to fill holes to make dispersed nanomaterials.

5 In ball milling, small balls are allowed to rotate around a drum and drop with gravity force on to a solid enclosed in the drum. Ball milling breaks down the structure into nanocrystallites. Ball milling is the preferred method of preparing nanometal oxides.

6 Nanoparticles can also be prepared by altering the pore spaces in phyllosilicates by means of various surfactants and then filling them if required.

7 Nanomaterials have found applications as insulators, batteries, machine tools, phosphors and magnets. They are used in aircraft, building structures and medicine.

3.10 EXERCISES

1 *Web search the following words: nanometals, electrochromic windows, imogolite, nanocomposites.*
2 *Select one reference in an area of interest, find it and read it.*
3 *Take a pack of cards and some carrots. Cut the carrots up and place them as pillars between each of the cards to form a pillared clay structure. Try building a conventional house of cards. Why is this less stable than a pillared clay structure?*
4 *Investigate the relationship between bulk and surface volume. Take a carrot and cut it into a cylinder. Measure the volume as* $\pi r^2 h$ *where* h *is the length of the pillar and* r *is its radius. Measure the surface circumference,* $2\pi rh$*. Then cut the carrot in half and do the same calculation but add the two circumferences together. The volume is the same of course. Plot circumference divided by volume versus number of cuts. Extrapolate your data until you find the number of slices to make the carrot only available in two-dimensional space. This experiment demonstrates the potential value in increasing surface to bulk space.*
5 *Find out more about zeolites or phyllosilicates, particularly kaolinite, illite and montmorillonite.*
6 *Find a structure of imogolite showing bond linkages on the Web or in a reference and see that one silicon oxygen is attached to one H and the other three are attached to Al.*
7 *For an interesting article on nanotubes read Hulteen JC and Martin CR (1997) A template-based method for the preparation of nanomaterials.* J. Mater. Chem. *7: 1075–87.*

3.11 REFERENCES

Most scientific journals in physical science do not use report titles of papers. Therefore as the reader should now have been able to find articles the title format is dropped except obviously for books.

1 Feynman RP (1960) *Engineering and science.* California Institute of Technology, February edn.
2 Drexler K E (1990) *Engines of Creation.* Fourth Estate, London, 296 pp.
3 Pang LSK, Vassallo AM & Wilson MA (1992) *Energy and Fuels* 6: 176–79.
4 Cassell AM, Raymakers JA, Kong J & Dai H (1999) *J. Phys. Chem. B* 103: 6484–92.
5 Attard GS, Glyde JC & Göltner C (1995) *Nature* 378: 366–68.
6 Attard GS, Edgar M, Emsley JW & Göltner CG (1996) *Materials Research Society Symp. Proc.* 425: 179–84.
7 Attard GS, Edgar M, Emsley JW & Göltner CG (1996) *Materials Research Society Symp. Proc.* 425: 185–89.
8 Frisch HL, West JM, Göltner CG & Attard GS (1996) *J. Polym. Sci. A, Polym. Chem.* 34: 1823–26.
9 Attard GS, Edgar M & Göltner CG (1998) *Acta Materialia* 46: 751–58.
10 Attard GS, Göltner CG, Corker JM, Henke S & Templer RH (1997) *Angewandte Chemie Intl. Edn. Engl.* 36: 1315–17.
11 Attard GS, Bartlett PN, Coleman NRB, Elliott JM, Owen JR & Wang JH (1997) *Science* 278: 838–40.
12 Chen Y, Fitzgerald J, Chadderton LT & Chaffron L (1999) *Journal of Metastable and Nanocrystalline Materials* 2–6: 375–80.
13 Pierard N, Fonesca A, Konya Z, Willems I, Van Tenderloo G & Nagy JB (2001) *Chem. Phys. Letters* 335: 1–8.
14 Chopra NG, Luyken RJ, Crespi VH, Cherrey K, Zettl A & Cohen ML (1995) *Science* 269: 966–70.
15 Mitchell IV (1990) *Pillared Layered Structures: Current Trends and Applications.* Elsevier Applied Science, 1990.
16 Zhu HY & Lu GQ (2001) *Langmuir* 17: 588–94.
17 Ben-Nissan B & Chai C (1995) Sol-gel derived bioactive hydroxyapatite coatings. In R Kossowsky & N Kossovsky (eds) *Advances in Materials Science and Implant Orthopaedic Surgery.* NATO ASI Series, Series E: Applied Sciences, Kluwer Academic Publishers, 294: 265–75.
18 Ben-Nissan B, Chai C & Evans LA (1999) Crystallographic and spectroscopic characterisation and morphology of biogenic and synthetic apatites. In Wise DL, Trantolo DJ, Altobelli DE, Yaszemski MY, Gresser JD & Schwartz ER (eds) *Encyclopedic Handbook of Biomaterials and Bioengineering B: Applications*, 5th edn. Marcel Dekker, New York: 191–221.

4

THE CARBON AGE

This chapter discusses some of the uses of rolls of graphitic sheets called carbon nanotubes. These sheets come in a variety of forms and have different properties. They may be valuable components for nanoelectronics or as storage devices.

4.1 NEW FORMS OF CARBON [1–4]

The carbon age is now with us. Many would argue that if diamond sheets could be made cheaply all objects that need to be hard and indestructible would be made from diamond. It is probably not too long before we will be able to do this. However, even more exciting forms of carbon are now available. In 1980 we knew of only three forms of carbon, namely diamond, graphite and amorphous (non-crystalline carbon). Today we know there are a whole family of other forms of carbon. The first of these to be discovered was buckminsterfullerene (also called buckyball and fullerene C_{60}). There are now thirty or more forms of fullerenes and also an extended family, their brothers and sisters, the nanotubes.

Fullerene C_{60}, (buckyball), is the first spherical carbon molecule with carbons arranged in a soccer ball shape (Figure 4.1a). In the structure there are sixty carbon atoms (hence C_{60}) and a number of five-membered rings isolated by six-membered rings. It may well be that these objects can be used as ball bearings in some of Drexler's mechanical devices. The second spherical carbon molecule in the same group is the rugby ball, C_{70}, whose structure has extra six-membered carbon rings (Figure 4.1b), but there are also a large number of other potential structures containing the same number of carbon atoms (isomers) depending on whether five-membered rings are isolated or not, or whether seven-membered rings are present. Many other forms of fullerenes up to and beyond C_{120} have been characterised and it is possible to draw lots of structures with five-membered rings in different positions and sometimes together. The important fact for nanotechnology is that atoms can be placed inside the fullerene ball. Atoms contained within the fullerene are said to be endohedral. Of course they can also be bound to fullerenes outside the ball as salts if the fullerene can gain electrons. The structure is then $M_x^+ C_{60}^{n-}$, where M^+ is a cation and x is the number to balance the charge on the fullerene. In this case the cation is said to be exahedral.

Figure 4.1

a) The first fullerene C_{60} or buckyball.

b) The second fullerene C_{70}, which is shaped like a rugby ball.

a

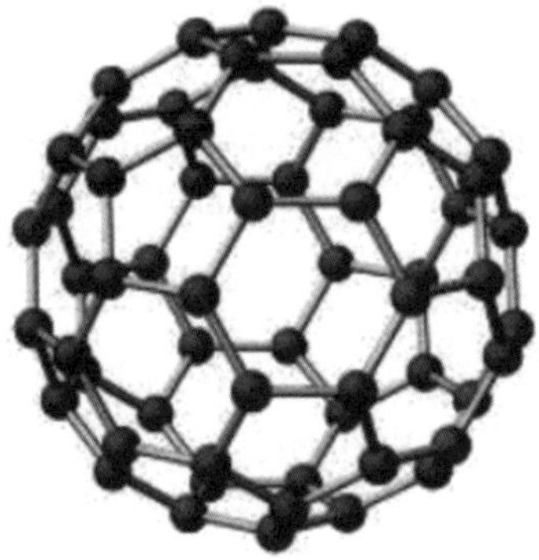

b

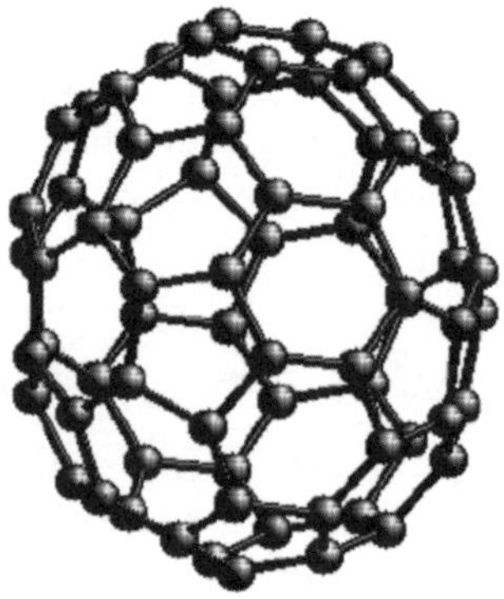

Bulk or thin films of pure C_{60} are only semi-conducting and have a room temperature resistivity of 108 Wcm. The electrical resistivity can be decreased by many orders of magnitude by adding alkali metals as salts. Here the metal ion is exahedral. As x in $M_x^+ C_{60}^{n-}$ increases, the resistivity decreases, reaching a minimum for the metallic x = 3 stoichiometry. By further increasing the alkali metal stoichiometry the resistivity increases again until an insulator is produced for x = 6. The metallic x = 3 state is superconducting, with a superconducting transition temperature of approximately -243°C for M = Rb, (rubidium). Recently, the addition of some simple organic compounds such as bromoform, $CHBr_3$, to the ball have been shown to increase conductivity.

Endohedral fullerenes can be produced in which metal atoms are captured within the fullerene cages. Up to three atoms have been trapped, but with bigger fullerenes this number can grow. Theory shows that the maximum conductivity is to be expected for metals, which will transfer three electrons to the fullerene.

Fullerenes can be dispersed on a surface as a monolayer. That is, there is only one layer of molecules, and they are said to be monodispersed. Provided fullerenes can be placed in specific locations they may be aligned to form a fullerene wire, like a line of marbles. Systems with appropriate material inside the fullerene ball are conducting and are of particular interest because they can be deposited to produce bead-like conducting circuits which look like a series of marbles in a row.

Combining endohedrally doped structures with non-doped structures changes the actual composition of a fullerene wire so that it may be tailored *in situ* during patterning. Hence within a single wire, insulating and conducting regions may be precisely defined. One-dimensional junction engineering becomes realistic. A real problem with building devices of a few atoms is the wear factor — it is easy to rub them off! Single atoms are easily removed from surfaces by friction, evaporation and just contact. Trapping metal atoms in place by encapsulating them in fullerenes and polymerising the carbon cages into a rigid framework makes sense. The barrier that must be overcome in order to escape the cage is quite large, about 6 eV. The inert fullerene cage and the novel additive nanofabrication technology allow for atomically neat lines and discrete, sharp junctions that are critical issues for nanoscale technologies.

Possibly more important than fullerenes are carbon nanotubes, which are related to graphite. The molecular structure of graphite resembles a sheet of chicken wire — a network of hexagonal rings of carbon. In conventional graphite the sheets of carbon are stacked on top of one another, allowing them to easily slide past each other. That is why graphite is not hard, feels greasy and can be used as a lubricant. When graphite sheets are coiled, they form carbon nanotubes. Only the tangents of the graphitic planes come into contact with each other, and hence the properties are more like those of a molecule. Nanotubes come in a variety of diameters and lengths. They may have different sized internal cylindrical cavities and may have more than one sheath. The end caps are half fullerene balls and these can differ. There are also possibilities in the arrangement of the hexagonal sheets, leading to left- and right-spiralled forms (chirality) and also folds and indentations in the sheets.

4.2 TYPES OF NANOTUBES [5–7]

A nanotube may consist of one tube of graphite (a single-walled nanotube, SWNT) or a number of concentric tubes, called multiwalled nanotubes (MWNTs). When viewed by transmission electron microscopy these tubes appear as planes. Whereas in SWNTs two planes are observed, representing the edges (Figure 4.2), in MWNTs more than two planes are observed, and these can be seen as a series of parallel lines (Figure 4.2).

Figure 4.2

Single-walled nanotube and multiwalled nanotubes. The two walls are shown as a) and b) respectively in the single-walled nanotube. Eighteen walls on each side are present in the multiwalled nanotube.

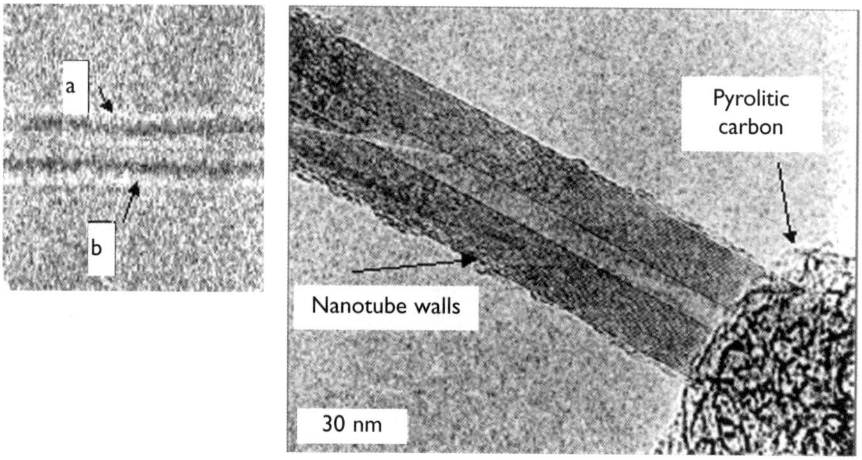

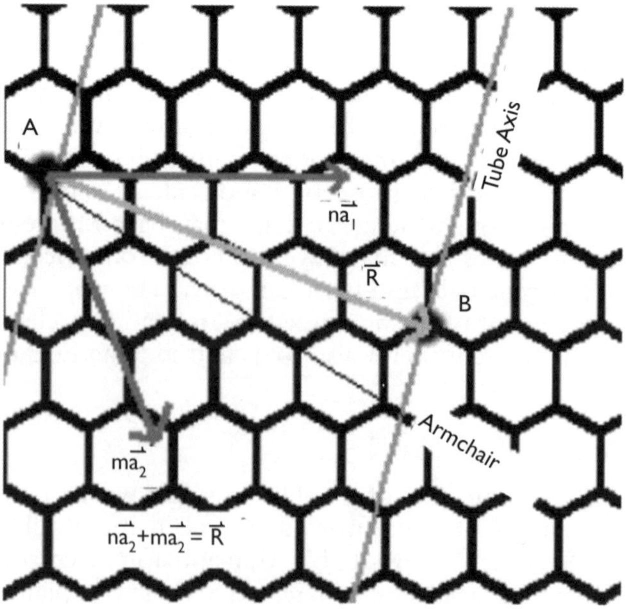

Figure 4.3

Naming nanotubes

There are different types of SWNTs because the graphitic sheet can be rolled in different ways. The conventional way to describe this is by looking at the unrolled sheet and expressing the rolling process by vectors $(\vec{a}_1,\vec{a}_2)$, where n and m are integers of the vector equation $\vec{R} = n\vec{a}_1 + m\vec{a}_2$. For example, a (10,11) nanotube has $n = 10$ and $m = 11$.

Figure 4.3 shows how this is done. This figure has two parallel lines labelled as axes where the graphite is rolled up to create the tube. A is any point on one of the lines that intersects one of the carbon atoms. One vector, $n\vec{a}_1$ is placed horizontally on the graphite sheet along the zigzag bonds. The other vector is placed in any other location. The armchair line (the thin line) is the one that travels across each hexagon, separating them into two equal halves. It is also shown in the figure. If the tube is rolled directly over this line, the resultant nanotube is an armchair nanotube. In this case, to ensure the vector sum lies over the armchair line, the second vector, $m\vec{a}_2$ is placed at an angle such that the armchair line bisects the angle made with the first vector, $n\vec{a}_1$ That is, the sum of the vectors $\vec{R}$ equates to rolling over the armchair line and $n\vec{a}_1 = m\vec{a}_2$. Armchair nanotubes thus have configurations (5,5: 6,6: 7,7 and so on) because the vectors must be the same to bring $\vec{R}$ over the armchair line. The numbers 5,5, 6,6 and so on reflect the diameter of the nanotube and the pairing of numbers in its stereochemistry.

There are other possible values of the second vector, $m\vec{a}_2$. If it is made zero then $\vec{R}$ lies along $n\vec{a}_1$, and the tube rolls along this line. Zigzag nanotubes have configurations (9,0; 10,0; 11,0 and so on), since one vector must be zero. Other possibilities where $n\vec{a}_1$ and $m\vec{a}_2$ are not zero or equal result in a chiral nanotube which has left- and right-handed forms.

Typical nanotube variants are shown in Figure 4.4. It is possible to recognise zigzag, chiral and armchair nanotubes just by following the pattern across the diameter of the tubes. The void space in these nanotubes varies with different values of n and m so that a range of voids is possible. Values between 1.2 nm and 2 nm are common.

Single-walled nanotubes can come as unaligned structures or bundles of ropes. Ropes consist of bundles of tubes packed together in an orderly manner [7, 8]. There are gaps between the nanotube bundles, and these are described as the lattice packing parameter. Calculations show that the packing depends on the type of nanotube. Armchair (10,10) tubes should have a lattice parameter of 1.678 nm. Zigzag (17, 0) tubes should have a lattice parameter of 1.652 nm and chiral (12,6) SWNTs should have a lattice parameter of 1.652 nm. There are also spaces between the layers of tubes, called interlayer spacings. The interlayer spacing value between the bundled tubes is also dependent on nanotube structure. For example, for tubes of the same diameter, armchair tubes should have a spacing of 0.338 nm, zigzag tubes 0.341 nm, and chiral tubes 0.339 nm. These spacings are not very different from

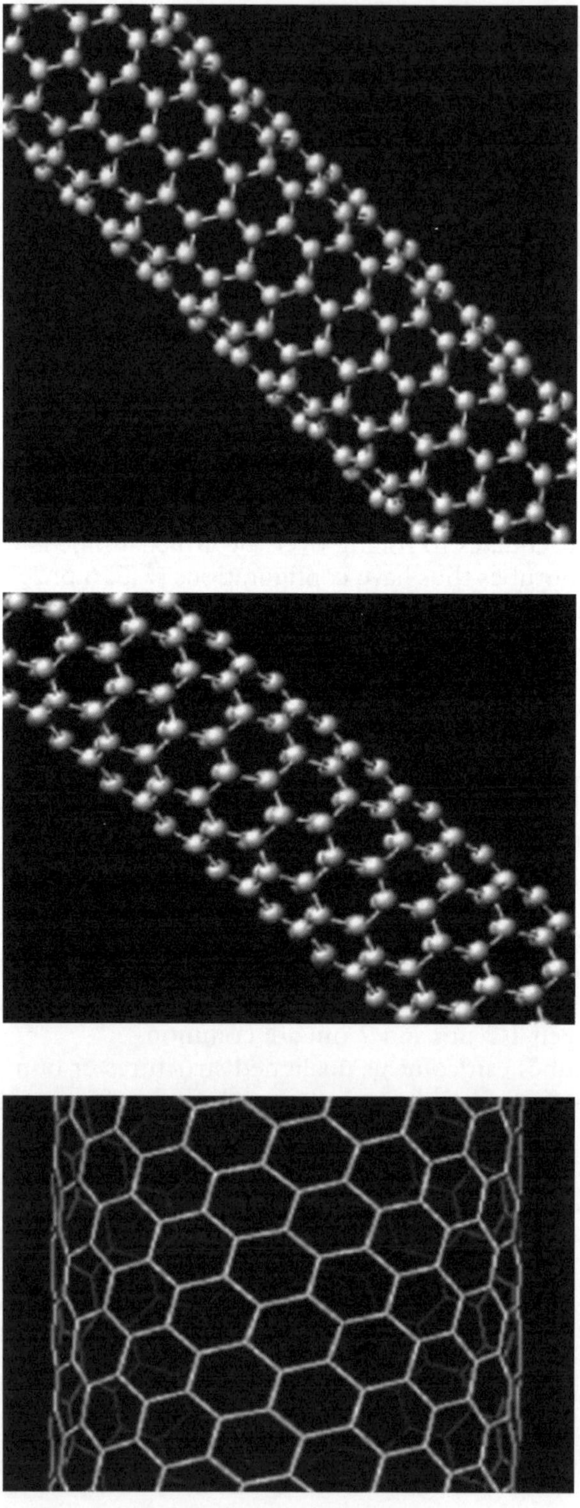

Figure 4.4

Types of nanotubes.
a) Zigzag (10,0).

b) Armchair (5,5).

c) Chiral (14,5).

those found for the spaces between the different layers in graphite. However, because of the different lattice packing parameters, the densities of the different tubular solids vary. It has been predicted that pure zigzag (17,0), armchair (10,10) and chiral (12,6) SWNTs should have densities of 1.33, 1.34 and 1.40 gcm^{-3} respectively. Preparing pure types of each nanotube is an area of ongoing research but a technique called Raman spectroscopy can detect differences in motion between the bonds in the different nanotubes and can be used to distinguish between them.

MWNTs can come in an even more complex array of forms because each concentric single-walled nanotube can have different structures and hence there are a variety of sequential arrangements. The simplest sequence is when the concentric layers are identical but different in diameter (such as sequence a,a,a,a,a for armchair nanotubes or zigzag nanotubes). However, mixed variants are possible consisting of two or more types of nanotubes arranged in different orders (for example, a,b). These can have regular layering (a,b,a,b,a,b,a) or random layering (such as a,a,b,a,b,b,b,a).

The structure of the nanotube influences its properties, including conductance, density and lattice structure. It is known that some nanotubes are conductors, that is, they are metallic, while some are semiconductors. The conductance of individual single-walled nanotubes have been measured to show that this is true, and connections to each individual type and diameter, such as zigzag, armchair, chiral, and so on, have been made. Both type and diameter are important. The wider the diameter of the nanotube the more it behaves like graphite. The narrower the diameter the more its intrinsic properties depend on type.

4.3 FORMATION OF NANOTUBES

METHODS AND REACTANTS

There are a number of methods of making nanotubes and fullerenes. Fullerenes were first observed by vaporising graphite with a laser (laser ablation) however this was initially not practical for making gram or greater quantities. Carbon nanotubes have probably been around for a lot longer and may have been seen during carbon vapour deposition but electron microscopy was not developed enough to distinguish them from other types of tubes. The first method for preparing both carbon nanotubes and fullerenes in reasonable quantities — was by putting an electric current across two carbonaceous electrodes in a helium or argon atmosphere (Figure 4.5). This method is called plasma arcing (Chapter 3.2). It involves the evaporation of one electrode (the anode) as cations followed by deposition at the other electrode. Fullerenes and nanotubes are formed by plasma arcing of carbonaceous materials, particularly graphite. The fullerenes appear in the soot that is formed, while the nanotubes are deposited on the opposing electrode, the cathode. A cheap method, which may be suitable for industrial application, is to make them from coal, especially since fullerenes can be purified on the coal from which they are made [9, 10].

Figure 4.5

Mechanism of carbon nanotube formation. Carbon is evaporated from the anode and deposited on the cathode.

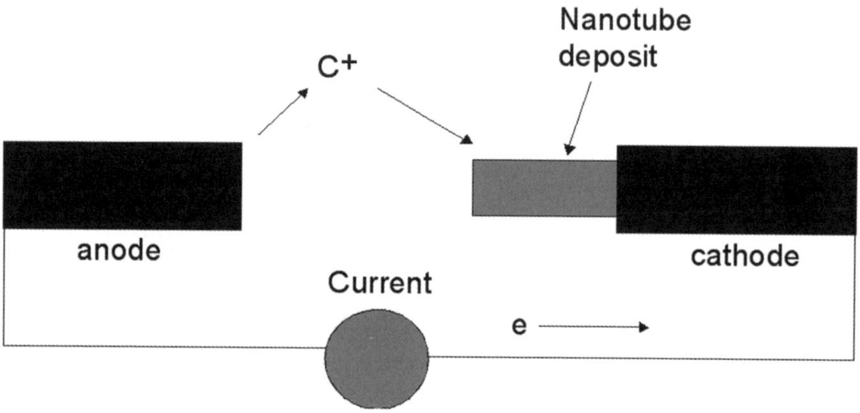

The products from coal differ from those obtained from graphite. While the ratio of C_{60}-fullerene and C_{70}-fullerene appears to be the same for coal as for graphite, other substances called polycyclic hydrocarbons are formed due to the hydrogen in the coal. The distribution of types of nanotubes appears also to depend on the various minor elements in the coal. They catalyse the growth of different types of nanotubes. Normally nanotubes formed by plasma arcing are multiwalled. However if the electrodes are bored out and cobalt or other metals are included, single-walled nanotubes are formed.

Like the soot from which fullerenes are extracted, the cathode deposit is heterogeneous. Under lowest resolution, cathode deposits consist of two phases. One phase is described here as 'feathers' and the other as 'matrix' [11]. The matrix is composed of nanotubes and nanoparticles (shortened scale-like nanotubes) while the feathers consist of the same mix but with amorphous carbon, called pyrolytic carbon. Depending on the arcing conditions and the nature of the anode, the ratio of feathers to matrix can change. For coal the ratio of feathers to matrix depends on its hydrogen content. More hydrogen appears to produce more feathers because the additional hydrogen promotes the formation of a disordered carbon material called pyrolytic carbon [12].

Experiments with coal also produce significant amounts of other types of structures that form on the cathode. These products are microfilaments rather than nanotubes and they come in a variety of forms [13, 14]. Some of them have a uniform central void (Figure 4.6) with graphitic planes aligned at an angle to a central void but not parallel as for nanotubes. The production of these filaments appears to be favoured when coal rather than graphite is used, and even more so when iron sulfides are present in the coal. Iron sulfide is observed at the growing end of the filament where it appears to act as a catalyst for growth. Some microfilaments have a worm-like shape with a nested cup structure (Figure 4.7) while others have collections of different types of voids (Figure 4.6) [13–15].

Figure 4.6

Void microfilaments

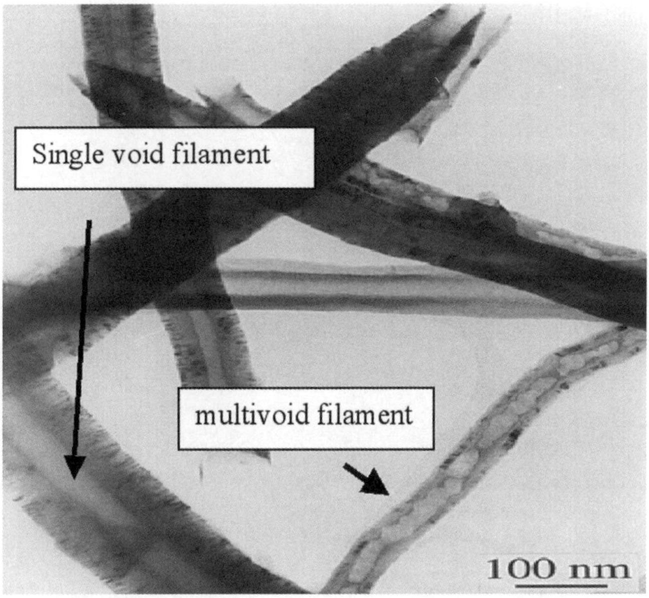

ARCING IN THE PRESENCE OF COBALT [16, 17]

The physical nature of the products formed from plasma arcing in the presence of cobalt in about three per cent concentration or more, is quite different to conventional arcing in which no cobalt is used. As noted above, the nanotube product is a compact cathode deposit of rod-like morphology. However when cobalt is added the product changes to a web with strands of 1 mm or so thickness that stretches from the cathode to the walls of the reaction vessel. The web consists of strings of SWNTs that can easily be recognised by Raman spectroscopy. The mechanism by which cobalt changes this process is unclear, however one possibility is that such metals affect the local electric fields and hence the formation of five-membered rings. If this is true, this process could be simulated in a different way to yield an easier method of preparing the same product or perhaps different products. Cobalt particles are found in independent globules and associated with the strands. They can often be found encapsulated in graphite layers. Bundles of SWNTs, globular cobalt particles and more amorphous bodies containing vesicles (Figure 4.7b) or bladders (Fig. 4.7c) are also observed [16, 17]. It should be noted that the bladders, vesicles and nodules formed could contain one or more cobalt particles (see the black dots in the bladders and vesicles). The trace amount of cobalt present in coal is not enough to enhance SWNT formation. However, similar processing with cobalt is possible when coal is used as the source of carbon.

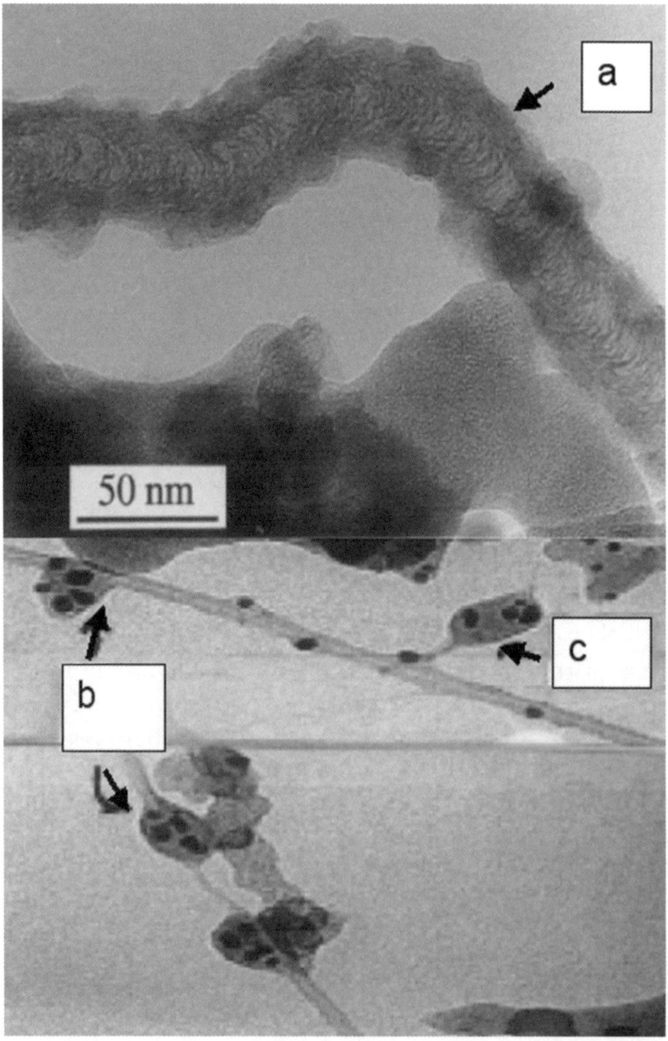

Figure 4.7

a) Nanoworms.

b) Single-walled nanotube bundles showing vesicles.

c) Bladders.

From a materials scientist's point of view, it is important to produce a uniform nanotube product. That is, a material consisting of nanotubes which all have the same diameter, are of similar length and have the same number of graphitic sheaths. It is clear that if an arcing process were to be used, cobalt or an element with similar function (such as nickel) would be necessary. However, elements such as sulfur and iron, which are present in coal, may also play some role here, because they alter the number of layers in the formation of multiwalled nanotubes. Similar work on graphite indicates the importance of sulfur and iron in these roles. Since the addition of organic substances such as naphthalene can affect multiwall formation, the organic structure of coal (as well as inorganic structures) may play some role. There seems to be scope for making more uniform products by a selective choice of coal.

Finding out how to do this depends on understanding the mechanism. The most important tool other than microscopy and spectroscopy is isotopic analysis. Where graphite is used as an anode, individual carbon atoms (C_1 units) are generated during arcing and migrate to the cathode, producing nanotube and fullerene soot. Fullerene soot contains a greater ratio of ^{13}C to ^{12}C (it is isotopically heavy) compared with the original graphite; and the cathode deposit is isotopically light [18]. The isotope fractionation can be interpreted in terms of the lower diffusion time and higher diffusion velocity of ^{12}C over ^{13}C species of C_1 or equivalent species in getting to the cathode because of the different charge to mass ratio. The slower ^{13}C species with less charge per unit mass will have a greater chance of undergoing reactions with other carbon species. Their products also have a smaller charge to mass ratio and may escape the electric field of the plasma arc to form soot or fullerenes.

In reactions involving species other than C_1, the magnitude of this isotope effect can be used to gauge whether molecular entities reach the soot or the cathode. For coal, there is strong evidence that much of the soot from the char and coal experiments is directly derived from molecular entities in the anode and not solely from C_1 units. The larger isotopic difference between anode and soot observed for graphite rather than coals is due to the absence of weak bonds which ensures that all the soot-forming process goes via C_1 or equivalent species.

During arcing, carbon experiences a temperature range from around 1700°C at the centre of the arc to about room temperature at its fringes. If big molecules such as naphthalene from coal are incorporated into fullerenes [19] it is most likely that fullerene synthesis occurs at lower temperatures at the edge of the arc. Likewise, if naphthalene can be incorporated in nanotube formation it must be possible for molecules surviving at the edge of the arc to tumble back into the electrode surface and become incorporated. In experiments where naphthalene has been added to the anode the average void space of the nanotubes in the cathode deposit increases by up to two nanometres compared with graphite. Presumably the larger void space occurs because there are more six-membered ring structures available when naphthalene is present in the arc and hence the half fullerene or related caps are wider. The wider tubes appear to be more predominant at the outer diameter regions of the anode deposit, possibly because naphthalene survives more readily here and can be incorporated into the structure. Moreover, other process conditions such as pressure can affect both fullerene and nanotube yields and structures, so there is real scope for further development.

LASER METHODS [7]

In 1996, a dual-pulsed laser vaporisation technique was used to optimise the laser method to produce SWNT in gram quantities and yields of > 70%. Samples were prepared by laser vaporisation of graphite rods with a 50:50 mixture of Co and Ni powder (particle size ~1 μm) at 1200°C in flowing argon, followed by heat treatment in vacuum at 1000°C to remove the C_{60} and other fullerenes. The initial laser vaporisation pulse was followed by a second pulse to vaporise the target more uniformly. The use of two successive laser pulses minimises the amount of carbon deposited as soot. The second laser pulse breaks up the larger particles ablated by the first one, and feeds them into the growing nanotube structure.

The material thus produced appears as a mat of 'ropes' 10–20 nm in diameter and up to 100 μm or more in length. Each rope is found to consist primarily of a bundle of SWNTs aligned along a common axis. By varying the growth temperature, the catalyst composition and other process parameters, the average nanotube diameter and distribution can be varied.

CHEMICAL VAPOUR DEPOSITION METHOD (CVD) [20, 21]

Arc-discharge and laser vaporisation are currently the principal methods for obtaining quantities of high quality carbon nanotubes. However, both methods suffer from some drawbacks. The first is that both methods involve evaporating the carbon source so that it has been unclear how to scale up nanotube production to the industrial level using these approaches. The second issue relates to the fact that vaporisation methods grow nanotubes in highly tangled forms mixed with unwanted forms of carbon or metal species. The nanotubes are difficult to purify, manipulate, and assemble for building nanotube-device architectures.

Chemical vapour deposition has been described in Chapter 3.4, and Figure 3.2 shows some pictures of aligned nanotubes. Chemical vapour deposition of hydrocarbons over metal catalysts is a classical method that has been used to produce various carbon materials such as carbon fibres, filaments, and whiskers for over twenty years [20, 21]. However, there was no evidence that this technique could be used to synthesise carbon nanotubes until relatively recently [20]. The method consists of the catalytic decomposition of acetylene over iron particles at 700°C. Using this method, four structural forms of carbon are formed: amorphous carbon layers on the surface of the catalyst, filaments of amorphous carbon, graphitic layers covering metal particles, and MWNTs. The MWNTs are usually covered with amorphous carbon on their outer layer. These carbon nanotubes have been studied by high-resolution TEM in both their as-grown form and after heat treatment. The as-grown nanotubes generally do not look fully formed. However, the structure is much improved after heat treatment to 2500–3000°C in argon.

Large amounts of carbon nanotubes can be formed by catalytic deposition of acetylene over Co and Fe catalysts supported on silica or zeolite [21]. (A zeolite is a caged structure with large voids into which molecules can enter). The carbon deposition activity seems to relate to the cobalt content of the catalysts, whereas the nanotubes' selectivity seems to be a function of the pH in catalyst preparation. Fullerenes and bundles of SWNTs were also found among the MWNTs produced on Co/zeolite catalysts.

Some researchers have studied the formation of nanotubes from ethylene [22]. Supported catalysts (Fe, Co, Ni) containing either a single metal or a mixture of metals seem to induce the growth of isolated SWNTs or SWNT bundles in the ethylene atmosphere. The production of SWNTs, including double-walled nanotubes (DWNTs) on Mo and Mo-Fe alloy has also been demonstrated. Chemical vapour deposition of carbon within the pores of a thin alumina template called a membrane with or without a Ni catalyst has been achieved. Ethylene was used with reaction temperatures of 545°C for Ni-catalysed chemical vapour deposition and 900°C for an uncatalysed process. The resultant carbon nanostructures have open ends. That is, they have no caps.

Methane has also been used as carbon feedstock. In particular it has been used to obtain 'nanotube chips' containing isolated SWNTs at controlled locations [23]. High yields (70–80%) of SWNTs have been obtained by catalytic decomposition of a H_2 / CH_4 mixture over well-dispersed metal particles (Co, Ni, Fe) on MgO at 1000°C. It has been reported that the synthesis of composite powders containing well-dispersed carbon nanotubes can be achieved by selective reduction in H_2 / CH_4 of oxide solid solutions between a non-reducible oxide such as Al_2O_3 or $MgAl_2O_4$ and one or more transition metal oxide(s). The reduction produces very small transition metal (Fe, Co, Ni and their alloys) nanoparticles at a temperature of usually > 800°C. The decomposition of CH_4 over the freshly formed nanoparticles prevents their further growth and thus results in a very high proportion of SWNTs and less MWNTs.

The disproportionation reaction of carbon monoxide has also been used to produce aligned carbon nanotubes.

$$2CO \rightarrow C + CO_2$$

A comprehensive study of the production of carbon nanotubes by the catalytic decomposition of various carbon-containing compounds over supported transition metal catalysts has been reported. It has been found that iron-containing compounds make good nanotube arrays [24]. For example, acetylene, ethylene, propylene, acetone, n-pentane, methanol, toluene, and methane were tested with iron-containing compounds and each resulted in the formation of carbon nanotubes.

BALL MILLING [21]

This has been described in Chapter 3.7. Ball milling and subsequent annealing is a simple method for the production of carbon nanotubes and could be the key to cheap methods of production industrially. Although it is well established that mechanical attrition of this type can lead to fully nanoporous microstructures, it was not until recently that nanotubes of carbon and boron nitride were produced from these powders by thermal annealing. Essentially the method consists of loading graphite powder (99.8% purity) into a stainless steel container along with four hardened steel balls. The container is purged and argon gas (300 kPa) is introduced. The milling is carried out at room temperature for up to 150 hours. Following milling, the powder is annealed under a nitrogen (or argon) gas flow at temperatures of 1400°C for six hours. The mechanism of this process is not known but it is thought that the ball milling process forms nanotube nuclei and the annealing process activates the nanotube growth. Work has shown that multiwalled nanotubes are formed but single-walled nanotubes are more difficult to prepare [21].

OTHER METHODS

Carbon nanotubes can also be produced by diffusion flame synthesis, electrolysis, solar energy, heat treatment of a polymer, and low temperature solid pyrolysis. In flame synthesis, combustion of a portion of the hydrocarbon gas provides the elevated temperature required, with the remaining fuel quite naturally serving as the hydrocarbon reagent. Hence the flame constitutes an efficient source of energy and hydrocarbon reactant. Furthermore, since combustion synthesis has been shown to be scalable for high-volume commercial production, the method holds promise in this regard.

4.4 ASSEMBLIES [21]

To make commercial use of nanotubes it is necessary to assemble them into useful structures. Ajayan and co-workers [25] developed a simple method to produce aligned arrays of carbon nanotubes by cutting a polymer resin-nanotube composite. However, the degree of orientation of the nanotubes in the composite is affected by the thickness of the slices, and the aligning effect becomes less pronounced with increasing slice thickness.

In some cases, for example in CVD experiments when the catalysts are prefabricated into patterned arrays, well-aligned nanotube assemblies are produced. Highly ordered, isolated long carbon nanotubes can be produced by iron nanoparticles embedded in mesoporous silica. The use of a patterned catalyst apparently encourages the formation of aligned nanotubes. The method offers control over length (up to 50 μm) and fairly uniform diameters (30–50 nm), as well as producing nanotubes in high yield, uncontaminated by polyhedral particles. Catalyst islands fabricated on silicon wafers using electron beam lithography can be used to grow (by CVD) SWNTs oriented in the plane of the substrate. Similarly, template based approaches are also used, where the aligned pores of a nanoporous membrane (such as electrodeposited porous alumina) are filled with carbon species through vapour deposition and later graphitised to produce nanotubes. The template membrane is then removed to obtain aligned nanotube arrays.

4.5 PURIFICATION OF CARBON NANOTUBES [21]

Purification of carbon nanotubes generally refers to the separation of carbon nanotubes from other entities. The classical chemical techniques for purification (such as filtering, chromatography, and centrifugation) have been tried, but they have not been found to be effective in removing the carbon nanoparticles, amorphous carbon, and other unwanted species. Three basic methods have been used with limited success for purification of the MWNTs, namely gas phase, liquid phase, and intercalation methods. The current purification procedure follows certain essential steps: preliminary filtration to get rid of large graphite particles; dissolution to remove fullerenes (in organic solvents) and catalyst particles (in concentrated acids); microfiltration; and chromatography to either separate MWNT and unwanted nanoparticles or SWNT and the amorphous carbon impurities. It is important to keep the nanotubes separated in solution, and nanotubes are typically dispersed using a surfactant prior to the last stage in separation.

Generally, a centrifugal separation is necessary to concentrate the SWNT in a low-yield soot before the microfiltration operation, since the nanoparticles easily contaminate the membrane filter. The advantage of this method is that unwanted nanoparticles and amorphous carbon are removed simultaneously and nanotubes are not chemically modified. However 2–3 M nitric acid is useful for chemically removing impurities. Gold particles can also act as catalytic sites for oxygen to attack organic impurities.

Efforts have been made to use separation techniques called size exclusion chromatography (SEC) and cascade filtration. These techniques have yielded good separation between nanotubes and nanoparticles in the samples. SEC is an effective and non-destructive method for the purification and size separation of carbon SWNTs. Both of these techniques rely on the fact that nanotubes have large molecular weights and do not enter small pores, but it is impossible to scale them up to obtain larger samples.

It is now possible to cut nanotubes (SWNT) into smaller segments by extended sonication (agitation using ultrasound) in concentrated acid mixtures. The resulting pieces of broken SWNT (open pipes that are typically a few hundred nanometres in length) form a colloid suspension in solvents. They can be deposited on substrates or further manipulated in solution and functionalised at the ends. It has been shown that sulfur functionalised nanotubes (nanotubes that have been reacted with thionyl chloride and octadecylamine) can be dissolved in organic solvents. Extensions of this technique may in the future be used to separate nanotubes using solvation to produce high-purity, well-defined samples. Functionalising nanotubes with sulfur is valuable since sulfur binds to gold, which is an ideal template for building nanostructures where nanotubes are orientated to form parts of machines or circuits.

4.6 THE PROPERTIES OF NANOTUBES [26–28]

CONDUCTIVITY

There has been considerable interest in the conductivity of nanotubes. As noted earlier, nanotubes with particular combinations of n and m are believed to be conducting and hence metallic. Conductivity has also been shown to be a function of diameter.

Conductivity in multiwalled nanotubes is quite complex. Frank and Poncharalp carefully contacted multiwalled nanotube fibres with a mercury surface [29]. The conductance of MWNTs jumped by increments as additional nanotubes were touched to the mercury surface. This quantised conductance was found in all sizes of nanotubes and is also observed in metal nanowires. Some types of armchair nanotubes appear to conduct better than other metallic nanotubes. Furthermore, interwall reactions of MWNTs were found to redistribute the current over individual tubes across the structure non-uniformly.

Atomic force microscopy tips (Chapter 2.5) have been used to investigate the electronic properties of individual single-walled nanotubes.

There is no change in current across different parts of metallic single-walled nanotubes. However, the behaviour of ropes of semi-conducting SWNTs is different in that the transport current changes abruptly at various positions on the nanotubes.

The conductivity and resistivity of ropes of SWNTs has been measured directly with a technique in which four electrodes have been placed at different parts of the nanotubes (Plate 3). The resistivity of these SWNT ropes was in the order of 10^{-4} ohms per cm at 27°C. This means that the ropes are the most highly conductive carbon fibres known. Measurements showed the current density in the tube was greater than 10^7 A/cm^2. However, this may have been underestimated as theory demonstrates nanotubes could sustain stable current densities as high as 10^{13} A/cm^2.

It has been reported that individual SWNTs may contain defects. Fortuitously these defects allow the SWNTs to act as transistors. Likewise, joining nanotubes together may form transistor-like devices [30]. A single nanotube with a natural junction (that is, where a straight section is joined to a chiral section (Figure 4.8), behaves as a rectifying diode — a half-transistor in a single molecule.

Figure 4.8

A nanotube transistor

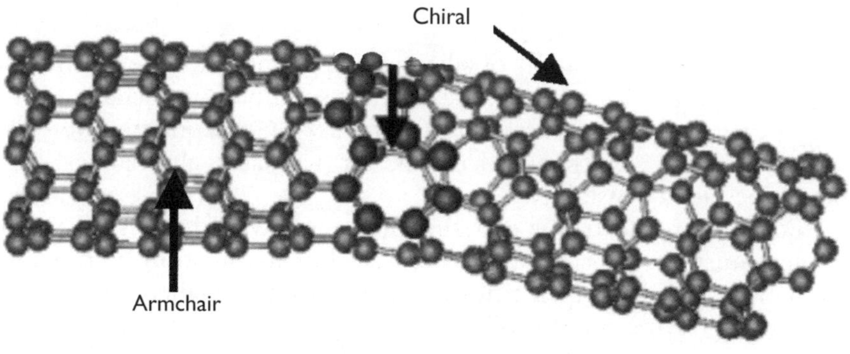

The properties of bent nanotubes have been explored. Nanotubes that are suspended and then deflected from an equilibrium position can be described as springs. Their conductivity is reduced when they are stressed.

STRENGTH AND ELASTICITY

SWNTs are stiffer than steel and are resistant to damage from physical forces. Pressing on the tip of the nanotube will cause it to bend without damage to the tip. When the force is removed, the tip of the nanotube will recover to its original state. Quantification of these effects, however, is rather difficult and an exact numerical value cannot be agreed upon. Using an AFM, unanchored ends of freestanding nanotubes can be pushed out of their equilibrium position and the force required to push the nanotube can be measured. The current Young's Modulus value (Chapter 3.9) of SWNTs is about 1 TPa but this value is disputed and a value as high as 1.8 TPa has been reported. Other workers have also reported different values. A (10,10) armchair nanotube had a Young's Modulus of 640.30 GPa; for a (17,0) zigzag tube it was 648.43 GPa, and a (12,6) gave a value of 673.94 GPa. The differences probably arise through different experimental measurement techniques. Others have shown theoretically that the Young's Modulus depends on the size and chirality of the SWNT, ranging from 1.22 TPa for the (10,0) and (6, 6) to 1.26 TPa for the large (20,0) SWNT. They calculated a value of 1.09 TPa for a generic nanotube. However, when working with different MWNTs, others have noted that modulus measurements of MWNTs using AFM do not strongly depend on the diameter. Instead, they argue that the modulus of MWNTs correlates to the amount of disorder in the nanotube walls. Not surprisingly when multiwalled nanotubes break, the outermost layers break first.

4.7 USES OF NANOTUBES

ELECTRONICS

Continuing miniaturisation of silicon components and fine control of electronic properties at smaller scales may soon pose intractable problems (This is discussed further in Chapter 9). Consequently the electronics industry has begun to look for workable alternatives, and nanotubes are a possibility.

For delicate electronic experiments, single-walled tubes of specific type must be painstakingly separated, but it will soon be possible to manufacture bulk quantities of identical single-walled nanotubes. Indeed, by relying on the ability of the metallic nanotubes to conduct electricity, high currents have been used to burn out the metallic nanotubes from mixtures. Some success has now been achieved in controlling the length of the nanotubes. Scientists have succeeded in hooking up transistor nanotubes and even made two transistor sections along the same nanotube. Circuits have been built by draping a single nanotube over three parallel gold electrodes, adding a polymer between the electrodes and sprinkling potassium atoms on top. The potassium atoms add electrons to the nanotube. The nanotube has been used in a computer circuit to make a logic circuit — a device that registers true or false, as in the binary number calculations of a computer.

Fortunately, not all electronic applications need to be so elegant. Field emission is the property that makes flat-panel displays work. Even mixtures of multiwalled nanotubes are good at field emission, as they emit electrons under the influence of an electrical field. Millions of nanotubes arranged just below the screen could provide each pixel, and several companies around the world are trying to exploit nanotubes in flat-panel displays. Researchers at Samsung in Suwon, South Korea, have a prototype that requires half the power of conventional liquid-crystal displays and the nanotubes appear to meet the 10 000-hour lifetime typically demanded of electronics components.

HYDROGEN STORAGE

Nanotubes may have a role in hydrogen storage [21]. Whilst its energy content on a mass-for-mass basis is better than petrol, hydrogen has difficulty competing with fossil fuel because it is a gas. The target for hydrogen capacity that would interest car manufacturers is about 6.5 percent by weight, regardless of storage medium. Nanotubes can also store helium.

Nanotubes can also be used to store other materials like oxides or metals such as copper. Hence they can be used as nano- test tubes, or the carbon can be removed to produce nano-copper wires for nano-electrical circuits. Nanotubes may also have a use in batteries. Graphite can store lithium ions, the charge carriers for some batteries, but six carbon atoms are needed for every lithium ion. The geometry inherent in bundles of nanotubes may allow them to accommodate more than one lithium ion for every six carbons.

MATERIALS

As noted above, nanotubes are useful as materials because of their high Young's Modulus. The value of carbon fibre is already proven in composite materials, and carbon nanotubes certainly have promise in the same market because of their exceptionally high length-to-diameter ratio, most notably in stress transmission. Nanotube fibres have now been produced, and although they are weak in strength they have flexible properties unlike traditional graphitic fibres [31]. Conventional carbon fibres need to be prepared at high temperatures in which the graphitic planes are aligned, but nanotube fibres can be prepared at room temperature from colloidal solution.

SWNTs can deform reversibly when electrochemically charged [32]. Thus the electrical properties of SWNTs can be used to generate mechanical motion from electrical energy. The extreme sensitivity of nanotube electronic properties to the presence of a trace element also enhances their potential as gas sensors, and for other types of sensors as well.

Figure 4.9

Nanogears a) and b) to effect different translational motions and
c) to act as a bearing or switch. Reprinted with permission of the NASA Ames Centre for Nanotechnology.

a

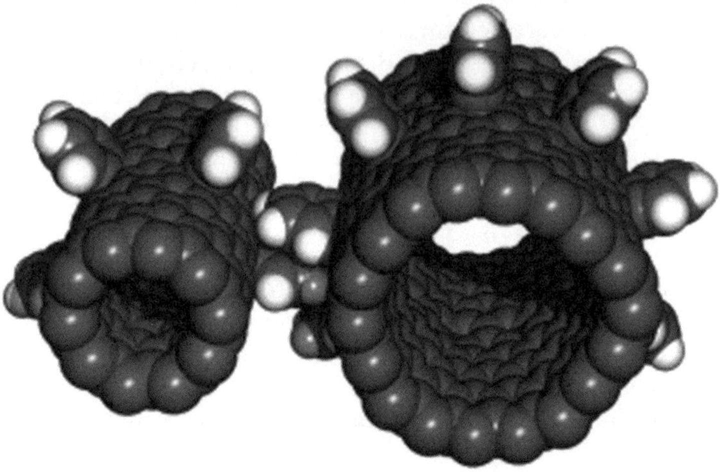

b

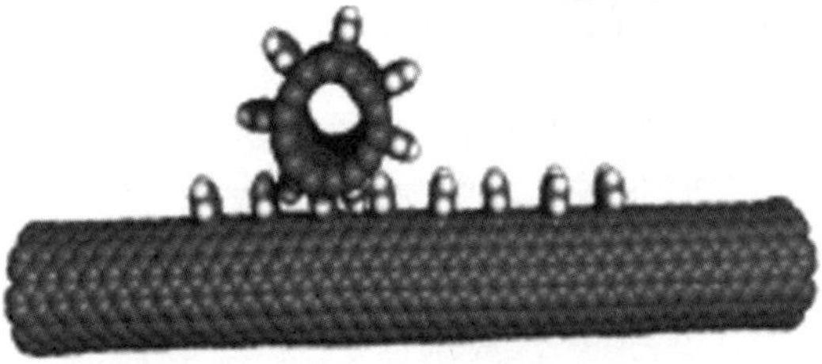

c

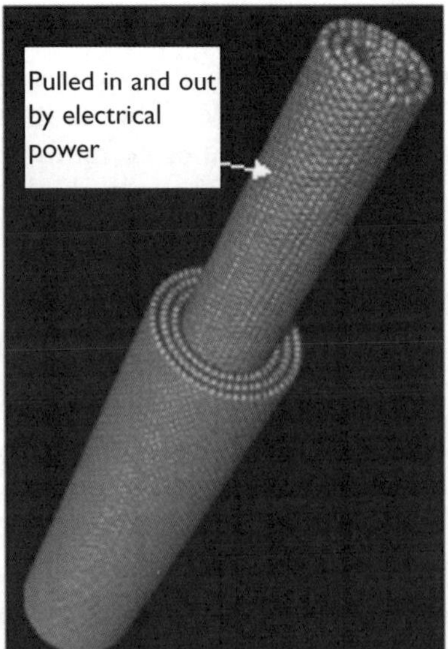

MECHANICAL MACHINES

Substituted nanotubes are the nanogears of Drexler's fanciful and inspirational mechanical nanomachines [33, 34]. Nanotubes suitably substituted with various structures can act as axles in nanomachines. It may be possible to gear different nanotubes together to translate different rotational motion (Figure 4.9) or change the direction of that motion (Figure 4.9). This could be done by building gear teeth (substituents) on the nanotubes. Combinations of nanotubes and fullerenes have been conceived as molecular pumps or pistons. This is not science fiction. The first pump has been made (Figure 4.9c). Researchers from the University of California, Berkeley have developed what can be called the first nano-bearings by attaching one end of a multiwalled carbon nanotube to a stationary gold electrode. They then used a nanomanipulator to pull the inner tubes out. Using a scanning electron microscope, the team could watch how the inner core was pulled back inside by intramolecular van der Waals forces, thus making the MWNTs act like a bearing [35]. Another extremely interesting application of the bearing is as a so-called nanoswitch. By applying a voltage to the bearing, researchers can very rapidly force the central tube to slide out. By moving the inner tube of a multiwalled nanotube, a piston is formed.

Drexler, Merke and collaborators [36, 37] have been modelling designs for mechanical systems at the nanoscale level (1 to 100 nm) that could in principle be manufactured using structures like the nanotubes discussed above. At the nanoscale, one can no longer think of the material as a continuum whose properties change continuously as it is cut and shaped. Rather one has to consider that it is formed from discrete atoms. This requires the analysis of the material science properties of atoms rather than wood, metal or plastic. They are related but the relationship is complex. Recall that in nanotechnology the surface to bulk ratio is large. In effect, the considerations are those in extrapolating from three-dimensional to two-dimensional space.

Models can be studied with all the precision that would be used to study conventional chemical models. They can be constructed with all the detail that would be used in mechanical engineering to create such a device as a macro gear. These tiny structures would have molecular weights and molecular volumes. A typical Drexler gear structure has 3557 atoms and a molecular weight of around 5000. The problem is that a careful balance must be achieved between having the gear and race atoms so close on the one hand that short-range repulsive van der Waals interactions cause the gear to freeze, and on the other hand having them too far apart so that the gear teeth slip past each other. It is possible to balance these in the static system, but the dynamic motions lead to fluctuations (vibrations) that allow slippage. This does not mean that these types of nanomachines will not work. It means that the forces and mechanical engineering need to be modelled using new principles. Indeed research so far shows how the gears will need to be fine-tuned for these effects. It is worth mentioning in passing that it is not possible to use a lubricant. The lubricant molecules are as big as the gears themselves!

Molecular dynamics simulations use simple engineering principles that one would use to understand operations as simple as a rotating drum or other dynamic phenomena. In the course of rotating a gear the angular momentum is slowly transferred to the output unit — the second gear. For nanotechnology the time units are very small, not microseconds but picoseconds (10^{-12} seconds). However the dynamics are the same. For a gear that rotates 360° in one picosecond, the other gear may only rotate by 180°. Over the cycle the initial angular kinetic energy partially dissipates through conversion into thermal energy via atomic collisions. However there are differences between this system and a simple macroscale system. The problem here is the discreteness of the atoms. A macroscopic system is treated as a continuum. To make a gear one can design the shape of the gear teeth and the shape of the opposing teeth of the face to have exactly the same spacing as the gears turn. It is then only necessary to design the casting or injection molding system with appropriate tolerances to achieve good performance. Tolerances of micrometres are adequate for macromachines. However the surfaces of atoms are round, soft, flexible and bumpy.

Thus, nanogears must be designed differently from macroscopic gears. The shape is envisaged to be a V design, because a gear tooth in the x,y plane cannot be atomically smooth in the z direction. The v-shaped gear tooth in the z direction nestles within a v-shaped notch in the race to maintain stability in the z direction as the teeth contact in the x,y plane. Pictorially this looks like a gear with round bites in which the atoms of the other gear fit. Other factors, such as the polarisability of the electron clouds, electronegativity, electron density and atomic compressibility may all be important; and van der Waals forces will make the gears stick together as they rotate. The researchers in this field say it's a bit like making gears out of toffee.

SPACE ELEVATORS

The overriding factor in the design of aircraft or spacecraft that need to enter the planet's atmosphere is the weight to power ratio. Smaller, lighter craft are cheaper to make air- or space-borne. Carbon nanotube structural materials can radically reduce structural mass, miniaturise electronics, and reduce power consumption. The use of atomically precise materials and components should shrink most other components (Chapter 3.9). Thermal protection of spacecraft is crucial for atmospheric re-entry and for other tasks involving high temperatures. Carbon nanotubes, like graphite, should withstand high temperatures.

As noted above, carbon nanotubes have a Young's modulus of at least one terapascal (pascals x 10^{12}), which is also of benefit in withstanding aeronautical strains, including the strains of atmospheric re-entry.

Some researchers have investigated the possibility of constructing a space elevator [38, 39]. This would consist of a cable extending from the Earth's surface into space with a centre of mass at a geosynchronous altitude so that it does not drag behind. The cable would be attached to a satellite. Goods and people could ascend and descend along the cable. Gravity would bring them down but other energy sources will be needed to send them up. But what happens if the cable breaks? Thousands of kilometres of cable will drop to Earth, causing serious damage.

The point of maximum stress occurs at an altitude where the angular velocity of the cable and satellite is greater than at the Earth, so the cable must be thickest there and taper as it approaches Earth. These taper factors have actually been measured, and for steel the pinnacle in space must be over 10 000 times wider than the base. Given that the base must be strong enough to support the structure, it is clear that steel is useless for such an application. For diamond, the taper factor is 21.9 [39]. However, diamond is brittle. Carbon nanotubes have a tensile strength similar to diamond, but bundles of these nanometre-scale radius tubes shouldn't propagate cracks nearly as well as the diamond tetrahedral lattice. Thus, if we can overcome the difficult problem of developing a molecular nanotechnology capable of making nearly perfect carbon nanotube systems approximately 70 000 kilometres long, the first serious problem of a transportation system capable of truly large-scale transfers of mass to orbit can be solved. To say this is a long way off is an understatement.

WHAT YOU SHOULD KNOW NOW

1 Carbon nanotubes come in three forms: chiral, zigzag and armchair. They come in a variety of diameters and in single and multiwalled forms. The different types of nanotubes are best represented using vectors, which describe the rolling process that occurs when a graphite sheet is transformed to a tube.

2 Carbon nanotubes can be made by a variety of processes. The most common are laser ablation, plasma arcing and chemical vapour deposition. Various materials can be used to make the nanotubes, including coal, graphite and various compounds containing transition metals. Cobalt is particularly useful as an additive for making single-walled nanotubes.

3 Carbon nanotubes with certain defects can behave as transistors. Others can conduct electricity and have been made into simple logic

circuits. Non-metallic nanotubes can be separated from metallic tubes using a high oxidising current.

4 Carbon nanotubes can store hydrogen and may also be useful with lithium as batteries. They have unusual tensile strength and may make valuable building materials if manufactured cheaply in quantity.

5 Carbon nanotubes have been proposed as nanomachines. They make good nanotweezer tips for electron microscopy and multi-walled nanotubes can be pulled in and out like pistons. Calculations have been made to understand their role as mechanical nanogears.

4.8 EXERCISES

1 *Find a soccer ball and count the number of five-membered rings. Count the intercepts to confirm there are 60 carbons in buckminsterfullerene. Count the number of bonds.*

2 *Use chicken wire to build the three types of nanotubes: armchair, chiral and zigzag. Determine the vector nomenclature of the particular structure you build. Alternatively try this experiment using a rolled transparency with chicken wire markings drawn on it.*

3 *Read the Nobel Laureates' accounts of their discovery of fullerenes:*

Curl RF (1997) Dawn of the Fullerenes: conjecture and experiment (Nobel Lecture). Angew. Chem. Int. Ed. Engl. *36(15): 1566–76.*

Kroto H (1997) Symmetry, space, stars, and C_{60} (Nobel Lecture). Angew. Chem. Int. Ed. Engl. *36(15): 1578–93.*

Smalley RE (1997) Discovering the Fullerenes (Nobel Lecture). Angew. Chem. Int. Ed. Engl. *36(15): 1594–1601.*

4 *Find out all you can about mechanical gearing and the equations that relate to angular momentum in gears. Change the time and dimensions in these equations to picoseconds and nanometres and do some calculations.*

5 *Think how you might go about separating a mixture of different geometry nanotubes. Collins and coworkers removed conducting nanotubes. Collins et al,* Science, *2001, 292, 706–709,* Phys Rev Letters, *2001, 86, 3128–3131. These references might give you some ideas.*

6 *Find out all you can about Raman spectroscopy.*

4.9 REFERENCES

1 Kroto HW, Allaf AW & Balm SP (1991) *Chemical Reviews* 91: 1213–35.

2 Kraetschmer W, Lamb LD, Fostiropoulos K & Huffman DR (1990) *Nature* 347: 354–58.

3 Haufler RE, Coceicao J, Chibante LPF, Chai Y, Byrne NE, Flanagan S, Haley MM, O'Brien SC, Pan C, Xiao Z, Billups WE, Ciufolini MA,

Hauge RH, Margrave JL, Wilson LJ, Curl RF & Smalley RE (1990) *Journal of Physical Chemistry* 94: 8634–40.
4 Dresselhaus MS, Dresselhaus G & Eklund PC (1996) In *Science of Fullerenes and Carbon Nanotubes.* Academic Press, San Diego CA, USA, p. 69.
5 Iijima S, Ichihashi T & Anodo Y (1992) *Nature* 356: 776–78.
6 Journet C, Maser WK, Bernier P, Loiseau A, Delachapelle ML, Lefrant S, Deniard P, Lee R & Fischer JE (1997) *Nature* 388: 756–58.
7 Thess A, Lee R, Nikolaev P, Dai HJ, Petit P, Robert J, Xu CH, Lee YH, Kim SG, Rinzler AG, Colbert DT, Scuseria GE, Tománek D, Fischer JE & Smalley RE (1996) *Science* 273: 483–87
8 Rao AM, Richter E, Bandow S, Chase B, Eklund PC, Williams KA, Fang S, Subbaswamy R, Melon M, Thess A, Smalley RE, Dresselhaus G & Dresselhaus MS (1997) *Science* 275: 187–91.
9 Pang LSK, Vassallo AM & Wilson MA (1991) *Nature* 352: 480.
10 Wilson MA, Pang LSK & Vassallo AM (1992) *Nature* 355: 117–18.
11 Taylor GH, Fitzgerald JD, Pang LSK & Wilson MA (1994) *Journal of Crystal Growth* 135: 157–64.
12 Pang LSK, Prochazka L, Quezada R, Wilson MA, Pallasser R, Fisher KJ, Fitzgerald JD, Taylor GH, Willett GD & Dance IG (1995) *Energy and Fuels* 9: 38–44.
13 Pang LSK, Wilson MA, Taylor GH, FitzGerald J & Brunckhorst L (1992) *Carbon* 30: 1130–32.
14 Fitzgerald JD, Taylor GH, Brunckhorst L, Pang LSK & Wilson MA (1993) *Carbon* 31: 240–44.
15 Wang Y (1994) *J. Amer. Chem. Soc.* 116: 397–98.
16 Ajayan PM, Lambert JM, Bernier P, Barbedette L, Colliex C & Planeix JM (1993) *Chem. Phys. Letters* 215: 509–14.
17 Kalman JF, Nordlund C, Patney HK, Evans LA & Wilson MA (2001) *Carbon* 39: 137–44.
18 Pallasser, R Pang LSK, Prochazka L, Rigby D & Wilson MA (1993) *J. Am. Chem. Soc.* 115: 11634–35.
19 Wilson MA, Moy A, Rose H, Kannangara GSK, Young BR, McCulloch DG & Cockayne DJH (2000) *Fuel* 79: 47–56.
20 Yacaman MJ, Yoshida MM, Rendon L & Santiesteban JG (1993) *Appl. Phys. Lett.* 62: 202–204 and 657–59.
21 Ding RG, Lu GQ, Yan ZF & Wilson MA (2001) *Journal of Nanoscience and Nanotechnology* 1: 1–23.
22 Fonseca A, Hernadi K, Nagy JB, Bernaerts D & Lucas AA (1996) *J. Mol. Catal. A: Chemical* 107: 159–68.
23 Colomer J-F, Bister G, Willems I, Konya Z, Fonseca A, Van Tendeloo G & Nagy JB (1999) *Chem. Commun.* 1999: 1343–44.
24 Kong J, Soh HT, Cassell AM, Quate CF & Dai HJ (1998) *Nature* 395: 878–81; see also L Dai et al. (1999) *J. Chem. Phys. B.* 103: 4223–27.
25 Ajayan PM, Stephan O, Colliex C & Trauth D (1994) *Science* 265: 1212–14.
26 Iijima S & Endo M (guest eds) (1995) Special issue on carbon nanotubes. *Carbon* 33.
27 Subramoney S (1998) *Advanced Materials* 10: 1157–73.
28 Yakobson BI & Smalley RE (1997) *American Scientist* 85: 324–30.

29 Frank S, Poncharalp S, Wang ZL & de Heer WA (1998) *Science* 280: 1744–48.
30 Fuhrer MS, Nygard J, Shih L, Forero M, Yoon Y-G, Mazzoni MSC, Choi HJ, Ihm J, Louie SG, Zettl A & McEuen PL (2000) *Science* 288: 494–97
31 Poulin P, Vigola B, Penicaud A & Coulin C (1999) CRNS French Patent 0002272.
32 Baughman RH, Cui CX, Zakhidov AA, Iqbal Z, Barisci JN, Spinks GM, Wallace GG, Mazzoldi A, De Rossi D, Rinzler AG, Jaschinski O, Roth S & Kertesz M (1999) *Science* 284: 1340–45.
33 Drexler KE (1981) *Proceedings of the National Academy of Sciences USA* 78 (September): 5275–78.
34 Drexler KE (1990) *Engines of Creation*. Fourth Estate, London, 296 pp.
35 Cumings J & Zettl A (2000) *Science* 289: 602–604; see also p. 505.
36 Han J, Globus A, Jaffe R & Deardorff G (1997) *Nanotechnology* 8: 95–102.
37 Globus A, Bauschlicher C, Han J, Jaffe R, Levit C & Srivastava D (1998) *Nanotechnology* 9: 192–99.
38 Isaacs JD, Vine AC, Bradner H & Bachus GE (1996) *Science* 151: 682–83.
39 Drexler KE (1992) *Journal of the British Interplanetary Society* 45: 401–405.

5

MOLECULAR MIMICS

This chapter describes the types of molecules chemists synthesise to use in nanotechnology. These molecules have a common purpose in that they seek to act as moving components of nanomachines. You will learn how they are anchored on surfaces. Some of these machines may be used to make chemical computers.

Many molecules bear an uncanny resemblance to things we have engineered at the macro scale or that occur in nature. You may have tried to solve a puzzle where two nails or chain links are tied together in some way, and they must be separated. Some molecules have similar properties and they may be useful as molecular machines in nanotechnology. A molecular level machine can be defined as an assembly of a distinct number of molecular components that are designed to perform machine-like movements as a result of some input [1]. In common with their macroscopic counterparts, a molecular machine is characterised by several criteria. These are:

- the type of energy needed to make it work
- the nature of the movement during which work is done or energy is transformed
- the way it can be controlled
- the ability to repeat its operation
- the time needed for a useful action.

5.1 CATENANES AND ROTAXANES [1]

In the last chapter we dealt with the possibility of building nanomachines from substituted single walled nanotubes and multiwalled nanotubes. Substituted single walled nanotubes mimic gears when intermeshed, while multiwalled nanotubes, when used as moving cylinders, mimic pistons. However there are a large number of other possible molecular nanomachines. Some useful structures are joined bracelets linked through each other. They are called catenanes, and their polymers, which consist of catenanes chained together like Christmas decorations, are called polycatenanes (Figure 5.1a). Just as Christmas chains must be ripped apart to separate them, these molecules cannot be broken into their constituent parts without physically breaking one of the macromolecules. The naming of these compounds is quite simple. In assigning nomenclature a number is placed in the front to describe the number of rings. Thus a [2] catenane has two rings and a [3] polycatenane has three rings. Two-bracelet catenanes ([2] catenanes) are useful because each bracelet can be rotated around the other.

Another useful structure that can be modelled at the molecular level is the axle and collar. In this structure a central rod holds the axle to the superstructure through a collar, and dumbbell structures are at each end to stop the collar falling off. The molecular equivalents of these structures are called rotaxanes (Figure 5.1a). Rotaxanes are of particular interest because the collar structure can often be moved by some external force and used as a molecular switch [1].

5.2 MOLECULAR SWITCHES

The principle behind using rotaxanes as molecular switches is the ability to move the collar under some external influence. If the collar can be switched back and forth (Figure 5.1b) from each site 'S' then it acts as a switch. The number of sites at which the collar can stop can be increased so that there are a number of stations where it may decide to stop. These structures then become molecular railway lines and the collar is in effect a train. It is possible to put more than one train on the line and also design trains that stop only at particular stations. The idea is to have collars that will stop only at certain sites depending on different environmental changes. The structures that can stop a train or collar, or make it move, are variable and depend on the nature of the collar or track. There are a variety of different types of collars but nitrogen-containing aromatic compounds and carbohydrates are popular, primarily because they are easier to make. There are a number of desired phenomena for collar movement, however one cannot always design a rotaxane for such purposes. Preferred train departure signals are biological materials, because these substances are sensitive detectors for cancer or biochemical imbalances, but the easiest responses are to

Figure 5.1

The structure of rotaxanes and catenanes (a) and the mechanism of rotaxane shuttling (b).

a

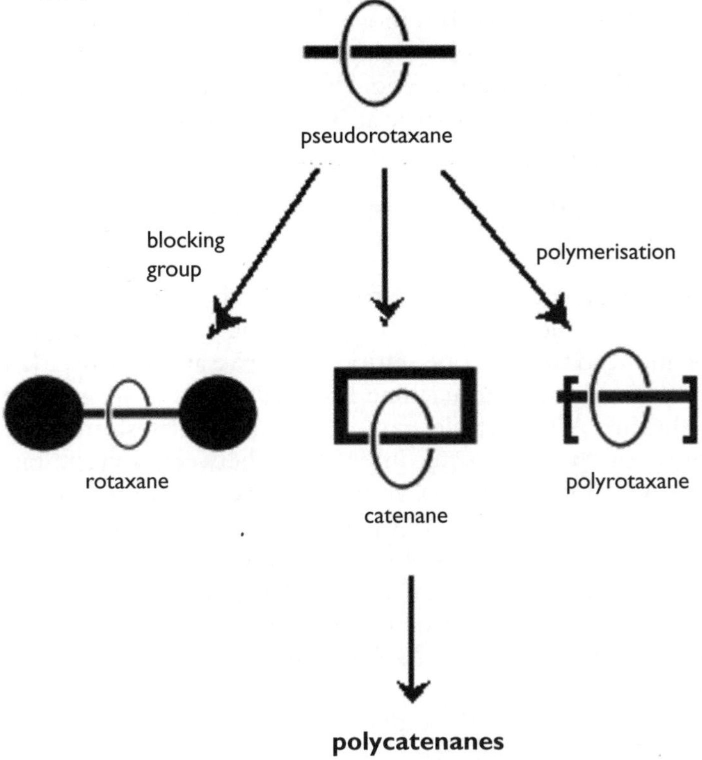

b

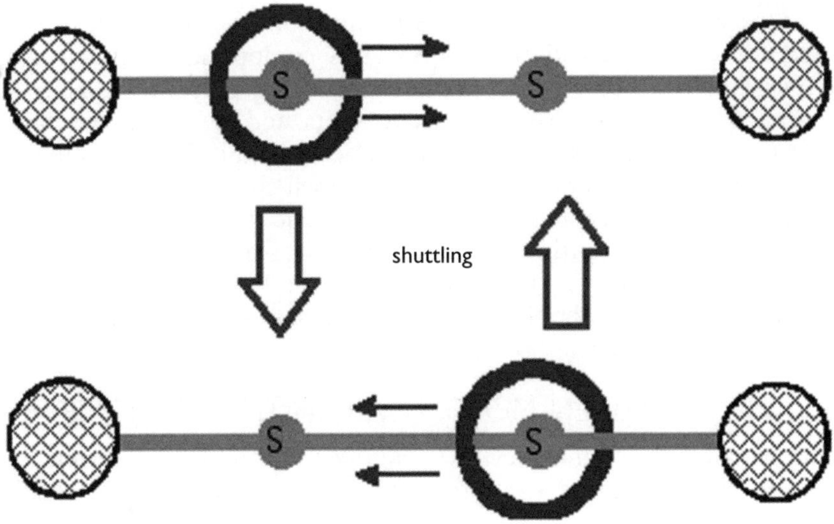

electrons, light or heat. A number of useful sensors can be generated in this way particularly if these switches can be made to function in tandem on a larger scale. If a large number of rotaxanes are aligned on a surface, then when all the collars move they can influence macroscopic properties. Thus, a molecular change can be transmitted to a measuring device that can detect the change. This then becomes a nanomachine that is easy to observe at the level at which human beings can observe and control.

5.3 THE ELECTRON DRIVEN MOLECULAR SHUTTLE SWITCH

If part of the rotaxane can be made to lose electrons (called oxidation) or gain electrons (called reduction), this can be an agent by which a train can be induced to leave one station and arrive at the next [2]. This is an electrically driven nanomachine because electrons are needed to move the shuttle. A typical electron driven molecular shuttle switch works because of charge transfer interactions between a cyclic bipyridinium unit, which forms the collar and the axle. Figure 5.2 shows an example of this type of switch. An oxidisable benzidine unit forms one station (station 1) and a biphenol group forms the other station (station 2) on a rotaxane rod. When the molecule is in its electron rich (unoxidised) state the train shuttle resides mainly on the benzidine station (middle structure Figure 5.2). On removal of an electron (oxidation) from the benzidine station, the train shuttle moves along the rotaxane rod to reside over the biphenol station (top structure Figure 5.2). The process is reversible, and when an electron is returned to the benzidine station the shuttle moves to the benzidine station again. Thus, if connected to a molecular wire the structure can act as a molecular switch operated by electrical current.

This structure is not unique. A number of other molecules can be used for oxidation and reduction. Thus compounds terminated by tetraarylmethane-based stoppers and possessing polyether chains and central 9,10 or 2,6-dioxyanthracene units and a 1,4-dioxybenzene structure have been synthesised and operated [1]. The rotaxane has been prepared by interlocking these dumbbell-shaped compounds with a bipyridinium-based tetracationic cyclophane collar [1]. The central 9,10 or a 2,6-dioxyanthracene units and a 1,4-dioxybenzenestructure are the train stations. The cyclophane can be displaced from the dioxyanthracene to the 1,4-dioxybenzene station by electrochemical removal of electrons.

5.4 THE pH DRIVEN MOLECULAR SHUTTLE SWITCH

The benzidine biphenol device in Figure 5.2 can also be driven by a change in pH because the nitrogens can be protonated. In alkaline solution, such as pyridine, the shuttle resides mainly on the benzidine station (Figure 5.2, middle position), but when an acid is added (TFA, trifluoroacetic acid) the shuttle moves along the rotaxane rod to reside over the biphenol station (Figure 5.2, bottom). This molecular switch could be very sensitive in detecting molecular acidity imbalances.

5.5 THE LIGHT DRIVEN MOLECULAR SHUTTLE SWITCH

Some double bond structures will transpose from *trans* to *cis* and *cis* to *trans* stereochemistry under the influence of light, sometimes in the visible range. Others change structure in the presence of ultraviolet light. Nakashima and coworkers [3] have invented a light driven molecular switch based on rotaxanes. The light driven molecular switch structure consists of an azobenzene molecule connected to dipyridyl units with methylene spacers and 2,4–dinitrobenzene moieties as end caps with a cyclodextrin sheath around the outside (Figure 5.3a). A cyclodextrin is a polymeric carbohydrate molecule that has the shape of a fat washer and hence makes an excellent collar.

The azobenzene molecule can be converted between two forms (*trans* and *cis*) with light. In the *trans* form the cyclodextrin sits at the azobenzene structure (Figure 5.3) but moves to surround the methylene structure in the *cis* form. This is between the N^+ and the O atoms in the molecular structure shown in Figure 5.3. The structure can thus operate as a light controlled molecular electric switch.

Like other molecular shuttles, light driven *cis-trans* shuttles are not unique. A device has been put together to form a rotaxane, where the active unit is a π electron-donating macrocycle polyether, namely bis–*p*-phenylene-34-crown ether and a dumbbell shaped component that contains a ruthenium polypyridine complex as one of its stoppers (Figure 5.3b). A *p*-terphenyl type ring system is used as a rigid spacer and two 4,4'-bipyridinium units are used as π electron accepting stations. A tetraarylmethane group is used as the second stopper. Here the switch operates between the two bipyridinium stations, because ruthenium 2^+ complexes of polypyridine ligands exhibit photo induced electron transfer processes and assist the transfer. In effect the train is remotely driven from the ruthenium control box.

Temperature controlled (-140 to 50°C) molecular shuttles based on rotaxanes can also be built if the bond strengths which hold the train are relatively stronger at one site than another. This is significant when bonds are weak. Important factors are hydrogen bonding, charge transfer interactions, electrostatic and π-π stacking. Ideally, if the collar can be made with two binding sites, these can be used to allow differential bonding at two different temperatures. It should therefore be possible in principle to produce a rotaxane that responds to at least three different stimuli: light, heat and pH.

Figure 5.2

Electrically driven molecular switch.

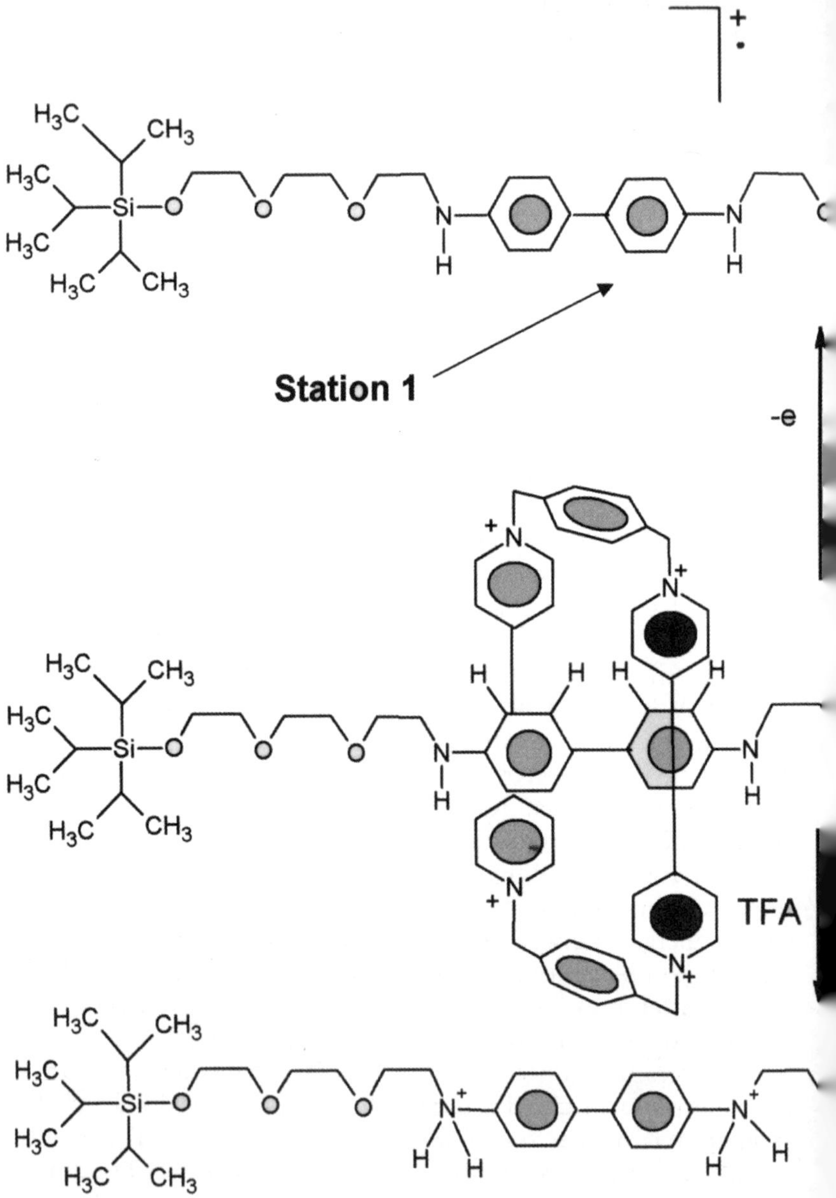

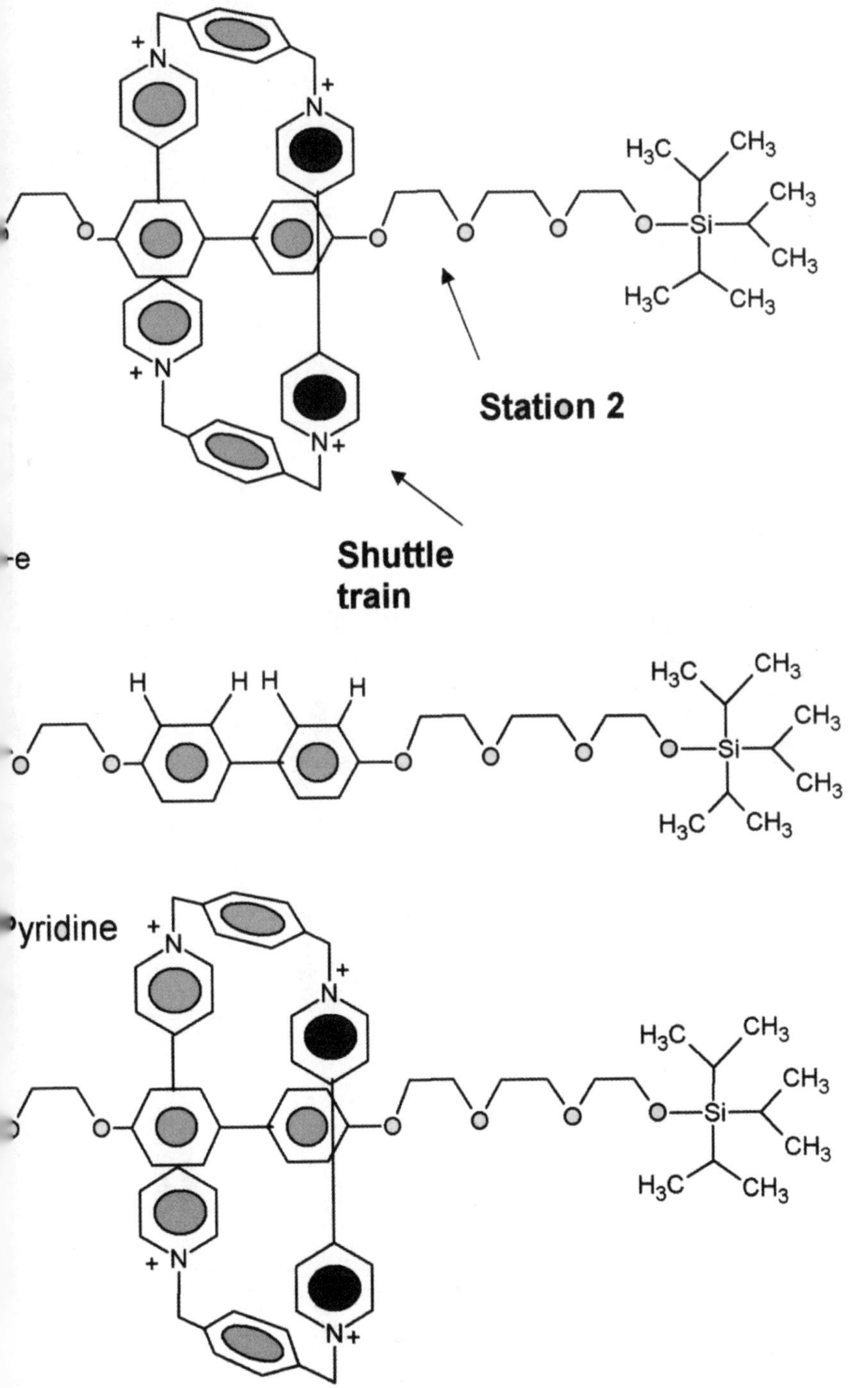

Station 2
Shuttle train
yridine

Figure 5.3

a) Azobenzene light driven molecular switch,

b) The active unit is a π-electron-donating macrocycle polyether, namely bis-*p*-phenylene-34-crown ether, and a dumbbell shaped component that contains a ruthenium polypyridine complex as one of its stoppers.

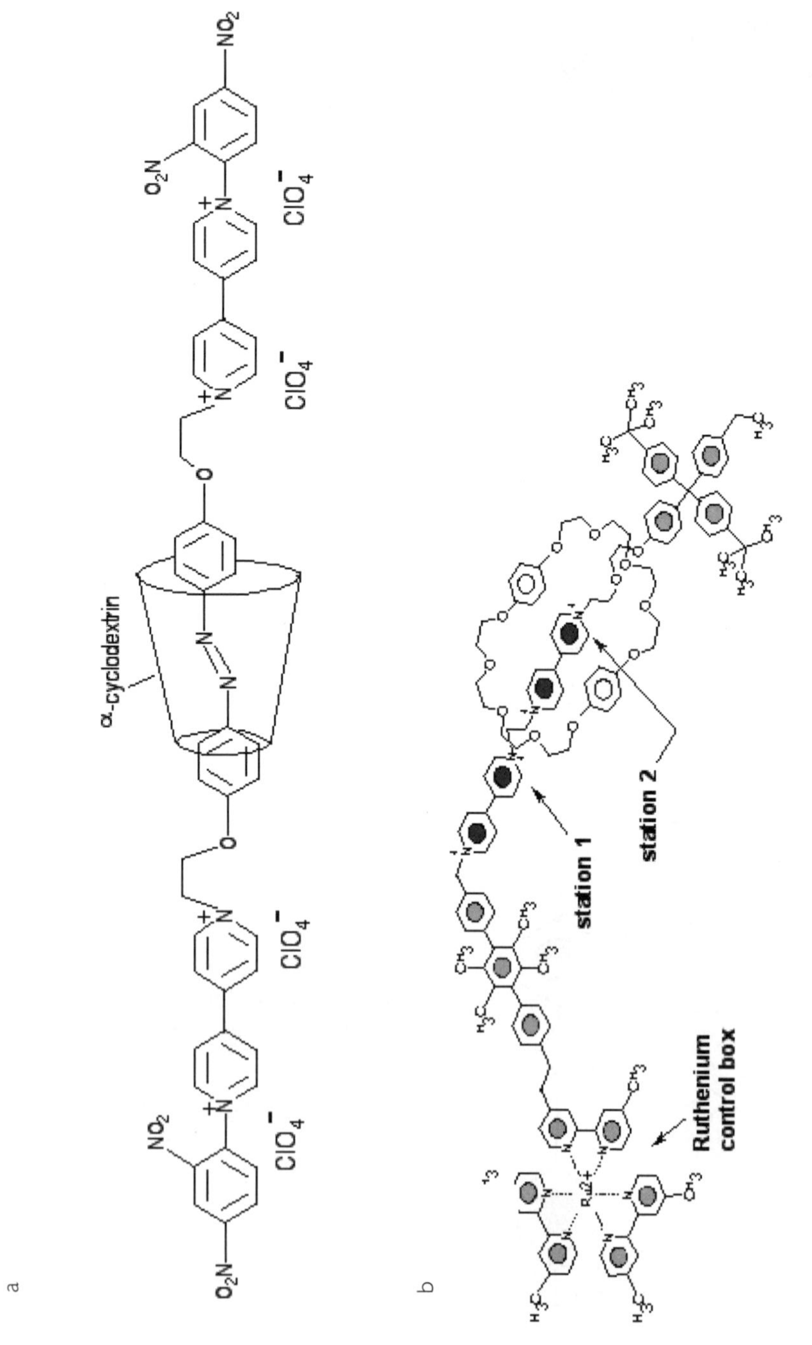

5.6 SYNTHESIS OF ROTAXANES AND CATENANES

Rotaxanes are often produced by the statistical approach. The statistical approach relies on the chance of forming a macromolecule by threading one molecule through another [1]. The yields of desirable product not surprisingly, are normally low. However, if a cyclodextrin is used as a collar in the production of rotaxanes, the structure self assembles, because the cyclodextrin readily encompasses and loosely binds a reactant to form the rod in its structure. Sometimes a blocking group is put on to stop the collar falling off or moving around before putting on the two end caps by binding one side to a resin (Figure 5.4a). However the end caps can either be joined afterwards, or sometimes at the same time as the collar is assembled with the rod.

Catenanes are made in a different way, some of which are rather clever and based on the principle of cutting up a pretzel. For example it is possible to break up a molecule that has a twist in it to produce a catenane. Figure 5.4b shows a molecule that is bound across the ring in several places by joining groups. By swapping chains during cleavage a twist can be generated, which on breaking of the initial bonds forms either a catenane or a trefoil knot (last structure Figure 5.4b), depending on the number of twists. Catenanes can also be produced by a directed method in which an initial link is installed to hold the parts together and the molecule is then built around this. Subsequently the template linkage is removed. Figure 5.5 is a diagram of the process. Figure 5.6 is a real example of how a catenane is formed and Figure 5.7 is a real example of how a rotaxane is formed.

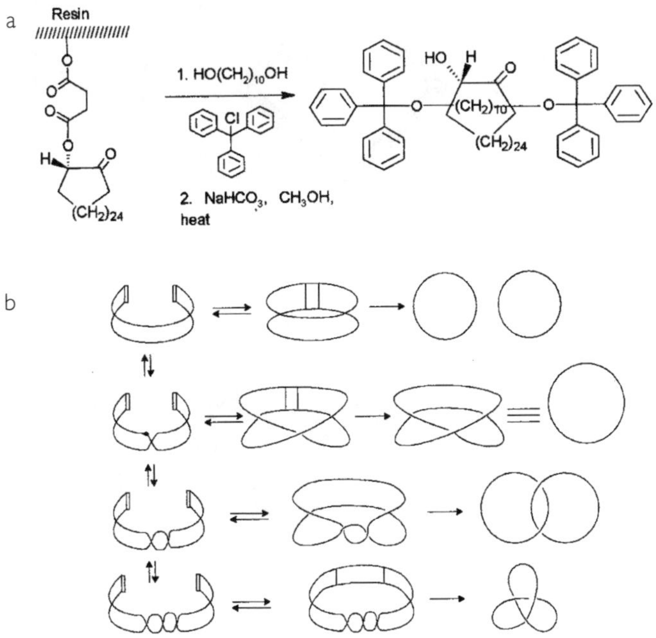

Figure 5.4

Synthetic routes

a) Binding synthesis route to rotaxanes.

b) Synthesis of [2] catenane and a trefoil knot by cleaving links in twisted strip-like molecules. Amabilino DB and Stoddart JF (1995) *Chemical Reviews* 95: 2725–828.

Figure 5.5

Diagrammatic representation of direct synthesis of catenanes involving template linkages.

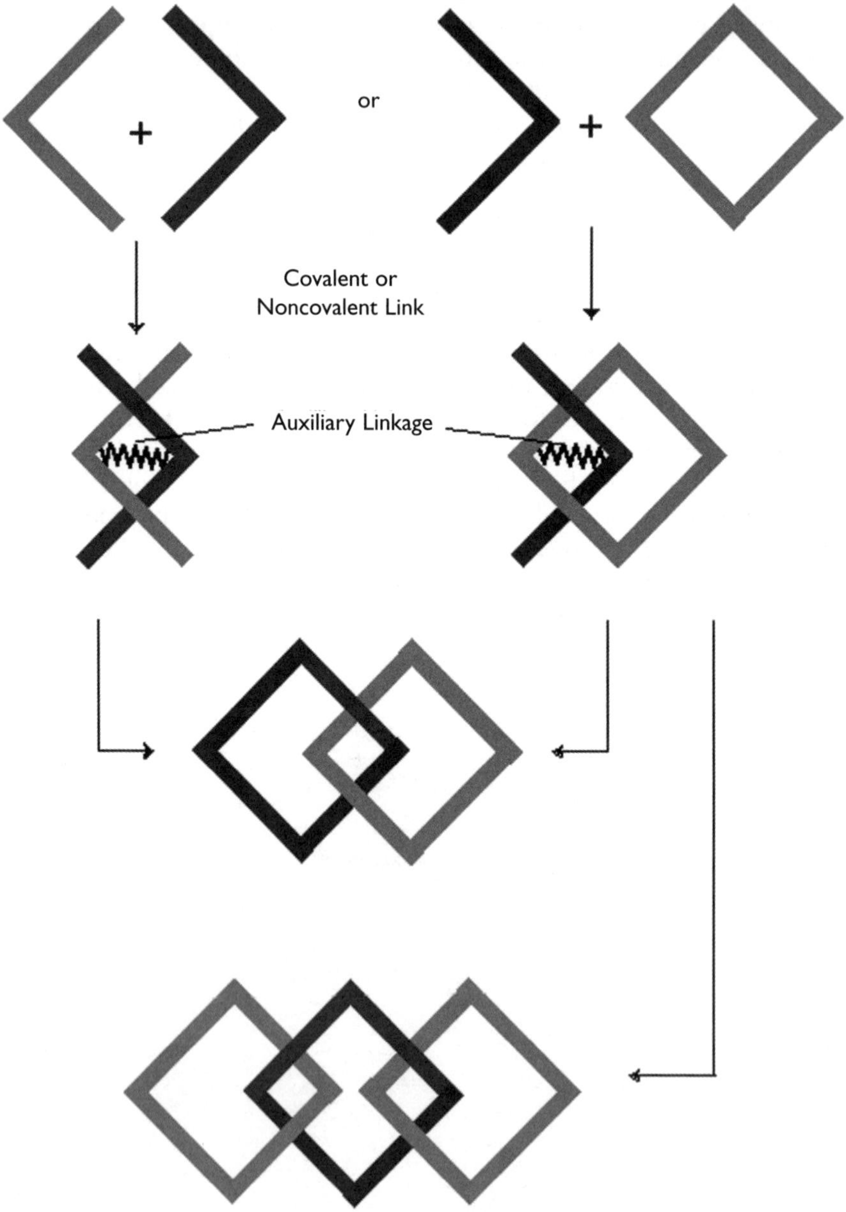

Figure 5.6

Actual representation of direct synthesis of catenanes involving template linkages. Amabilino DB and Stoddart JF (1995) *Chemical Reviews* 95: 2725–828.

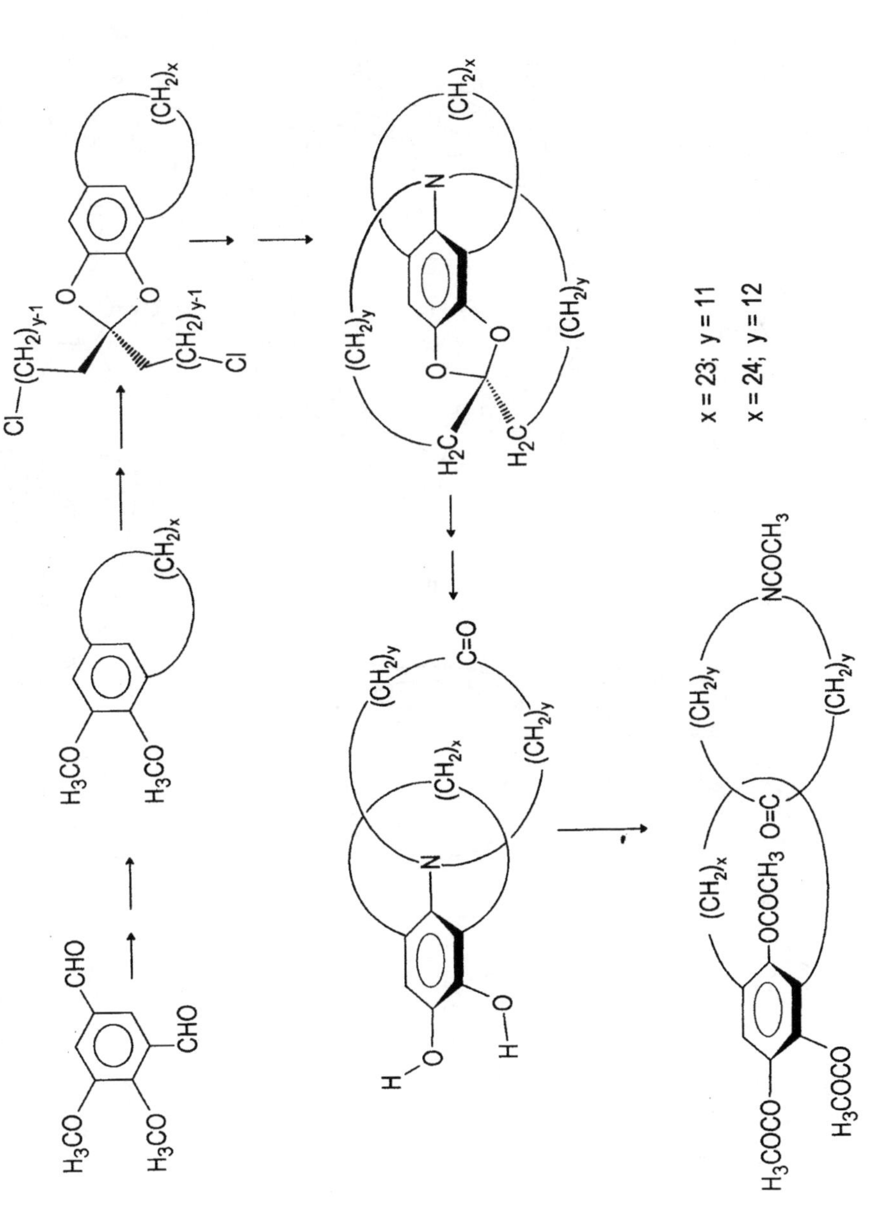

Figure 5.7

The self-assembly by a threading procedure of a [2] rotaxane incorporating a cyclodextrin. Amabilino DB and Stoddart JF (1995) *Chemical Reviews* 95: 2725–828.

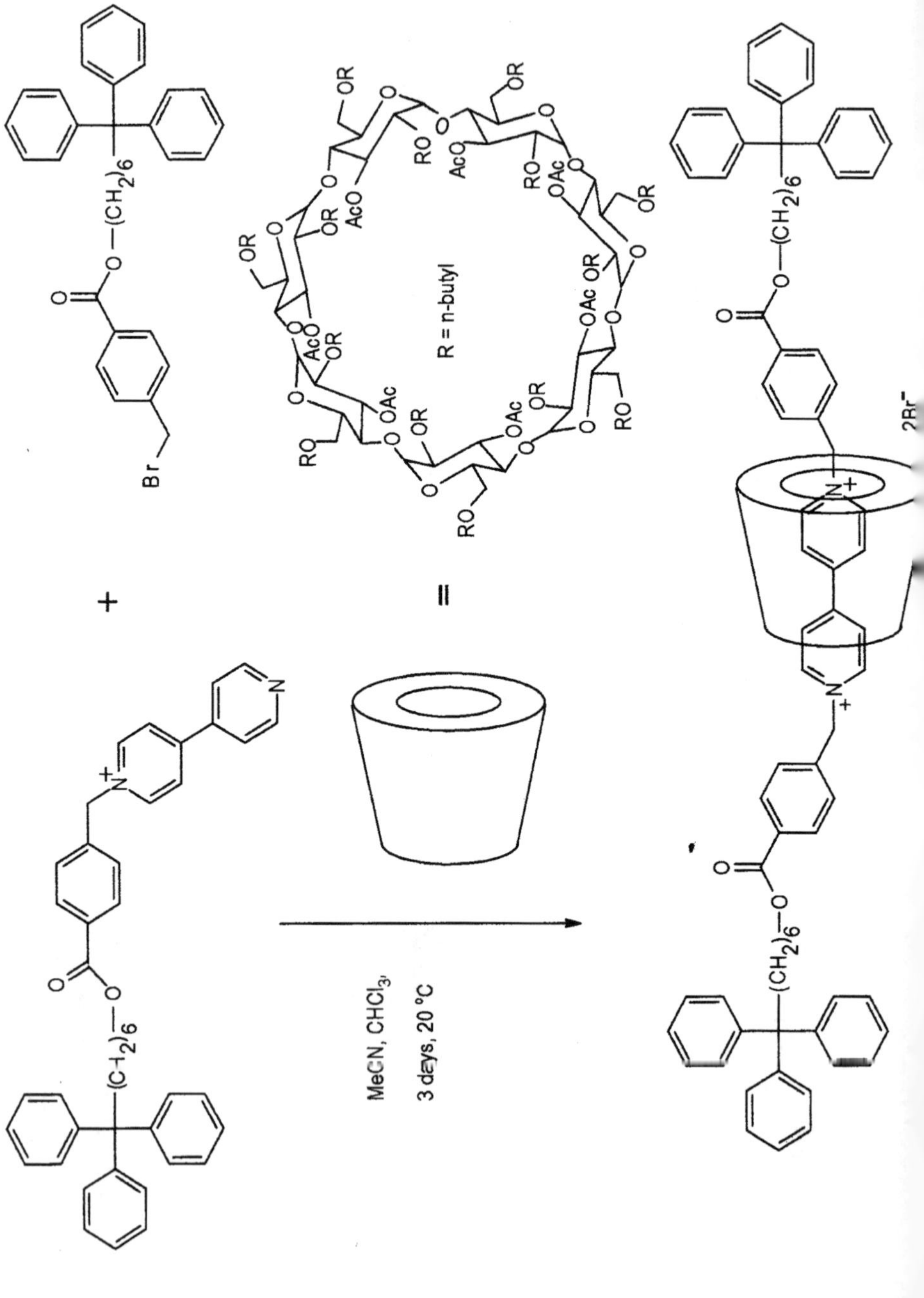

5.7 ROTAXANES AND MOLECULAR COMPUTERS

Logic gates are the counting devices for computers. They are based on microsilicon devices called silicon chips. Silicon chips are dealt with in more detail in Chapter 8. It does not take much imagination to see that a row of rotaxanes could be used as an abacus and hence as a molecular-based logic gate. Indeed it has been shown that they can do exactly the same task as silicon. Rotaxane counting devices are called chemical computers. They do not suffer from the packing defects that exist in surface based technology. Unlike silicon technology each molecule is a counting device. In a mole of rotaxane there are Avogadro's number of counting units; thus even a thousandth of a mole of rotaxane can undergo computations at a level undreamed of by conventional devices. Molecular computers hold the promise of being far less expensive and much smaller and faster than today's silicon-based computers. In principle, it is possible that these machines will perform 100 billion times better than a current old fashioned chip machine. While silicon based chips are not going to improve by much more than a factor of ten, with molecular computers we can potentially get the computational power of 100 workstations on the size of a grain of sand. Molecular electronic based computers would also have vastly reduced power consumption. In addition, such computers would contain vast amounts of memory resources. This implies that all data could be securely hardwired into the machine, and that nothing would ever need to be erased, making such machines immune to disruptions such as those caused by computer viruses.

To make one type of molecular computer, a set of wires is arranged in one direction, above a layer of molecular switches, and a second set of wires is aligned in the opposite direction below. At the junction of the wires is a single layer of rotaxanes. Each rotaxane collar is moved by electrical input which is controlled by the wire circuits. The apparatus then operates as a logic circuit as in a conventional computer. Quantum computers (Chapter 8) are similar except they rely on single atom orientations to count and have the advantage that they can use the power of the quantum behaviour of atoms and electrons to provide additional counting methodologies.

5.8 CHEMICAL ROTORS [4]

Bulky groups on crowded molecules have similar morphology to a form of rotors called propellers. A molecule comprised of two bulky aromatic rings attached to a focal atom can be regarded as the molecular equivalent of a macroscopic two bladed propeller. The structure shown in the top trace of Figure 5.8 incorporates two identical aromatic rings linked to the same atom. Rotation of one ring in one direction about the single bond linking it to the focal carbon group in the middle (called a methine carbon atom) forces the other ring to rotate in the opposite direction. Thus when one ring rotates clockwise the other must rotate anticlockwise. This arrangement can be used in a traditional macroscopic manner as paddles.

There are other examples in which molecules resemble everyday macroscopic rotors. The coupled motions of two structures have been exploited in the design of molecular gears incorporating 9-triptycyl ring systems. For example, the gear shown in the middle trace of Figure 5.8 incorporates two 9-triptycyl ring systems bridged by a methylene group. The aryl rings of the two triptycyl ring systems interdigitate in a manner reminiscent of the notches of a pair of meshed gears, much like those in models for carbon based nanotube gears (Chapter 4). As a result, the rotations of the two 9-triptycyl ring systems about a single bond linked to a common group are coupled. When one 9-triptycyl ring system rotates clockwise the other rotates anticlockwise and vice versa.

Figure 5.8

Parts of molecular machines: top trace, paddle; middle trace, gears; and bottom trace, an armature. The structures assigned as R are also shown. Modified from Balzani V, Credi A, Raymo FM and Stoddart JF (2000) *Angew. Chemie Int. Ed.* 39: 3348–91.

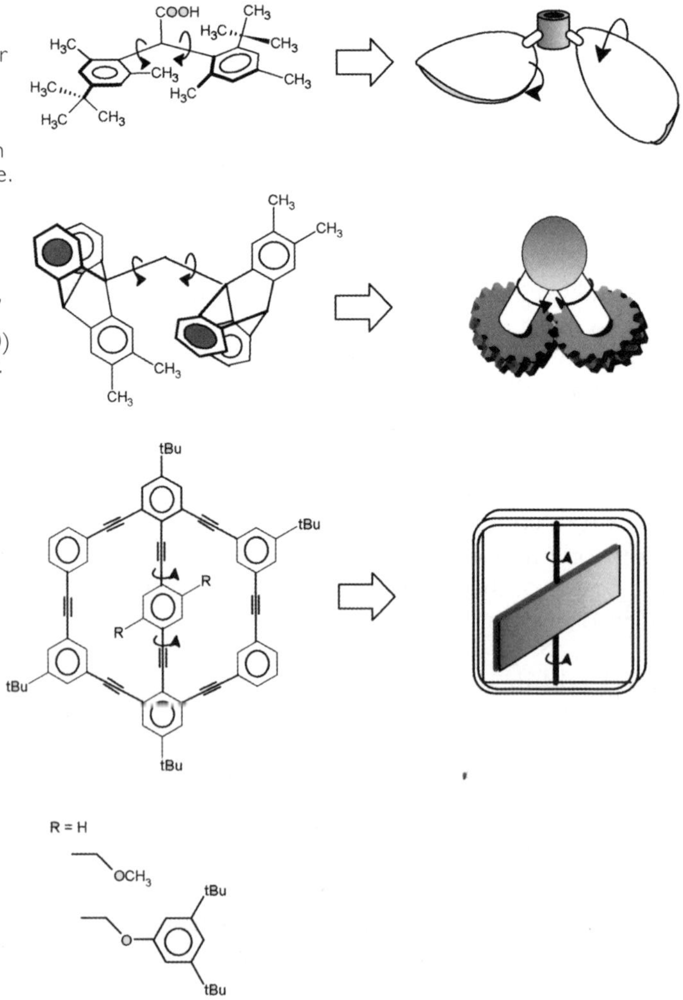

In Figure 5.8, the bottom trace depicts a rotor in a framework. The groups labelled 'R' can be modified. Thus when R is small, such as a proton, the rotor rotates freely. However when R is a bulky group it does not rotate at all. If the central molecule is made magnetic and magnets are arranged at each side of the framework we have the essence of an electric dynamo. Thus electricity could be generated and passed down a conducting carbon nanotube wire of a few nanometres thick or a copper wire prepared by filling a non conducting carbon nanotube with copper (Chapter 4). It could then be used to drive some other device, such as a molecular syringe (see below); or provide current for a molecular computer.

Figure 5.9a depicts a 9-triptycyl ring system attached to a 2,2'-bipyridine unit. The rotation about the single bond is fast. The structure can be operated with a brake. Upon addition of mercuric triflouroacetate, $Hg(O_2 CCF_3)_2$, the metal ion is coordinated by the bipyridine ligand. As a result the conformation is locked. Thus the locked conformation of the 2,2'-bypyridine unit brakes rotation of the 9-triptycyl ring system. The molecular brake can be released by adding ethylene diamine tetra acetic acid (EDTA), which reacts with the mercury ions and disengages the brake.

In order to build a better machine the design of the molecular brake has been modified. A molecule with a flexible end has been used (Figure 5.9b). A small blocking unit is attached to the main rotational unit. Because the flexible unit bends freely without interaction more easily in one direction than another, the structure only rotates in one direction. However, the rotation is slow and conformational stability is reached after six hours, beyond which time the machine does not work. The principle is good though, and it should be possible to synthesise light charged or other energetically charged rotors that rotate in only one direction. It should also be possible to store energy up in the propeller, hold it with a brake and release it when required.

Another example of chemically controlled braking comes from the catenanes. The [2] catenane compound, Figure 5.10, incorporates two identical large ring components called macrocycles. There is a phenanthroline unit and a polyether chain connected by two *p*-phenylene rings. The central Cu^+ ion is embraced by two phenanthroline aromatic structures while the two polyether chains are separated from each other. However, the demetallation of Cu^+ using a solution of potassium cyanide (KCN) gives rise to a co-conformational change, which involves the circumrotation of both macrocycles through the cavity of the other (Figure 5.10a). This conformation, where the phenanthroline structures are separated from each other, can be reversed with the addition of Cu^+ ions to the medium.

Figure 5.9

a) Unidirectional propeller and

b) propeller with brake. Modified from Balzani V, Credi A, Raymo FM and Stoddart JF (2000) *Angew. Chemie Int. Ed.* 39: 3348–91.

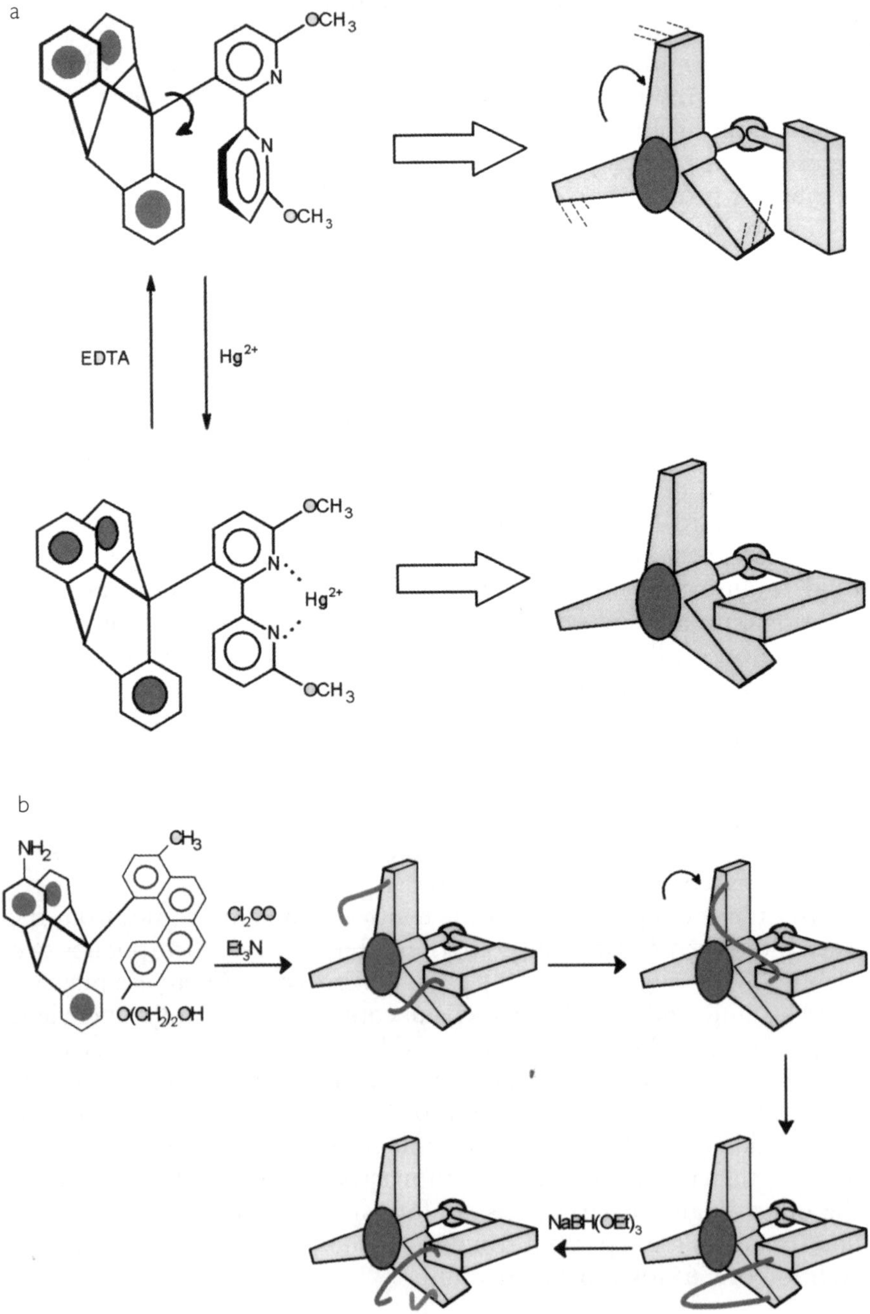

Figure 5.10

a) Copper, Cu+ on rotating catenane rings

b) Light driven molecular ball and socket molecule. Modified from Balzani V, Credi A, Raymo FM and Stoddart JF (2000) *Angew. Chemie Int. Ed.* 39: 3348–91.

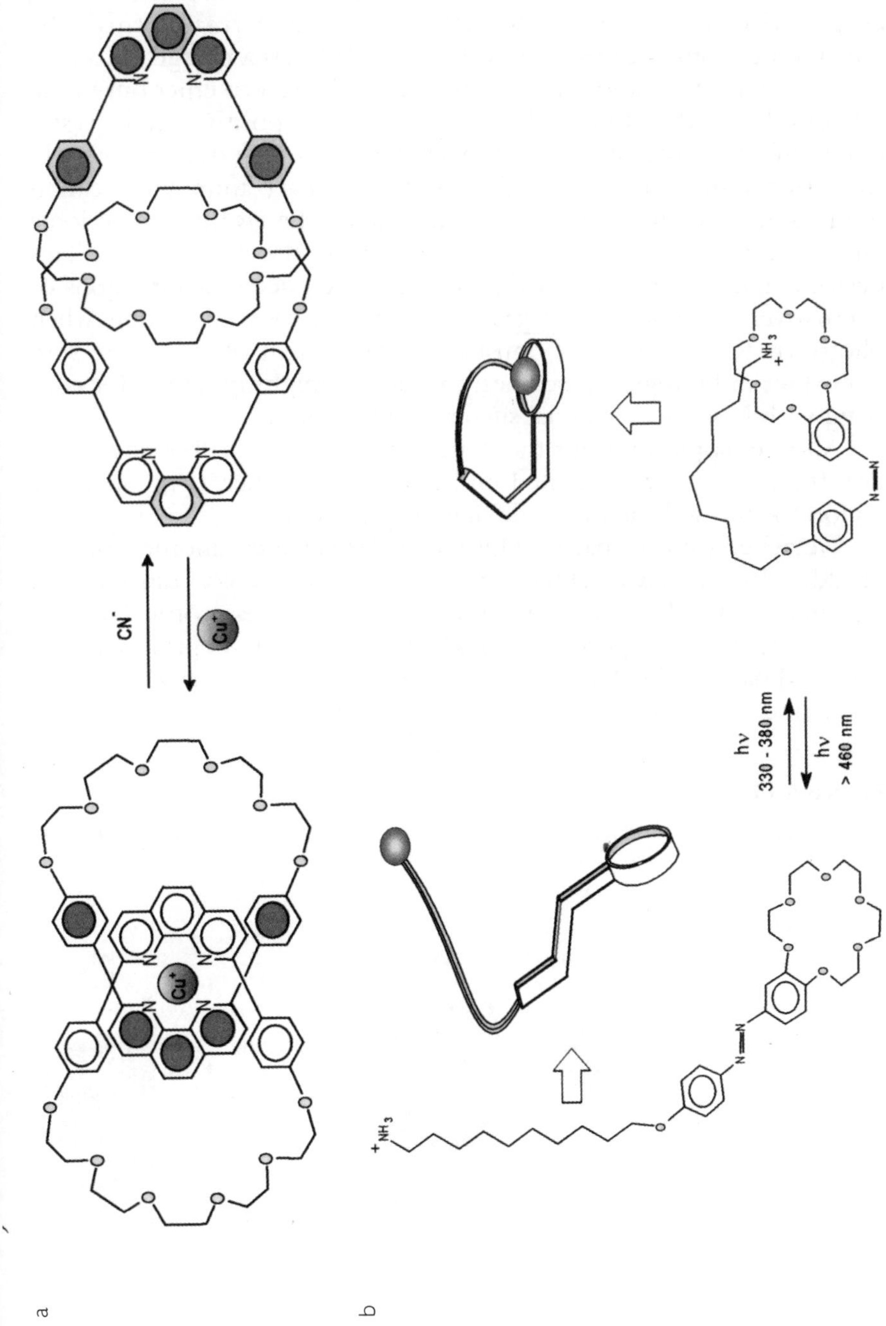

5.9 PRODDERS

We have noted that the azobenzene structure can adopt *cis* and *trans* configurations, and they can be interconverted by irradiation at appropriate wavelengths. Thus the geometry is altered by interconversion from *cis* to *trans*. The cyclic structure in Figure 5.10b is called a crown ether and this particular crown ether has *cis* and *trans* forms. The crown ether compound shown in Figure 5.10b has a charged head. In the *trans* form the cation head is away from the crown ether but in the *cis* form the head is close to the crown ether. The positive charge is stabilised by intramolecular bonding to the electron rich oxygen atoms in the crown ether. In the *trans* form the quaternary ammonium cation head forms ionic linkages with anions, but in the *cis* form these ionic linkages cannot be formed when stuck inside the crown ether. This device could therefore be used as a light driven detector for anions or alternatively it could act as a gate, which will release anions only when photo irradiated. Since the transfer and movement of anions are very important in biological machines, this device might be useful for coupling with biological material such as membranes.

Another application for the structure in Figure 5.10b is as a prodder. If a molecule was situated under the crown ether, then when the *cis* form is formed the molecule may be pushed away.

Detached cations make useful molecular syringes, another form of prodder. In this device (Figure 5.11) a silver ion is coordinated to a nitrogen atom. However, when the nitrogen atom is protonated with acid the silver ion is pushed through the tube; upon deprotonation it is sucked back again. This device could be used to move any free material in and below the silver ion. It also acts like a ball valve.

Figure 5.11

A molecular syringe. Modified from Balzani V, Credi A, Raymo FM and Stoddart JF (2000) *Angew. Chemie Int. Ed.* 39: 3348–91.

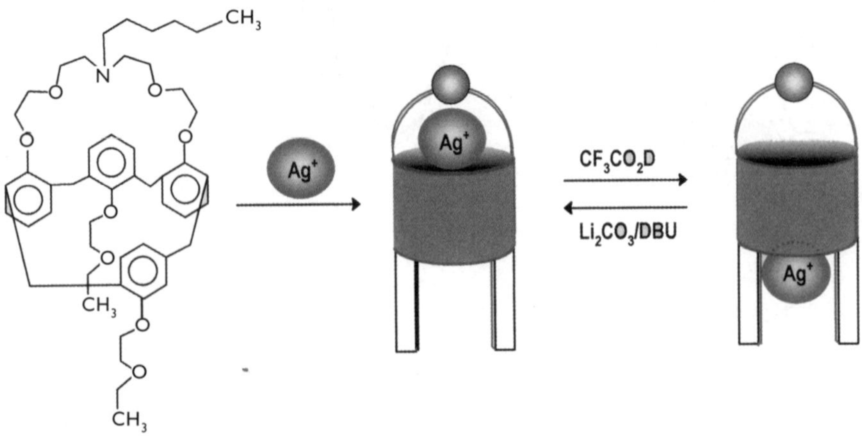

The structure in Figure 5.12 shows a cyclodextrin with a large hole into which aromatic rings can fit. It is relevant that molecules containing elemental iron, and which are magnetic, can fit in. An example is ferrocene. When iron is oxidised to Fe^{2+}, the iron-containing molecule moves out of the cyclodextrin. This is significant because it means that iron metal complexes can be included. With modification this can become the basis of an electric piston motor. Oxidation, reduction or protonation can dethread and thread the inclusion unit. Because a magnet is involved, the molecule could also be an electric motor.

Figure 5.12

The electrically driven ferrocene plunger. Modified from Balzani V, Credi A, Raymo FM and Stoddart JF (2000) *Angew. Chemie Int. Ed.* 39: 3348–91.

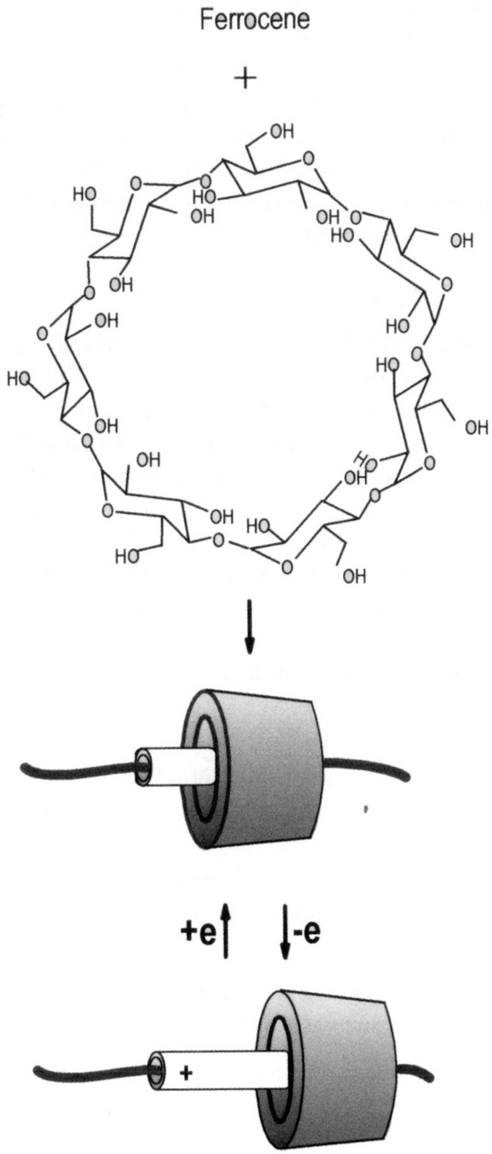

5.10 FLIPPERS [4]

Polycyclic compounds that are sterically hindered can act as flippers because one bulky group cannot rotate freely past another. In Figure 5.13a the two stereoisomers are interconverted by light at different wavelengths. By changing the frequency rapidly, rapid flipping can be induced. However the flipping is not unidirectional. To do this, the thioxanthylidene ring system is replaced with another tetrahydrophenanthrene unit (Figure 5.13b). Each of the helical subunits can have right (P) or left (M) handed geometry. Thus four stereoisomers are possible. The *cis-trans* isomerisations are reversed by irradiation, but second inversions are not possible using light but can be induced by thermal means.

This is therefore a two-mechanism switch. On irradiation with light, PP *trans* changes to *cis*. On heating, *cis* changes to a different *cis* compound. Further irradiation converts this different compound to a *trans* form. More heat converts this *trans* form back to the original *trans* form. Thus a series of temperature and light induced transformations can be used to move this molecular flipper in only one direction.

5.11 ATOM SHUTTLES

When a metal ion is held by a molecular train it can be carried from station to station. Figure 5.14 shows the shuttling of copper. The rotaxane has a phenanthroline and a terpyridine in its structure and it incorporates copper, Cu^{+} by coordinating with the phenanthroline structure in a tetrahedral geometry. Material coordinated to a metal ion in this way is called a ligand. The Cu^{+} ion can be converted to Cu^{2+} by electrolysis, however the Cu^{2+} now prefers penta-coordinate geometry. Hence it moves to the trinitrogen station. An important use for this machine could be in moving copper from one nanosite to another. It could be used to move a stockpile of atoms to a site where they can be used individually.

5.12 ACTUATORS

Actuators are machines which convert electrical motion to mechanical energy or vice versa. Nanoactuators may be formed when:

1 the behaviour of nanostructures under the influence of electrical stress such as changes in surrounding ion charges causes some molecular movement;

2 the transfer of photons to single molecules causes some conformational change such as *cis* to *trans*;

3 electrical currents are formed in single molecules by molecular motion.

Figure 5.13

Molecular flippers. a) Light driven and b) light and heat driven. Modified from Balzani V, Credi A, Raymo FM and Stoddart JF (2000) *Angew. Chemie Int. Ed.* 39: 3348–91.

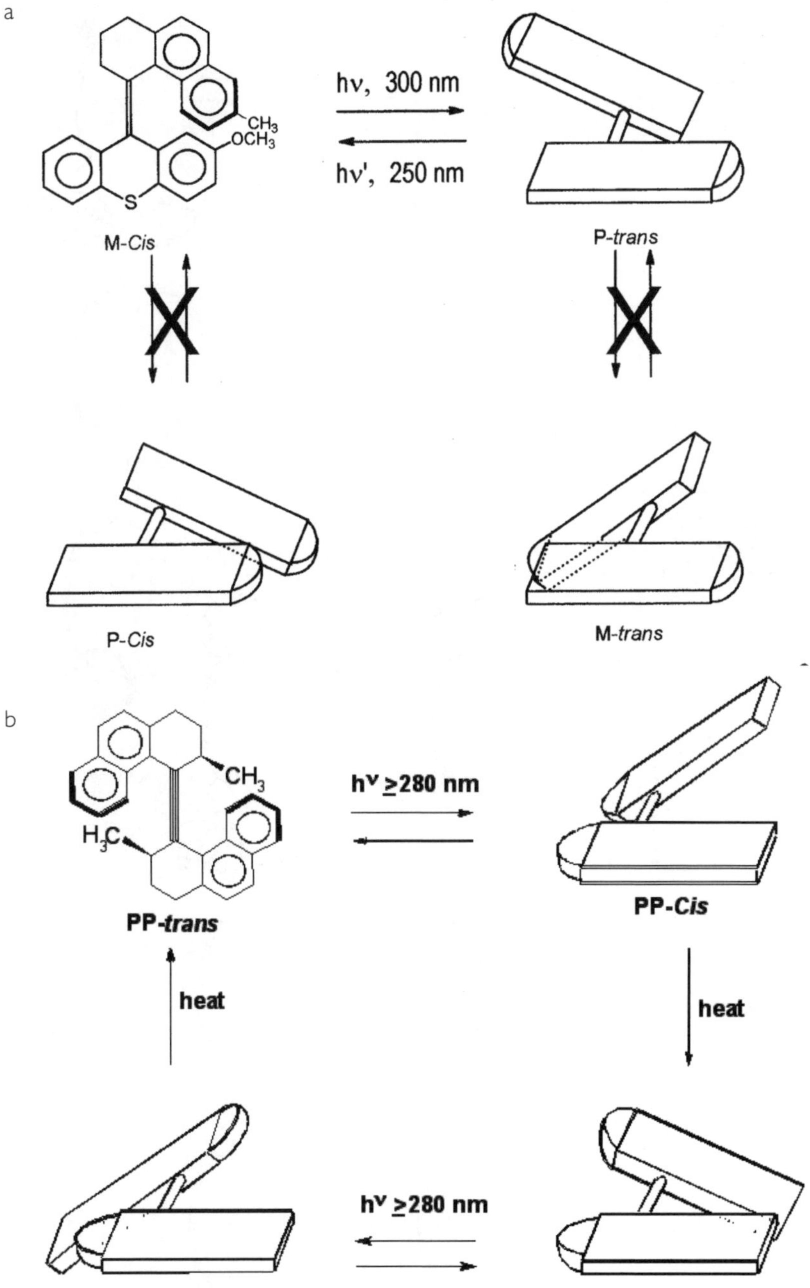

Figure 5.14

Copper-carrying molecular train at different stations. Modified from Balzani V, Credi A, Raymo FM and Stoddart JF (2000) *Angew. Chemie Int. Ed.* 39: 3348–91.

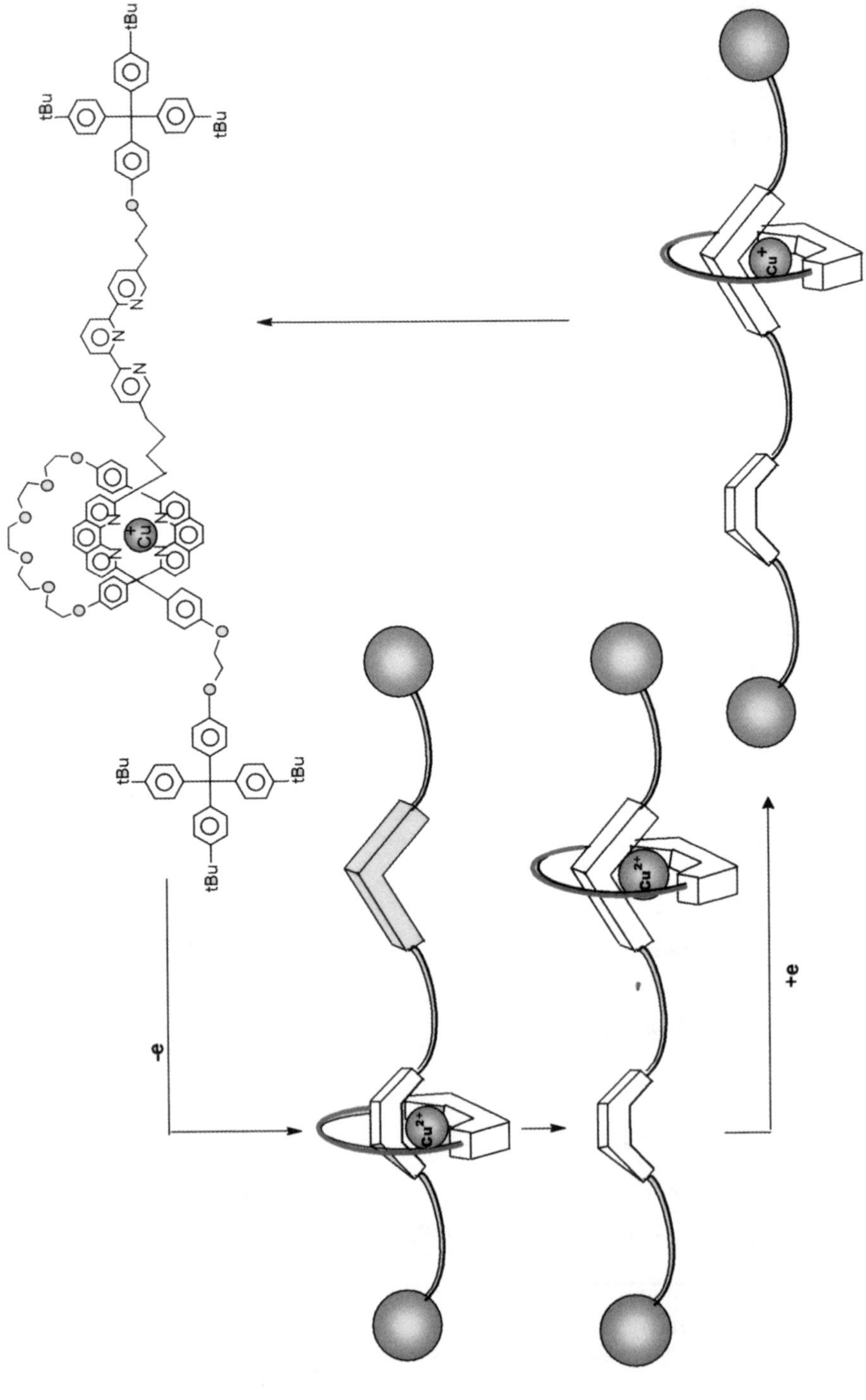

We have seen the possibility of all these types of nanomachines in the examples given above.

Electromechanical macroscopic actuators built on sheets of single-walled nanotubes have been produced from bundles of single-walled nanotubes [5]. The macroscopic actuators operate on a cantilever system, where bundles of single-walled nanotubes are adhered to sticky tape or a polyvinyl chloride strip. To stabilise charge, the device is placed in sodium chloride solution so that a double layer is formed. On applying current to one side of the actuator, the nanotubes on the surface are charged differently to those on the uncharged side due to the double layer, and hence they elongate. Thus the bundle bends.

This process can be achieved on a nanoscale so that tiny flagella, like those that push microorganisms through water and sperm through vaginal fluids, are formed. They could operate to push tiny nanomachines through media such as blood. On the macroscale, this method could be used to produce artificial muscles. When nanotubes are layered (Chapter 4) by chemical vapour deposition (Chapter 3) to form forests, changing the voltage to a single nanotube causes the layers to move just like natural muscle fibres. Just like the bundles described above, the performance of artificial muscle depends on the ability to extract or inject charge.

5.13 CONTACTS

The components of nanostructures, such as those compounds discussed above, will still need to be assembled into a useful structure. A very good structure is the biological membrane, which is discussed in Chapter 6. However, there are other structures on which machines can be built. A very flat surface, such as gold, is preferred since it is very difficult to assemble a mechanism accurately on an uneven surface, such as the copper surface in Chapter 2, Figure 2.15.

The nanomachines are not assembled molecule by molecule, as this would take too long. Rather they are deposited by various means such as chemical vapour deposition, from a liquid, or electrochemically to form structures and then contacted.

This involves four steps. In the first step, the molecules of interest are adsorbed onto the isolating substrate in their millions. Since gold is a good electrode it can be used as the substrate and then masked (Chapter 3.4 and in more detail in Chapter 8.4) after the molecules are added. Whether molecules are covered or destroyed in masking does not matter since there is a huge inbuilt redundancy. What is important is that there are clear areas where the molecules are free.

The second step includes the visualisation of the molecule distribution, mostly by STM. In the third step the molecule is contacted, either through the scanning tip, which adds molecules (for example, dip pen technology) or by structuring of electrodes. This can be done by adding electrode material, by demasking certain areas, or by addition. All of this is very slow and the images from STM are usually used for localising the molecules of interest. Arrays of electrode pairs are usually structured from ~100 nm gold (primed with a ~5 nm titanium layer) on a thermally oxidised silicon wafer using lithographic techniques (Chapter 3.4 and Chapter 8.4). After this, in the fourth step, electrical characterisation must be achieved. All this needs to be sped up for practical utilisation. Numerous circuits may be made (as with silicon chip technology) and only the best kept. These may have largely non-working systems, but this does not matter since only a few structures need to work per 100 nm^2.

Figure 5.15

Fixing a molecular component to a frame for wiring. Units for wiring are shown in the table. Modified from Balzani V, Credi A, Raymo FM and Stoddart JF (2000) *Angew. Chemie Int. Ed.* 39: 3348–91.

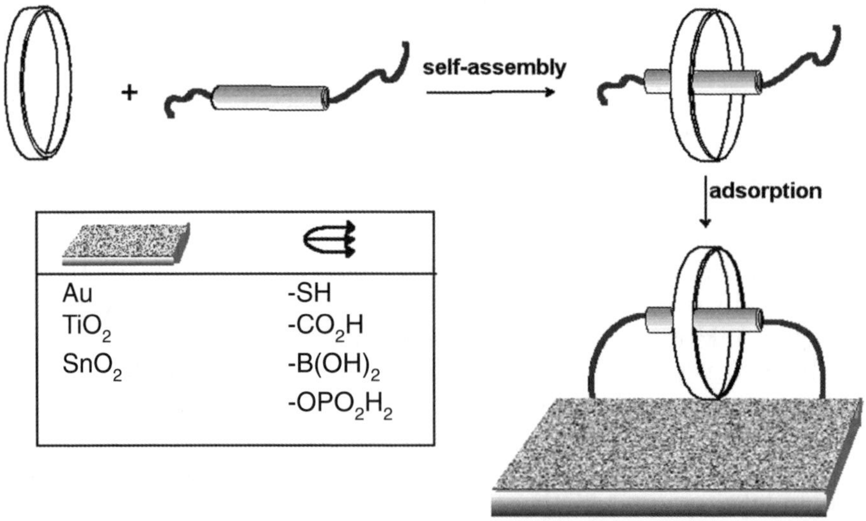

For optimal electrical contacting, a parallel orientation of the adsorbed molecules referred to the pre-structured gold or other electrodes is needed. A simple method that has been shown to work in inducing a preferred molecule orientation is flow adsorption. As the molecules hit the surface they self-align. If flow is controlled, they can be systematically placed at one end of a surface and then laid down to yield a high percentage of molecules aligned along the flow direction. Although the results from aligned molecules are promising, there is still the need for a technique that allows the contacting of single molecules out of an ensemble deposited at the substrate surface, such as to adjust various electrode gaps, or to choose special molecules for contacting purposes. Many of the structures could use thiols, which bind readily to gold surfaces. A thiol at each end of a structure could be used to 'solder in' an electronic part such as a molecular switch (Figure 5.15). However, other solders are useful, such as carboxylic acid, boron hydroxide or phosphate groups. Other 'circuit board' substrates include titanium oxide, TiO_2 and stannic oxide, SnO_2. These structures can already be organised as a single layer (monolayer) on a surface [6] or congregated in defined environments [7, 8] or between electrodes [9–13].

One useful method would be to create a 'club sandwich' with gold thiol and operational molecules with masks in between so the gold only contacts the molecules and does not 'short circuit'. By carefully assembling each section of the sandwich, a structure with all electrical parts in place could be built up. This is not science fiction; an electrode covered with a superstructure made of gold nanoparticles cross-linked by molecules has been developed for sensor applications [13]. Artificial machines such as those described here have been able to complete computing operations [15–23] and are the forerunners of molecular computers. However, for the time being they have strong competition from their other nanoscale counterparts, the quantum computers. These are covered in chapter eight.

If everything described above can be achieved, these machines need to be put in context. For solid objects, machines may be oriented in banks in small useful devices the size of a pinhead or smaller. However for many applications, such as active agents in nanomedicine, they will need to be in solution. This requires that all structures are very small and surrounded by some kind of solubilising medium. This medium will probably be biochemical, as Chapter 6 will describe. Whatever the outcome, the great challenge will be in assembling all the useful parts of these nanomachines together. Recently, a solid-state electronically accessible [2] catenane device was fabricated from a single monolayer of the [2] catenane, anchored with phospholipid counter ions, and sandwiched between a polysilicon bottom electrode and a titanium/aluminium top electrode. The switch can be opened at 2 V and closed at -2 V, read at 0.2 V and may be recycled [19]. Such a device should easily be made in a form that can enter blood vessels.

WHAT YOU SHOULD KNOW NOW

1 The difference in structure between rotaxanes and catenanes. Catenanes are molecular chains and therefore can be used to chain up structures so they cannot be moved. Rotaxanes are macromolecules with a movable collar structure held around an axle, like a bracelet around an arm.

2 How these compounds are synthesised.

3 How rotaxanes are used as molecular switches. The principle behind using rotaxanes as molecular switches is in being able to move the collar under some influence.

4 What is meant by a chemical computer. Logic gates are the counting devices for computers and are based on microsilicon devices called silicon chips. Rotaxanes are molecular-based logic gates. It has been shown that they can perform exactly the same task as silicon. A solid-state, electronically accessible [2] catenane device fabricated from a single monolayer of the [2] catenane, anchored with phospholipid counter ions, and sandwiched between a polysilicon bottom electrode and a titanium/ aluminium top electrode can be opened at 2 V and closed at -2 V, read at 0.2 V and may be recycled.

5 The structure of a number of other molecular machines, including, prodders, rotors, molecular brakes, atomic shuttles and trains.

6 How these machines are bound to surfaces using thiols and gold.

5.14 EXERCISES

1 *Build four models from Figure 5.13. Use the filled in structures rather than the molecular structures. Use matchsticks as axles and shape the structures from pieces of apple or other material. Show that there are four isomers*

2 *Build a model of Figure 5.8 top trace from tape, paper and matchsticks and convince yourself that the diagram shows clockwise and anticlockwise rotation.*

3 *Read Balzani V, Credi A, Raymo F and Soddart JF (2000)* Angew. Chem. Int. Ed. *39: 3348–3391. This paper is a brilliant review of the field.*

4 *Web search cyclodextrin, nanojunction, ethylene diamine tetra acetic acid, actuator, and ferrocene.*

5 *Look up the definition of polycyclic aromatic compounds.*

6 *Take some string and build the structures shown in Figure 5.4b. Demonstrate how these structures produce the products shown.*

5.15 REFERENCES

1 Amabilino DB & Stoddart JF (1995) *Chemical Reviews* 95: 2725–828.

2 Bissell RA, Cordova E, Kaifer AE & Stoddart JF (1994) *Nature* 369: 133.

3 Murakami H, Kawabushi A, Kotoo K, Kunitake M & Nakashima N (1997) *J. Amer. Chem. Soc.* 119: 7605–606.

4 Balzani V, Credi A, Raymo FM & Stoddart JF (2000) *Angew. Chemie Int. Ed.* 39: 3348–91.

5 Baughaman RH, Cui C, Zakhidov AA, Iqbal Z, Barisci JN, Spinks GM,

Mazzoldi A, De Rossi D, Rinzler AG, Jaschinski O, Roth S, Kertesz M & Wallace GG (1999) *Science* 284: 1340–44.

6 Asakawa A, Higuchi M, Mattersteig G, Nakarumura T, Pease AR, Raymo FM, Shimizu T & Stoddart JF (2000) *Advanced Materials* 12: 1099–102.

7 Lynch DE, Hamiltion DG, Calos NJ, Wood B & Sanders JKM (1999) *Langmuir* 15: 5600–605.

8 Brown CL, Jonas U, Preece JA, Ringsdorf H, Seitz M & Stoddart JF (2000) *Langmuir* 16: 1924–30.

9 Buey J & Swager TM (2000) *Angew. Chemie* 112: 622–26.

10 Buey J & Swager TM (2000) *Angew. Chemie Int. Ed.* 39: 608–12.

11 Bidan G, Billon M, Divisia-Blohorn B, Kern J-M, Raehm L & Sauvage J-P (1998) *New J. Chem.* 22: 1139–41.

12 Shipway AN, Lahav M & Willner I (2000) *Adv. Materials* 12: 993–98.

13 Lahav M, Shipway AN & Willner I (1999) *J. Chem. Soc. Perkin Trans.* 2: 1925–31.

14 Will G, Boschloo G, Hoyle R, Rao SN & Fitzmaurice D (1998) *J. Phys. Chem.* 102: 10272–78.

15 Rouvray DH (1998) *Chemistry in Britain* 34: 26–29.

16 Muller DA, Sorsch T, Moccio S, Baumann FH, Evans-Lutterod K & Timp G (1999) *Nature* 399: 758–61.

17 Nalwa HS (2000) *Handbook of Nanostructured Materials and Nanotechnnology*. Academic Press, San Diego CA.

18 Adleman LM (1994) *Science* 266: 1021–24.

19 Collier CP, Matterstei G, Wong EW, Lou Y, Beverly K, Sampaio J, Raymo FM, Stoddart JF & Heath JR (2000) *Science* 289: 1172–75.

20 Wong EW, Collier CP, Belohradsky M, Raymo FM, Stoddart JF & Heath JR (2000) *J. Amer. Chem. Soc.* 122: 5831–40.

21 Collier CP, Wong EW, Belohradsky M, Raymo FM, Stoddart JF, Kuekes PJ, Williams RS & Heath JR (1999) *Science* 285: 391–94.

22 Ball P (2000) *Nature* 406: 118–20.

23 Reed MA & Tour JM (2000) *Scientific American* 282: 86–93.

6

NANOBIOMETRICS

Nanotechnology is making enormous breakthroughs using biology. In this chapter you will find out about the parts that compose nature's nanomachines: lipids, DNA and proteins. We will also discuss biological computing, which uses a protein-based 3D optical memory based on bacteriorhodopsin. Biological material such as DNA can be used as sensors and for making nanohinges, nanowires and nanoglue.

6.1 INTRODUCTION

If you want to be convinced of the power and possibilities of nanotechnology then you only have to look at nature to see the staggering possibilities. Nature is the ultimate nanotechnologist — for the moment!

Skeptics may ask whether nanotechnology can ever be made to work, whether it is even possible to build structures at the nanoscale that can perform complex tasks. The answer is that of course it is possible — nature has already produced functional examples of virtually every nanotechnology device mentioned in this book. The fact is however, that we're only just able to understand the language of biology. Learning, recognising and applying the lessons that biology can teach us is where things get interesting.

Nanorobots have existed in nature since the beginning of life. They are called bacteria and viruses. Cells contain many sorts of nanomachines including 12 nm diameter rotating motors, called ATPase, which we will look at in this chapter. However Chapter 7 covers some other examples from biology that rely on nanoscale organisation: optical structures on butterfly wings that are highly metallic in appearance, yet contain no pigments; exquisitely sensitive chemical sensors based on 4 nm-sized ion channels; composite nanomaterials, such as spiders' silk or abalone shells, that are tougher or stronger than the best synthetic materials; and methods to convert sunlight into chemical energy. All of this is done without the use of toxic solvents, billion dollar factories and endless pollution.

Humans are inveterate tinkerers. We want these materials, but we want to have control and to make them suit our purpose, not nature's. Our needs are different to those of an abalone or a butterfly. We need to be able to make standard materials, since Western society is not very good at dealing with products that are not identical. We can also produce materials that biology cannot make and so we may be able to develop new hybrid materials and structures to extend biology's properties. So, rightly or wrongly, for various reasons, we are not content with what nature provides. We want to develop our own nanotechnology.

How can we learn from biology? Firstly, just knowing something is possible is half the battle. If something works, even if we don't know how, we at least know that the physics is possible and that we are not trying to violate some basic scientific laws. Secondly, we can glean some of the basic principles behind how nature achieves her goals. What are the physics, chemistry, and materials science principles that make sense in the nanoworld? How should we approach building nanostructures and how can they be connected to the macro world that we inhabit? Is there a more subtle way of mining iron ore than blowing it out of the earth with dynamite? Could we use the same principles that enable bacteria to sequester minerals and excrete metal?

These lessons may be more important to learn than we think — to date our record of living in harmony with nature has been poor.

Biology shows us that only a few basic building blocks can self-assemble into more complex structures. These structures in turn can self-assemble into more complex hierarchical structures from which you can build devices ranging from the nanoscopic (such as nanoscale sensors based on proteins) to the gargantuan (for example, the Great Barrier Reef in Australia).

The rest of the chapter will describe three basic structural units found in biology — lipids, proteins and DNA — and how these units can self-assemble into structures and devices that can be used to develop areas of nanotechnology. It is important to remember that although we may want to mimic and to some extent imitate nature, we do not want to slavishly copy biology — hence the term biomimetic nanotechnology.

6.2 LIPIDS AS NANO- BRICKS AND MORTAR

Lipids, one of the most versatile building blocks in biology, are used to form three-dimensional structures and as a matrix to place other active nanomachines into [1].

If you want to build architecturally exciting nanostructures, then lipids are the bricks. The mortar that holds them together is a range of non-covalent bonds. Of course it is possible to include a range of features, such as protein windows and revolving doors, and with mod cons such as a DNA library in the annexe. It's far too easy to get carried away with this sort of analogy and one of the dangers is that you try and force totally new concepts into accepted paradigms.

Although the analogy that lipids are bricks is conceptually intriguing the important point that is missing from the analogy is 'who are the bricklayers?' And that's where biology excels — the 'bricklayers' are built into each brick. The structure of the individual bricks is such that they have the ability to self-assemble into the appropriate higher-order structure. To put it another way, the blueprint of the final structure is programmed into the structure of the individual nanoscale building blocks, so that when you activate the lipids (for example by adding water to the dry lipid) they automatically self-assemble into the desired structure.

LIPID STRUCTURE

The term lipid has been used in a variety of ways. Here the word lipid is generally used to refer to compounds that are extracted from, or are synthesised to mimic, the naturally occurring compounds. So what are lipids? Lipids are molecules that possess a hydrocarbon tail that is hydrophobic, or insoluble in water (just like oil is insoluble in water) and a polar head group that is hydrophilic, or water soluble (Figure 6.1a). This is a rule of thumb for solubility: like dissolves like. Generally the total length of the lipid molecules is 2–4 nm. From these molecules we can build supramolecular structures that are hundreds of nanometres to hundreds of microns in size [1, 2].

These lipids, as well as synthetic compounds where the hydrophilic head-groups and hydrophobic tails are made up of groups that may never occur in nature, are also known as amphiphilic compounds or amphiphiles (literally *amphi* – both, *phile* – to love, that is, to be happy in both oil and water). It is possible to either extract naturally occurring lipids or to synthesise lipids using standard organic chemistry techniques. Literally thousands of different amphiphiles have been extracted, synthesised and studied.

SELF-ORGANISING SUPRAMOLECULAR STRUCTURES

When water is added to a lipid, all the polar head groups will dissolve in the water. However, the hydrocarbon tails will not dissolve in the water but will stick to each other *via* weak van der Waals bonding interactions. An individual lipid molecule with a long hydrocarbon tail will not be soluble in water, but will align itself with a number of other lipid molecules to form structures where the hydrocarbon tails all line up

Plate 1

Corral of iron atoms in a circle on copper showing electron wave patterns. Reproduced with permission of IBM. Crommie MF, Lutz CP & Eigler DM (1995) Waves on a metal surface and quantum corrals. *Surface Review Letters* 2: 127–31. STM rounds up electron waves at the QM corral. *Physics Today* 46(11): 17–19.

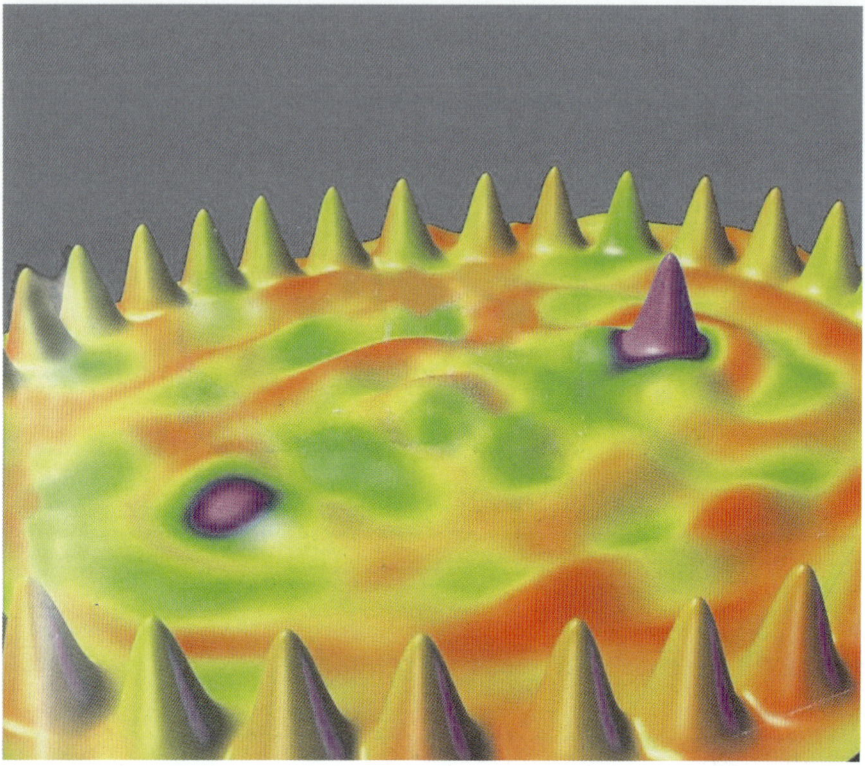

Plate 2

Ghost mirage atom caused by refocussing. The purple points are the introduced atom and its ghost in an electron sea. Reprinted from the cover of *Nature*, February 2000.

Plate 3

An individual carbon nanotube on seven electrodes. Courtesy of Cees Dekker, Delft University of Technology, The Netherlands.

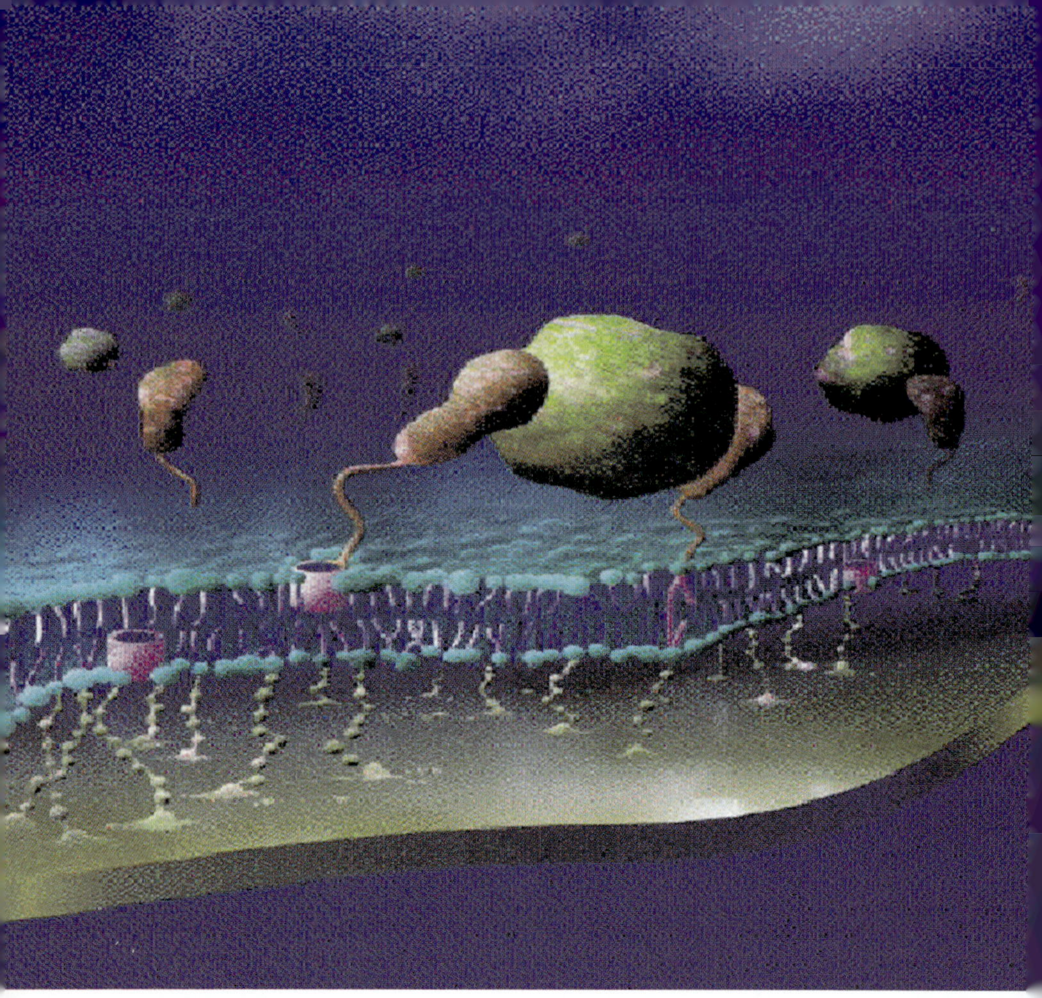

Plate 4
The AMBRI biosensor. The yellow object is the analyte. The red-brown objects are the antibodies, and the ion channels are shown in purple.

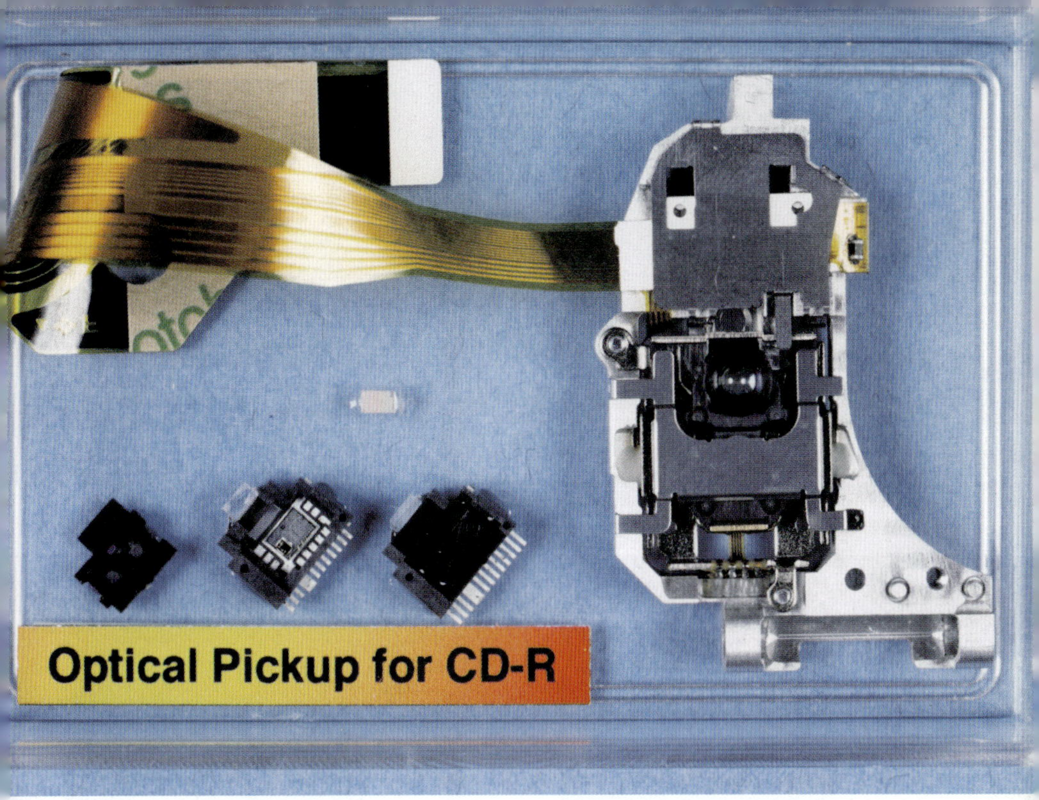

Plate 5

A holographic wave plate in an optical CD pickup for noise limitation. The thin film plates containing zigzag nanostructures are retarding or polarising sensitive elements. They are the faint objects in the centre of the image and at the end of two of the black plug-in elements. Courtesy of Toyota Central R&D Labs Inc.

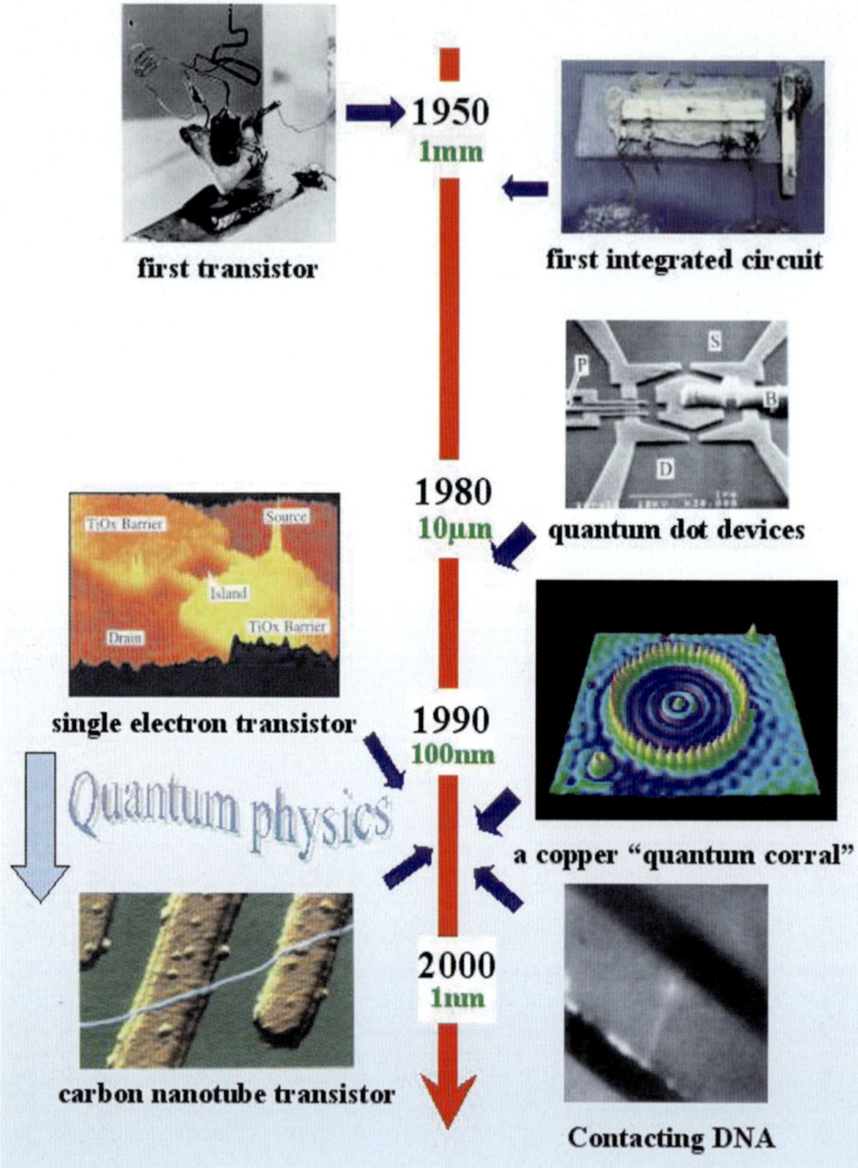

Plate 6

A brief history of electronic devices from the first transistor through to interconnects made from DNA.

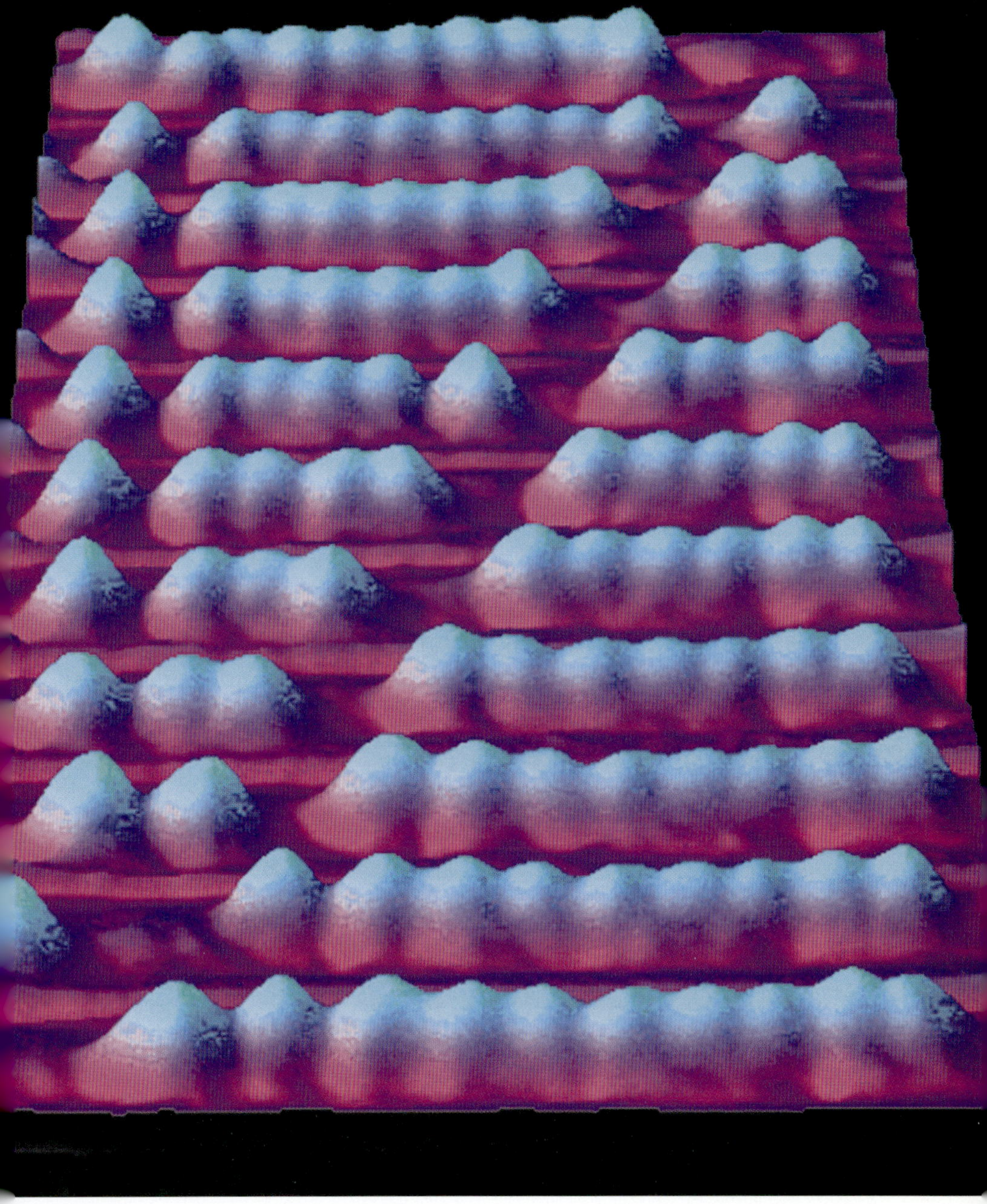

Plate 7

A molecular abacus, a new way of counting! Each atom can be moved like the beads of an old-fashioned abacus.

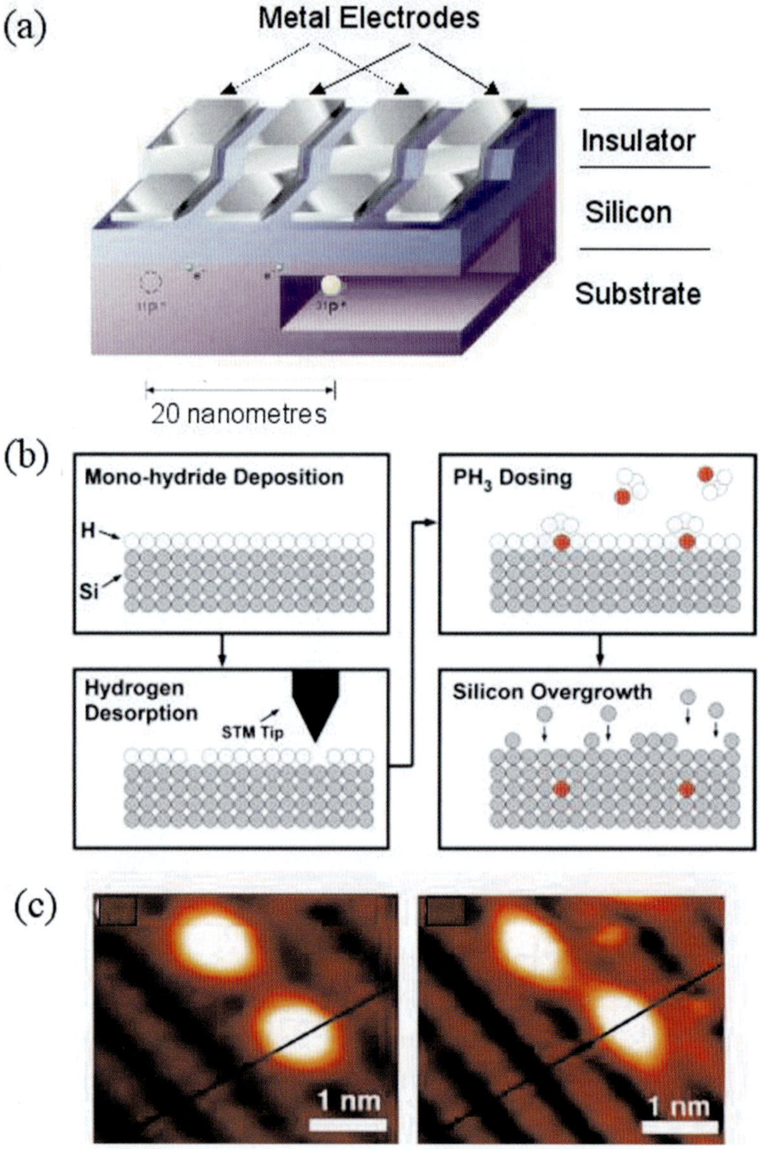

Plate 8

a) An illustration of the Kane architecture for a solid-state quantum computer based on 31^{P} qubits in silicon.

b) A schematic of a possible atomic fabrication route to the realisation of Kane's design. A low defect density silicon surface is passivated with a monolayer of hydrogen. An STM tip is used to selectively desorb hydrogen, exposing silicon on an atomic scale, permitting only one phosphine molecule to adsorb at each of the required sites. Low temperature silicon overgrowth encapsulates the phosphorus.

c) STM images of two desorption sites before (left) and after (right) phosphine deposition. The different shapes demonstrate single PH_3 adsorption through a hydrogen patterned resist.

a

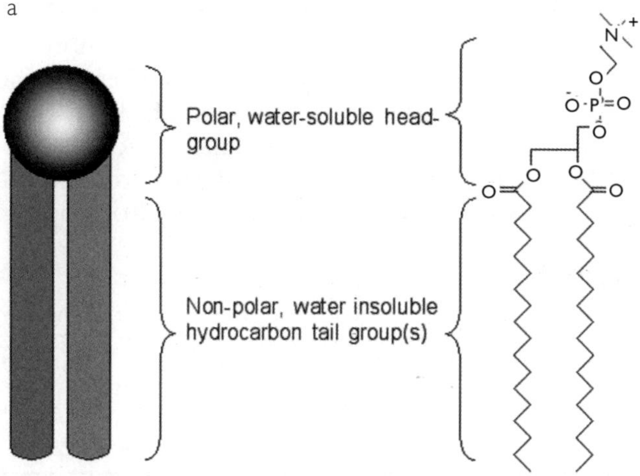

b

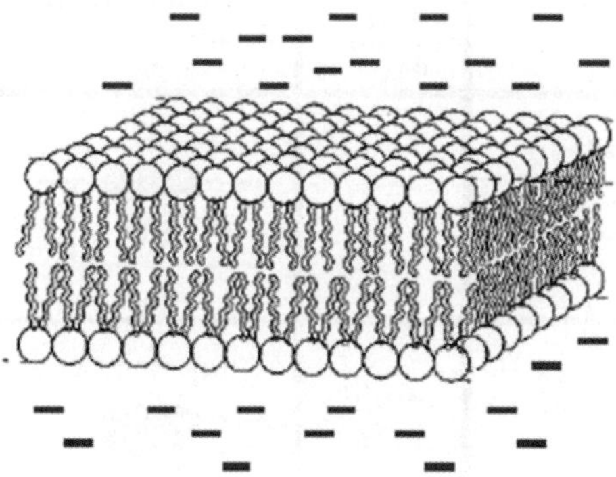

Figure 6.1

a) General structure of a lipid showing the hydrophilic head group and the hydrophobic tail.

b) Structure of a lipid bilayer in water. The hydrophobic hydrocarbon groups line up and the hydrophilic head groups are exposed to the water.

together on the inside of the structure and all the polar, water soluble headgroups coat the surface of the structure (Figure 6.1b).

We know that if we add water to lipids, they will self-assemble into ordered structures with the hydrocarbon tails on the inside and the head group projecting into the water. How can we control the shape of these higher hierarchical structures? Well, it turns out that at a first approximation, it is a conceptually relatively easy matter. We 'simply' have to adjust the size of the head group and the size of the hydrocarbon tail (Figure 6.2) [3]. As Figure 6.2 shows, if we make a compound that has a single hydrocarbon chain and a large water soluble headgroup, micelle structures form automatically. Typical examples of these compounds are the detergents found in all soaps and dishwashing liquids. They are effective at cleaning because any oily or greasy residue is dissolved in the hydrocarbon centre of the micelle. These micelles are suspended (emulsified) in the water and can be washed away easily. If the size of the head group is decreased it is possible to form structures called 'hexagonal phases'. These structures are basically long tubes of lipids with the hydrocarbon tails on the inside and water on the outside. When the headgroup size matches that of the hydrocarbon tail, planar bilayer structures, such as those that form cell membranes or vesicles (Figure 6.1b), are formed. By using molecules where the head group is smaller than the hydrocarbon tail group we can form the 'inverted hexagonal phase'. This is where long tubular structures are again formed but where the headgroups are pointing towards the inside, so that we end up with long tubes of water inside a hydrocarbon matrix.

Figure 6.2

Table showing the different structures that can be formed by adjusting the geometric size of the head group and hydrocarbon tail. A 'packing parameter' can be calculated from the volume of the hydrocarbon chains (V), the area per molecule at the hydrocarbon-water interface (A_0) and the length of the hydrocarbon region (l_c). The value of the packing parameter is a good indicator of the type of structure that the lipids can form.

Lipid	Critical Packing Parameter: V/A_0l_c	Critical Packing Shape	Self-assembled Structure Formed
Single chained lipids (e.g. detergents) with large head -groups	< 1/3	Cone 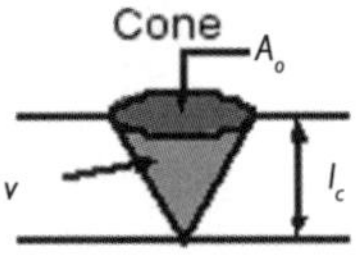	Spherical Micelles 
Single -chained lipids with small head-group areas	1/3–1/2	Truncated Cone 	Globular or Cylindrical Micelle 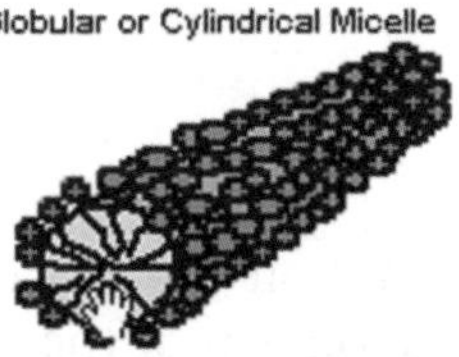
Double -chained lipids with large head -group areas and fluid chains	~ 1	Cylinder 	Bilayer Membranes 
Double-chained lipids with small head-groups	> 1	Inverted Truncated Cone 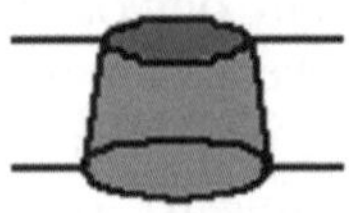	Inverted Micelles, tubules of water surrounded by lipid 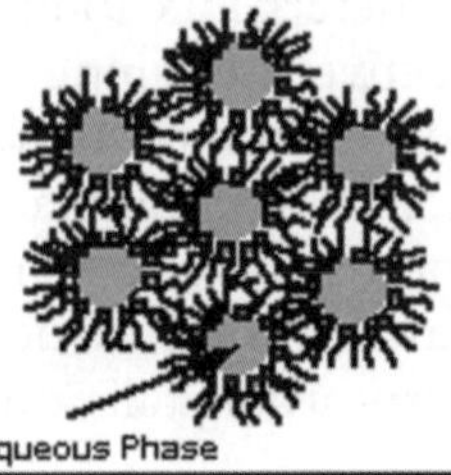

Because the molecules that make up these structures are defined precisely (for example, we can make the hydrocarbon tail an exact number of carbons long) the subsequent supramolecular structures are precisely defined as well. Given the same molecule, the same temperature and the same distance of water between one headgroup and the next, the thickness of the bilayer membrane, the diameter of the hexagonal phase tubule (or other structure) will be the same within a fraction of a nanometre each time the supramolecular structure is made. This is precisely the sort of control we need in order to build our nanoarchitecture.

THINGS TO DO WITH LIPIDS — TEMPLATES

Lipids are of interest in their own right when working with nanoarchitectures. We will return to some examples later on when we use them for incorporating protein nanomachines, but they are also of interest as templates to enable us to 'boot-strap' the formation of other non-biological nanomaterials [4, 5]. (Boot-strapping comes from the entrepreneurial Baron von Münchhausen who rescued himself from drowning by pulling himself out of the water by his own bootstraps). In this case the inherent self-assembly properties of the biological molecules are used to produce other non-biological structures.

As explained in Chapter 3, nanoparticles are of interest because in nanodimensions materials do not behave the same way as they do in bulk materials. Quantum properties of small ensembles of atoms become important. For instance, although gold is a bright shiny metal, at 20 nm in diameter it's no longer metallic golden but a beautiful burgundy red colour. When synthesising these materials it is easy to make roughly spherical nanomaterials, as this is the lowest energy structure. It is much more difficult to make other shapes, such as rods. When dealing with the ultra-small not only does size matter but shape also matters in determining the properties of the material.

In order to make nanorods we can make use of lipids that form the 'inverted hexagonal' water-in-oil structures as shown in Figure 6.2. If a water-soluble metal salt and a reducing agent are added, the metal nanoparticle will form in the water tubules formed by the lipids. The structure of the lipid template forces the nanoparticle to grow along the water tubule, forming nanorods rather than spherical nanoparticles (Figure 6.3) [6, 7]. This is similar to the principles used with sol-gels and with phyllosilicate particles (Chapters 3.5 and 3.8).

6.3 SAME BUT DIFFERENT: SELF-ASSEMBLED MONOLAYERS

A related area that is worth briefly mentioning is that of self-assembled monolayers (SAMs). SAMs are monolayers that are formed on surfaces by a self-assembly process. However, instead of using hydrophobic/

Figure 6.3

a) Transmission electron micrograph of copper nanoparticles and nanorods formed by reduction of Cu^{2+} salt to copper metal using tubular micelle structures as templates, and

b) some of the strange forms that can be created from the nanorods.

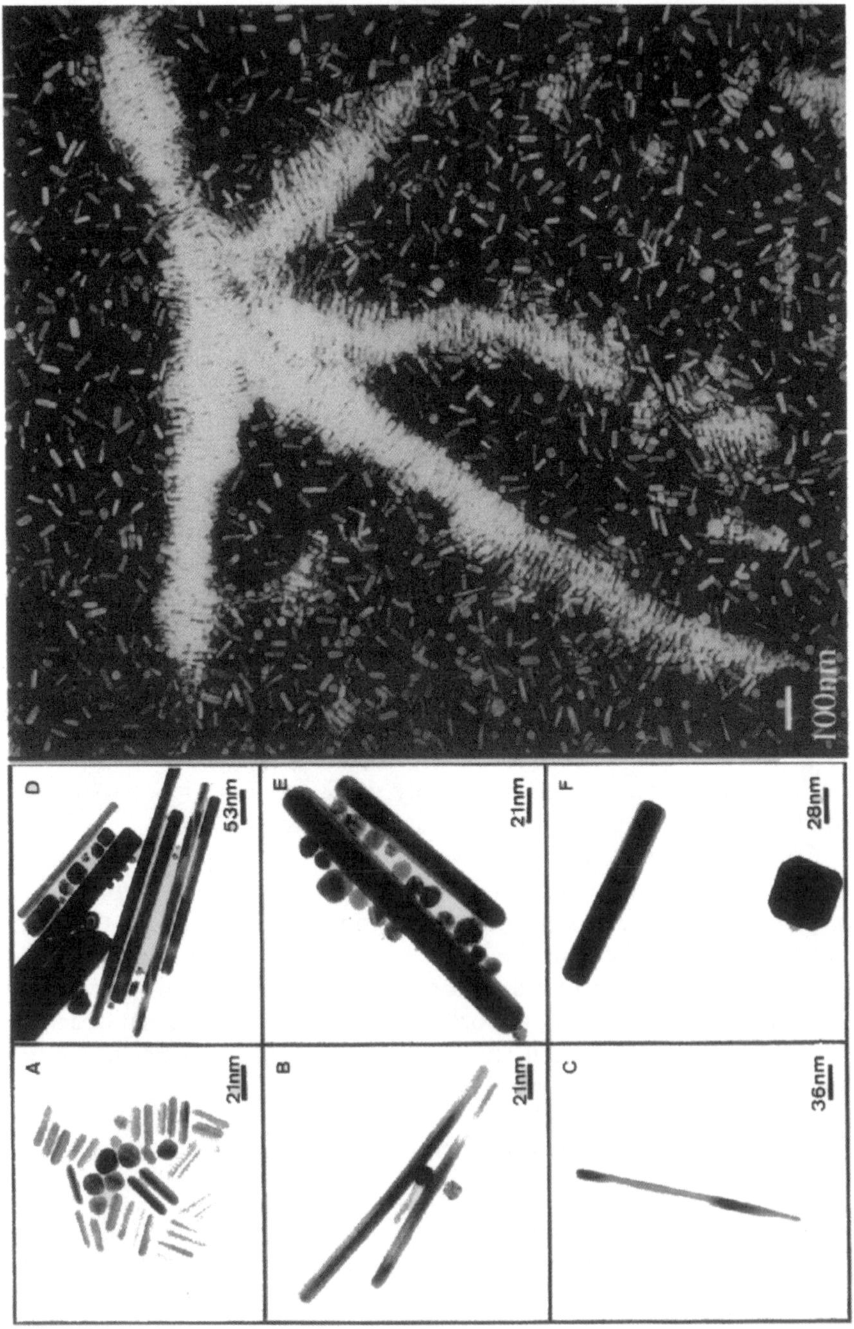

hydrophilic groups to drive the self-assembly process, in the case of SAMs, self-assembly is mainly driven by the strong interaction between a solid substrate and a specific functional group built into the molecule. As described in Chapters 3 and 5, the most widely studied SAM system is the monolayer formation of alkanethiols on gold surfaces [6].

Alkanethiols are similar to lipids because they possess a hydrocarbon tail group that is 1–4 nm in length, but instead of containing a headgroup designed to be water soluble, they have a head group designed to specifically bond to a gold surface, such as a thiol group (a sulfur-containing functional group, –SH). Quite often the alkanethiols also contain additional active groups at the other end of the thiol group such as positively or negatively charged groups, electroactive groups that can be oxidised or reduced, or groups that can be used to attach other molecules onto the SAM by further chemical modification.

Because of the strong interaction between thiols and gold, when a gold surface is placed in a dilute solution containing the alkanethiol (Figure 6.4), a monolayer of alkanethiols attaches itself to the gold surface *via* the thiol group. The hydrocarbon tails line up and point away from the surface. The driving force is the bond formation between the head group and the surface atoms, therefore once the surface is covered with a monolayer of molecules further alkanethiols will not adsorb. The gold substrate can be removed from the solution and rinsed to remove excess alkanethiol, leaving a surface that is coated with a monolayer of molecules.

This technique allows us to do a number of things. The really interesting thing, however, is that it allows us to control the surface properties of materials. As previously mentioned, when dealing with nanoscale structures, the ratio of surface to bulk material rises dramatically the smaller the structure becomes. Enrico Fermi (Nobel Prize winner for physics) is reputed to have said that 'God made the bulk but the Devil created the surface'. This is because the orbitals of the surface atoms are generally not as fully occupied as the atoms in the bulk of the material, and they are exposed to all the reactive chemicals present in the atmosphere or solvent (oxygen, nitrogen, carbon dioxide, water vapour, hydrocarbons and so on). Sometimes these surface effects are useful in optical devices (Chapter 7) or for producing other properties (Chapter 3). This leads to messy, ill-defined mixtures of stuff covering the surface. By coating the surface with a SAM, the properties of the surface can be uniformly controlled.

The power of the SAM formation technique is extended when one considers that it is possible to incorporate a wide range of functional groups into the individual molecules and that these functional groups can be used to further modify the properties of the surface. For instance, by making the functional group a methyl group, the surface becomes hydrophobic, so that water beads and runs off the surface.

Figure 6.4

Method of formation of a self-assembled monolayer. A gold surface is placed in a solution of hexadecanethiol (typically 1 mM in ethanol) and left for several minutes to hours. The thiol group will interact and bond to the gold surface, forming a monolayer. Excess alkanethiol is removed by rinsing, leaving only the monolayer of alkanethiol on the surface.

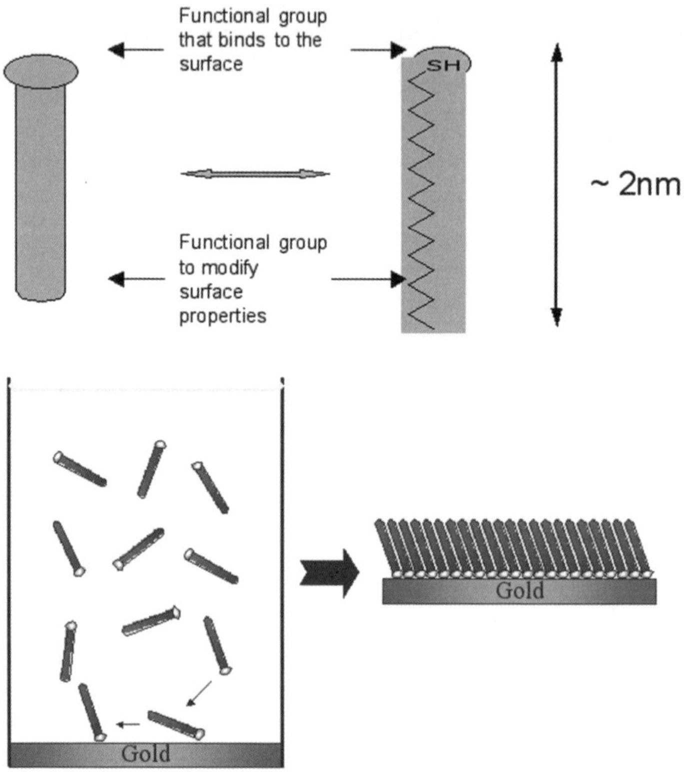

Replacing the methyl group with a negatively charged phosphate group makes the surface hydrophilic and completely wettable by water. The surface is sensitive to the acidity of the solution and thus can be used to measure pH changes. By incorporating electron withdrawing or electron donating groups into the SAM-forming molecules, it is even possible to influence the electronic nature of the substrate material.

6.4 THE BITS THAT DO THINGS — PROTEINS

THREE-DIMENSIONAL STRUCTURES USING A 20 AMINO ACID ALPHABET

In biology, a group of polymers called proteins are used to switch something on or off, move something, sense something, taste, smell, produce energy, convert sunlight to sugar, nitrogen to ammonia fertiliser, fight disease and so on. Proteins are made up of amino acids (Figure 6.5a) and there are 20 common amino acids that are linked using amide bond formation (Figure 6.5b). The amino acids basically only differ by the type of side chain functional group (the 'R' group in Figure 6.5a). This side chain can be hydrophobic, hydrophilic, positively or negatively charged. Biological systems synthesise linear polymers of these amino acids (Figure 6.5b). The amazing thing, however, is that out of all the possible structures that the polymer could adapt, during the synthesis the amino acid polymer folds into one specific stable three-dimensional structure This is no simple trick, and so far even using supercomputers we haven't been able to figure out how it is done. In other words, given a sequence of amino acids, we can't predict what structure it will adapt or how to force it to adopt a specific structure.

Globular proteins generally have a diameter of 2–8 nm. There are approximately 80 000 different proteins in the human body. Proteins can be made to function as very specific catalysts, as ion conductors in nerve cells to propagate nerve impulses, as the machinery that converts sunlight into chemical energy, as receptors that bind specifically to foreign material in the body and even to bind to certain crystal faces of ice, thus preventing fish in arctic waters from freezing.

But this is not a biology text, so how can these materials be used in nanotechnology?

NANOSCALE MOTORS [8–10]

People tend to think that biology doesn't use the wheel, that rotating motors and devices are foreign concepts in biology. The real answer is that biology likes to use legs or fins for locomotion but wheels for motors. Virtually every living cell is powered by a myriad of tiny rotating nanoturbines called ATPase (Figure 6.6). These nanoturbines are embedded in a lipid membrane (much like electrical turbines are embedded into the walls of dams) and when there is an excess of protons on one side of the membrane then the protons flow through the ATPase and literally turn a small shaft of the nanoturbine. At the same time, the turning of the shaft allows the ATPase to convert ADP (adenosine diphosphate) to ATP (adenosine triphosphate). Without going into details, the ATP synthesised during this process is a high energy molecule that is used to drive the chemistry that keeps cells alive. What's interesting from our perspective is that the ATPase complex (Figure 6.6) is truly a nanosized motor. It's roughly 12 nm in diameter and rotates at a calculated no-load velocity of 17 rotations per second. In the presence of a proton gradient the shaft rotates clockwise, whereas if there is no proton gradient but in the presence of ATP the reverse reactions occur and the shaft rotates anti-clockwise. In quite a staggering development, researchers at Cornell University (the

Figure 6.5

a) Picture of individual amino acids.

b) Linking amino acids together to form a polypeptide.

The tertiary structure of a protein is formed by folding a polypeptide chain into a three-dimensional structure.

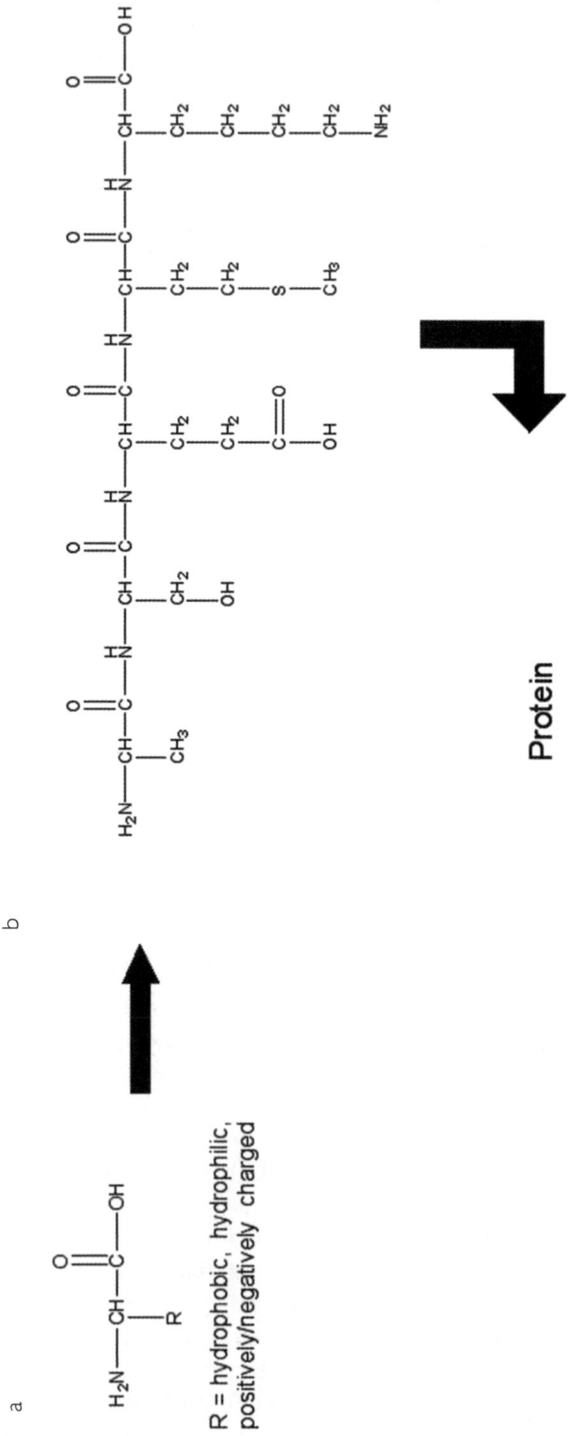

Montemagno group [9]) managed to isolate individual ATPases and transfer them onto a solid substrate. They then attached large micron-sized polystyrene beads or small silicon propellers to the shaft of the 12 nm diameter ATPase and observed the rotation of the polystyrene bead or silicon propeller.

Figure 6.6

Schematic picture of ATPase with a rotor attached.

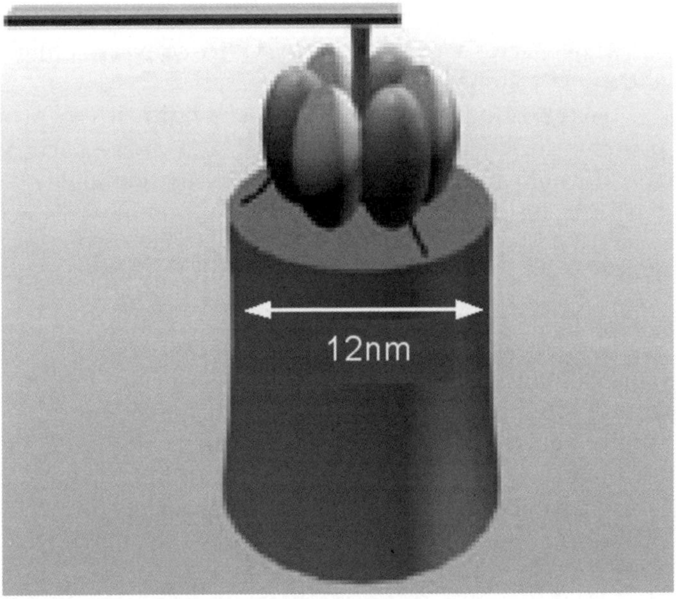

Whether this particular molecular motor is the one that has the appropriate stability, lifetime, and load capacity, or whether other genetically engineered or chemically modified versions are needed for practical nanomachines, is not known at this stage. However, the proof that it is possible to build functional nanomotors has been given and maybe these tiny motors could eventually be used to propel tiny bionic machines.

BIOLOGICAL COMPUTING — A PROTEIN-BASED 3D OPTICAL MEMORY BASED ON BACTERIORHODOPSIN

Bacteriorhodopsin is another interesting example of using something from biology for our own needs. Bacteriorhodopsin is a purple-coloured protein that is found in the lipid cell membrane of a bacterium (*Halobacterium salinarum*) that lives under extreme conditions in salt marshes, at high temperatures and under near-anaerobic conditions (that is, in the absence of oxygen). Because it is derived from a bacterium living in such harsh conditions, the bacteriorhodopsin is an extremely tough protein. For example, it is stable at temperatures up to 140°C while dry, it is not affected by exposure to sun and air for years, it tolerates pH values from 0 to 12 or high salt concentrations (3 M) and can even be incorporated into plastics without destroying its structure. This just goes to show that not all things biological are fragile. It is also a relatively simple protein compared to the chlorophyll-based photosynthetic complex used by plants. The protein contains a chemical that is very similar in structure and properties to those of the eye pigment rhodopsin, which enables us to sense light and to see. The bacteria use the pigment to convert sunlight into electrical and chemical energy.

Bacteriorhodopsin is essentially a light-driven proton pump. The proteins are lined up inside the cell wall of the bacterium and when it absorbs light it transports a proton across the lipid cell wall. This transport sets up a proton imbalance, or gradient. This proton gradient is now used by the ATPase proteins, which were discussed in the last section, to generate ATP. The bacteriorhodopsin sitting in the cell wall thus acts like a miniature solar cell.

The bacteriorhodopsin has other properties that are of interest to nanotechnology [8]. It can be used as an ultra fast bi-stable red/green photoswitch for making three-dimensional optical memories. In its simple version, the bacteriorhodopsin is dispersed in a polymer gel, forming a clear, purple cube. A pulse of light from a green laser beam is used to activate the bacteriorhodopsin (Figure 6.7a). The bacteriorhodopsin absorbs a photon and uses the energy from this photon to alter its structure from its initial ground state to an active state normally referred to as the *O* state. A second pulse of light from a red laser is used to drive the *O* into an intermediate *P* state, which quickly relaxes to a stable *Q* state. So, by using two sets of lasers, we have driven the bacteriorhodopsin from one stable state to another stable state. By placing the two sets of lasers (green and red) at right angles to each other on two sides of the cube, we can write information anywhere in the three-dimensional space of the cube, and only where the two lasers intersect will the bacteriorhodopsin be driven through all possible states.

To read back the stored information, low intensity light is used so that there is no interconversion, but the absorption can be monitored. If we shine the green laser at the cube, the ground state bacteriorhodopsin will again absorb a photon and be driven into its red light-absorbing *O* state. The bacteriorhodopsin that is already in the *Q* state will not be affected. If we now shine a very low intensity red light laser through the bacteriorhodopsin, those areas that are in the *O* state will absorb the red light, while those areas in the *Q* state will allow the red light to pass through, which can then be read by a high sensitivity light detector. The intensity of the red interrogating laser must be low enough so that it does not convert significant bacteriorhodopsin numbers from the *O* state to the *Q* state.

Figure 6.7

a) Schematic of the write cycle of a bacteriorhodopsin optical memory. In the first step (i) a green laser activates a line of bacteriorhodopsin, converting it to the O state. A red laser is then used to convert a patch of the bacteriorhodopsin to the P state (ii) from which it relaxes (iii) to the stable Q state.

b) Schematic of the read cycle of a bacteriorhodopsin optical memory. The green laser is again used to activate the bacteriorhodopsin (iv) to the red light-absorbing O state. The Q state does not absorb red light. In step (v) the bacteriorhodopsin-containing gel is interrogated with a low intensity red laser. The red light will pass through those areas that were activated in steps (i) to (iii). This can be readily measured using a photo-detector and with a value of either 1 or 0 assigned to the output.

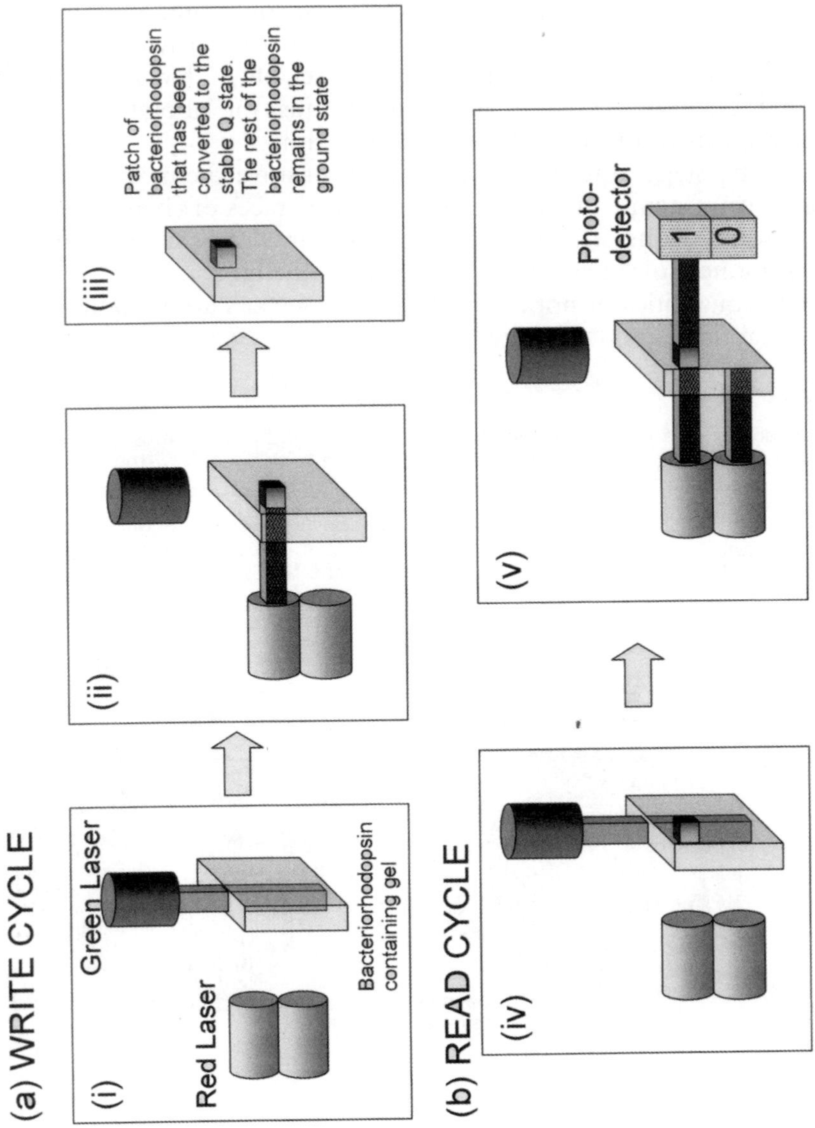

There are at least three advantages of this type of memory device. Firstly, the read-write operations can be carried out in parallel. Secondly, it is an optical technique and photons are faster than electrons. Thirdly, because the memory is three-dimensional it can store data at densities of 10^{11} to 10^{13} bits per cubic centimetre compared to two-dimensional optical memories, which are limited to approximately 10^8 bits per square centimetre.

ION CHANNELS AS SENSORS

One of the first practical applications in the area of biological nanotechnology will probably be chemical sensors [11–14]. To make chemical sensors, a number of issues must be confronted. For instance, the sensor should be highly specific towards the analyte to be measured, perhaps so sensitive that it can detect only a few molecules. It should function in a real environment, not just the laboratory, and the sensor mechanism should function in real time.

At this stage of technology, the most sensitive chemical sensors are biological systems. Bloodhounds can detect traces of chemicals left by a person walking in the woods, moths can detect a few thousand pheromone molecules per cubic metre, and the human body can detect minute quantities of hormones. Chemical sensors are ubiquitous and essential throughout biology.

Figure 6.8

a) Schematic of a cell bilayer membrane containing lipids and various membrane proteins, including an ion-channel. Ion channels are basically hollow tubes through which ions can traverse the insulating lipid membrane.

b) The ion flow through the channels can often be switched on or off depending on external stimuli, such as an analyte binding to the mouth of the channel or the presence of an electrical potential difference on the two sides of the membrane (as in the case of nerve cells).

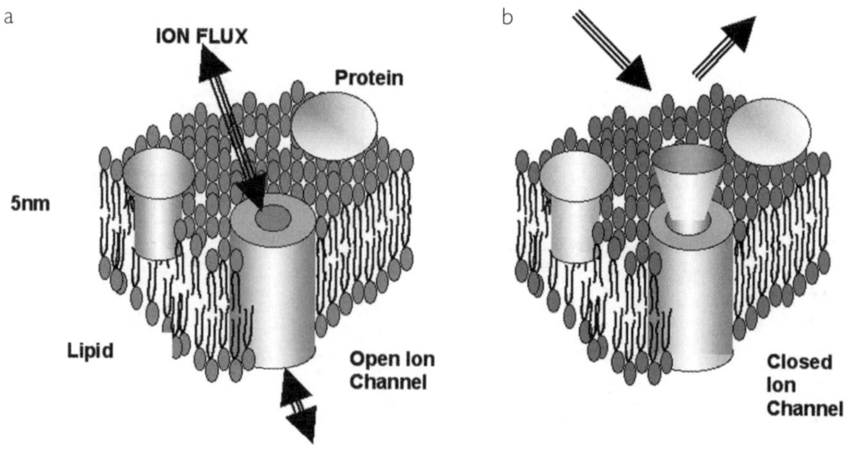

The essential sensing element in biology is a nanometer-sized, protein-based on/off switch. At its simplest, the switch is an ion channel stuck inside a lipid membrane (Figure 6.8a). As discussed in the previous section, lipid membranes are good insulators to aqueous ions such as sodium and chloride. The ion channel can be thought of as a hollow tube that penetrates from one side of the membrane to the other and normally allows ions to flow through its pore. The on/off switch is made by switching the ion channels on or off using an external stimulus. For instance, the ion channel may have a receptor for a particular chemical at the mouth of the tube. If the chemical binds to the mouth of the channel, it may simply physically block the channel so that ions can no longer flow through it (Figure 6.8b).

The passage of ions through an otherwise insulating membrane can be measured electrically because a current (ionic not electric) flows through the aqueous solution and through the membrane. Typically, between one and ten million sodium ions may pass through an ion channel in a second. The membrane conductivity can be measured in the same way as a million electrons passing along a copper wire.

The clever thing is that an ion channel can be made to act as a chemical amplifier. Measuring the presence of a single molecule directly is not easily done with our current technology. However, measuring the ion flow through a single ion channel is possible: in fact it is a standard technique in biophysics. Currents flowing through single ion channels are typically in the nano- to pico-amp range (10^{-9}–10^{-12} amps) when it is in the on state. If a single molecule now comes along and blocks the ion channel, the ion flow drops to zero. A single molecular event can effectively reduce a 1–10 million-fold flow of ions through the channel. Since current flows of pico-amps can easily be measured, it is simple to monitor an event that is caused by one molecule.

Ion channel conduction may be modulated by a number of mechanisms. For instance, rather than a simple blocking of the ion channel, the binding of molecules near the mouth of the ion channel may cause a conformational change to occur elsewhere in the channel; or the ion channels may be turned on or off by the presence of an electrical potential, as in the case of nerve cells.

For practical considerations the problem with building an ion channel based sensor is that most ion channels used by biology are large, relatively fragile proteins that are specific towards only a certain molecule. The exact structure of most ion channels is not even known yet and it would require enormous work to adapt the structure of the ion channel for every analyte we want to detect. In order to make a practical chemical sensor, at least three or four things are required.

Firstly, a universal molecular recognition element is needed. This is relatively simple, because such elements already exist and are widely used in biotechnology. They are called antibodies. Antibodies are Y-shaped proteins with regions at the top of the Y shape that bind to specific molecules (see Figure 6.5c). They are used by the body as the first line of defense in the immune system. If a foreign molecule gets into the body it responds by producing antibodies that bind to the foreign molecule. Once these antibodies are attached, the other defense mechanisms in the body take over to eliminate the intruder. Antibodies can be produced by modern biological methods to specifically bind to virtually any molecule. They are used as the basis of most modern pathology tests.

The second thing that is needed is an insulating lipid membrane, preferably attached to an electrode. In natural systems, the ion channels are attached to the cell membrane connected to a three-dimensional body. In a sensor, we need to be able to connect the sensor to the sensor electronics and read-out device. However, because we're interested in the flow of ions through the ion channel, both sides of the lipid membrane must be able to accommodate an ion flow. So, in a sensor device that has a lipid membrane attached to a solid electrode, an ionic reservoir between the electrode and the lipid membrane must be provided.

The third thing that is required is a general sensing mechanism that doesn't rely on changing the structure of the ion channel each time it tests for a new analyte. This is difficult, but it may be possible to use antibodies to bind to the specific analytes, as discussed above. If a simple, rugged ion channel could be found or developed, that would be an added bonus.

All of this sounds complex and at first glance fairly far-fetched. One has to take nanosized biological units such as antibodies, ion channels and lipids, chemically modify or create synthetically stable analogues of these units, then come up with a generic mechanism for sensing. By understanding the biophysics of the individual units, the sensor must be made to self-assemble so that it is manufacturable and will function with the right sensitivity and response time in the real world. Finally, the engineering must be right to enable the sensor electrode to connect to the world of electronics!

The power of combining modern biology, chemistry and physics and working across the disciplines was demonstrated by the fact that such a device has been successfully constructed. A schematic diagram of the sensor structure and the mechanism are shown in Figure 6.9.

How does it work in detail? The ion channel that it uses is a special bacterial ion channel called gramicidin. Gramicidin is a relatively small ion channel that is made up of two polypeptide tubes. Each tube is capable of punching a hole in one half of the lipid membrane. The gramicidin tubes can freely and independently diffuse through the inner and outer half of a lipid membrane. Cations can only flow through the membrane when the two halves line up so that a continuous hole is formed through the membrane. When they're not lined up

a

b

c

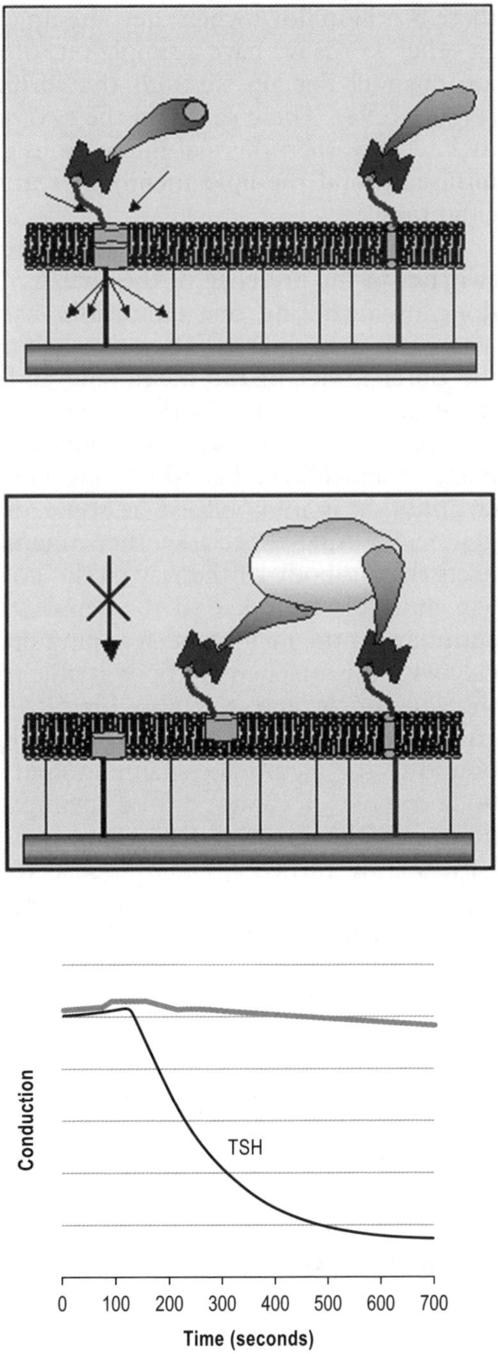

Figure 6.9

Ion channel switch biosensor. a) The lipid bilayer (4 nm thick) is tethered onto a gold electrode using hydrophilic tethering groups. The tethering groups form a 4 nm thick aqueous layer into which ions can flow. Gramicidin ion channels are incorporated into the lipid bilayer structure. One half of the ion channel is immobilised on the gold electrode, the other half is attached to a receptor molecule and is free to diffuse in the outer layer of the lipid membrane. In the absence of analyte, the two halves of the ion channel line up and ions flow freely through the membrane. That is, the membrane is ionically quite leaky and conductive. A second receptor is attached to a special lipid that spans the whole bilayer membrane and the reservoir, and is immobilised on the gold electrode.

b) If an analyte molecule is added to the biosensor the two receptors bind to it. This physically disrupts the two halves of the ion channel, preventing them from aligning. This stops the flow of ions through the ion channel, so that the membrane becomes less conductive.

c) Typical response curve on addition of analyte (top trace is a control sample without analyte, bottom trace is the response in the presence of analyte). The ion channels are switched off and the conduction through the membrane decreases.

there is no ion flow; when they line up on top of each other, ions flow. In other words we have a simple on/off action. The two halves of the ion channel line up through the formation of six hydrogen bonds (Figure 6.9a). These can hold the two halves together for roughly 0.1 to 1 second, then thermal fluctuations cause them to break apart and diffuse around the lipid membranes until they meet the next gramicidin tube.

The next trick is to be able to control this on/off switch so that it switches in the presence of the analyte that we want to detect. This is done by anchoring one gramicidin half onto the electrode surface, where it is immobilised. The other half of the gramicidin is placed into the outer leaflet of the membrane and is connected to an antibody receptor group. Normally the upper gramicidin can diffuse in the outer membrane leaflet and will every now and then sit on top of the immobilised gramicidin and conduct ions. For the sensor to function, another antibody is immobilised at some distance from the immobilised gramicidin using a special synthetic membrane-spanning lipid that connects the antibody all the way to the gold electrode surface. If an analyte molecule is added to the sensor membrane it will bind to the antibody on the membrane-spanning lipid. Now the gramicidin molecule with the attached antibody is able to diffuse in the outer leaflet of the membrane and will also bind to the analyte molecule that is attached to the membrane-spanning lipid. Once the gramicidin is bound to the membrane-spanning lipid *via* the antibody-analyte-antibody complex, it cannot diffuse any more and it cannot form ion channels with the inner leaflet gramicidin; that is, the ion channel is permanently turned off in response to an analyte molecule (Figure 6.9b). Figure 6.9c shows a typical response of the sensor towards a large protein analyte called TSH (a hormone). You can see that once a blood sample containing the analyte is added, the conduction of ions through the ion channel decreases with time as the ion channels are switched off.

Note that in the design of the sensor a 4 nm reservoir is formed between the lipid and the gold electrode. These reservoirs are formed from hydrophilic oligo-ethylene glycol groups that act as an ionic reservoir for the ions that go through the ion channels and, at the same time, are used to tether and immobilise the various ion channel and lipids to the gold electrode. Because this 4 nm ionic reservoir has only a limited capacity before it fills up, a sinusoidal voltage that swings from positive to negative is used to see if the ion channels are conducting. In other words, we measure the rush of ions going into and out of the reservoir through the ion channels. If a direct current source were used to measure the ion flow, then the ionic reservoir would be filled up very rapidly. Once filled, no more current would flow.

One of the reasons for going into a detailed description of this device is to demonstrate that it is possible to build very complex nanoscopic devices based on principles derived from biology — devices that have moving parts and that are robust enough to function in real world applications. The other reason is that this device also demonstrates that the individual units that make up the sensor can be designed so that the whole sensor basically self-assembles.

All of the pieces that make up the bottom part of the membrane, that is the immobilised gramicidin, the reservoir, the membrane-spanning lipid, and the inner half of the lipid membrane are formed simply by immersing a gold electrode into a solution of these components. The components all possess a sulfur group at one end. Using the mechanism discussed in the first section on self-assembled monolayers, all of the components spontaneously form a monolayer on the gold electrode. The electrode is rinsed, leaving this monolayer behind. A second solution containing a special mix of lipids and gramicidin is then added, and after addition of water, this lipid mixture again spontaneously self-assembles to form a lipid bilayer membrane. Why? Because the lipid structures were developed so that the same physics of cohesive and dispersive forces that make the lipids assume the structures shown in Figure 6.2 caused them to form a bilayer membrane on the surface of the self-assembled monolayer in the presence of water. They had no choice. Given defined conditions of temperature and component concentration, the final sensor structure was already built into the structure of the individual components.

The inherent power of this method of constructing complex nanosystems using self-assembly is one of the most important lessons that we need to learn from biology and apply to other areas of nanotechnology.

6.5 STRUCTURE IS INFORMATION — DNA

WHAT IS DNA?

Throughout this book we have assumed that the reader has a general knowledge of deoxyribonucleic acid, better known as DNA. Here we need to go into a little more detail. Out of the three groups of biological building blocks that we have considered so far (lipids, proteins and DNA), the DNA molecule is chemically the simplest. It is made up of a five-carbon sugar, deoxyribose, a phosphate and a linear polymer composed of four subunits of nitrogen-containing compounds (adenine, thymine, guanine and cytosine, which are more commonly represented by the letters A, T, G, C: Figure 6.10a). The characteristics of the subunits are such that adenine will bind to thymine, and guanine will bind to cytosine (Figure 6.10b). Therefore, if the two DNA polymers with the appropriate complementary sequence of nucleotides are made, then the two strands of the polymer will bind together. The structure that most people are familiar with is the famous double helix. Biology uses DNA to store information. In fact, the whole blueprint of the structure of each cell in your body is stored in the DNA — the information that tells certain cells to become liver, muscle or nerve cells, where and how to grow bones, which proteins to manufacture and much else. DNA has so much information-carrying capacity that much of the information is probably redundant.

Figure 6.10

The double-stranded DNA molecule is held together by chemical components called bases.

a) The chemical structure of the four bases. The bases are attached to the deoxyribose at the R group.

b) The bases are adenine (A) which bonds with thymine (T), and cytosine (C) which bonds with guanine (G).

a

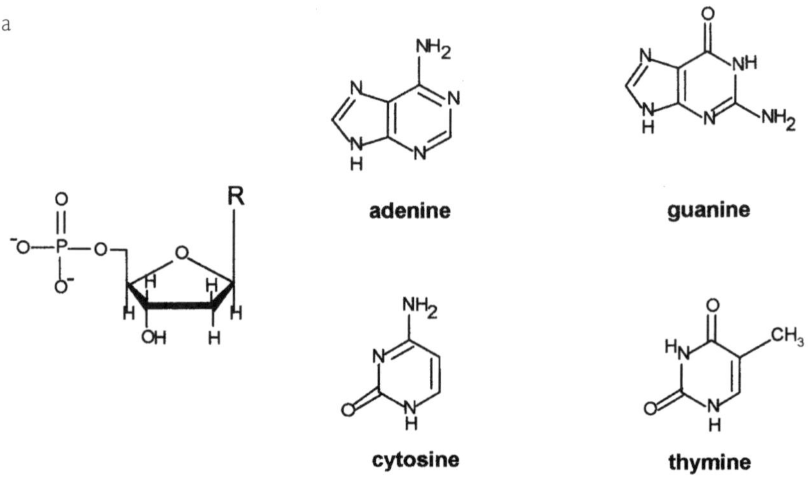

b

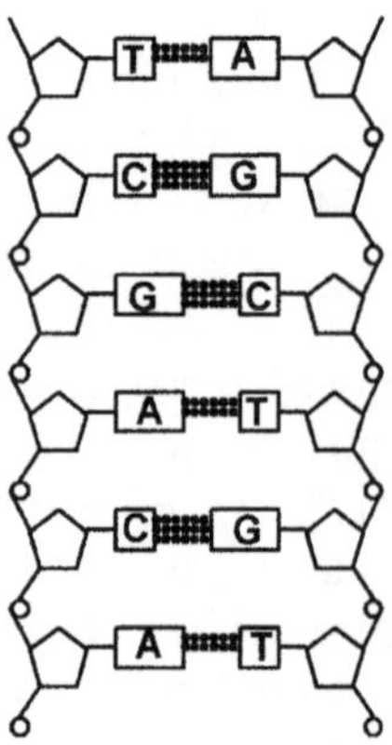

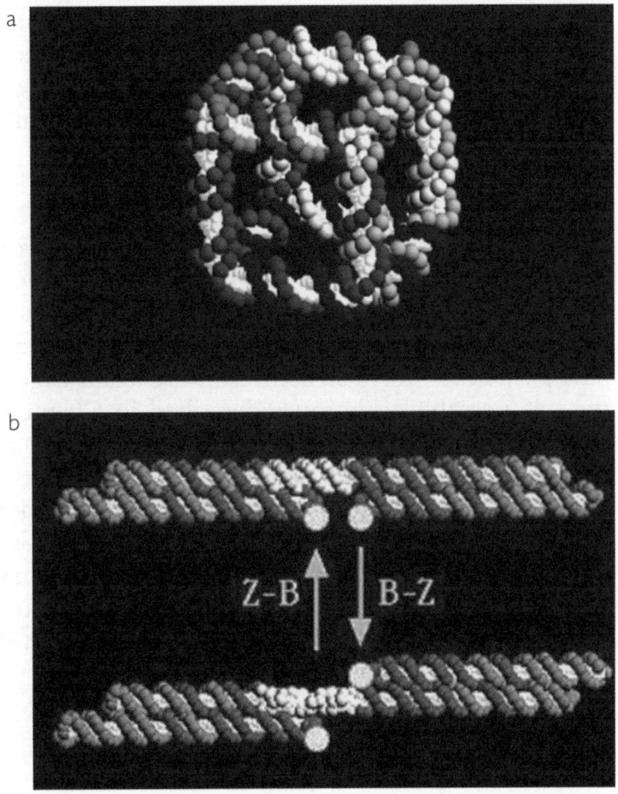

Figure 6.11

Molecular models of DNA structures.

a) A DNA cube.

b) A chemically driven molecular hinge.

Another useful substance is the related compound, ribonucleic acid or RNA. RNA and DNA differ in three ways:

1 the sugar in RNA is ribose

2 thiamine is replaced by uracil

3 RNA is usually single stranded.

USING DNA TO BUILD NANO-CUBES AND HINGES

Although we commonly think of DNA existing only as the double stranded helix that can store genetic information, it can also be used to make structures such as nanoscale knots, squares, and hinges.

For instance, Seeman and colleagues [15, 16] have found a way to produce branched Y-shaped and cross-shaped junctions rather than just the linear double helix structure. Using these structures it is possible to construct squares, cubes (Figure 6.11a) and even octahedra made just from DNA.

Recently it was shown that moveable hinge structures could be constructed from DNA (Figure 6.11b). In this structure the DNA rotates from one shape to the other in the presence of a chemical trigger, here labelled Z-B and B-Z. This device is being explored to see if it could function as another type of nanosized motor.

DNA AS SMART GLUE

Because one DNA sequence can be designed to bind specifically to another DNA sequence and essentially to none other, DNA can function very effectively as a 'smart' molecular glue that can be used to bond nanoscale objects.

For instance, as shown in Figure 6.12, DNA can be used to link gold nanoparticles to form structures of use against bioterrorism [17]. Using the self-assembly techniques described in Chapter 6.3, it is possible to coat gold nanoparticles with DNA strands that have a thiol attached to one end. Two different types of DNA are linked to the nanoparticles, and they are mixed together. As the DNA strands are not complementary, that is, they do not cross-link, the nanoparticles remain separate and stay in solution. When a third strand of complementary DNA is added to the mixture, the complementary strand binds to the DNA strands on the nanoparticles and cross-links them into arrays (Figure 6.12b)

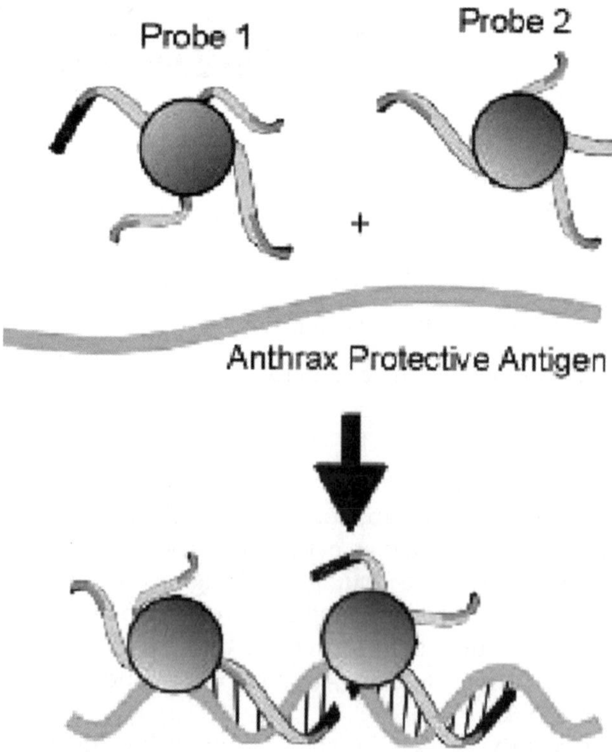

Figure 6.12
Using DNA as clever glue. Gold nanoparticles (circles) are first functionalised using two different, non-complementary types of DNA. Because the DNA strands do not bind to each other the gold nanoparticles remain separate. On addition of a strand of complementary DNA the nanoparticles bind to the DNA. Because of their close proximity their optical properties change and a red to blue colour change is observed.

An interesting optical effect arises when the distance separating the individual gold nanoparticles varies. When gold nanoparticles are well separated and dispersed in solution they range in colour from deep burgundy red to oranges and browns, depending on the size of the individual nanoparticles. When these nanoparticles are in relatively close contact, or if the nanoparticles aggregate, the colour changes to a blue-black. In the present case, where the DNA strand causes a cross-linking reaction, not only is a metal-DNA nanocomposite material created, but there is also a colour change due to the aggregation in the presence of the DNA strand. This effect is being exploited in the development of a sensitive optical DNA test.

The same metal-DNA nanocomposite material is being used to explore the ability of DNA to act as an electronic conductor. Although still a controversial area, various researchers have noticed that DNA can act as a conductor, semiconductor or insulator of electrons depending on the structure. The details are being intensively studied. If they can be proved, then this would add still another property to the utility of DNA as a nanomaterial.

WIRING UP THE NANOWORLD: DNA AS WIRE TEMPLATE

Even if it turns out that DNA does not behave as an electrically conducting molecule, it can still be used to form nanosized wires [18]. In this case the structure of DNA is used to link two electrodes. The DNA is then used as a template that can be metal plated to construct nanosize metal wires.

In this clever piece of work, use is made of the fact that it is relatively simple to obtain long strands of DNA and that the cross linking process can be used to glue the DNA strand in specific places. As shown schematically in Figure 6.13a, two electrodes were formed with a gap between them. The two electrodes were separately coated with two different short DNA sections called oligonucleotides (Figure 6.13a (a)), which were 'sticky' towards the two ends of a long piece of a type of DNA called λ-DNA. The electrodes were then immersed in a solution containing the long-strand λ-DNA. The ends of the λ-DNA bound to the oligonucleotides on the electrodes and formed a bridge (Figure 6.13a (b)).

DNA is negatively charged because of the phosphate groups along the backbone, and it is possible to exchange the sodium counter-ion with silver ions (Figure 6.13a (c)). The silver ions can be chemically reduced to silver metal (Figure 6.13a (d)). More metal can then be deposited onto the silver metal by standard silver deposition techniques until a conductive silver wire is formed (Figure 6.13a (e)). The final wire is approximately 100 nm in diameter and 12 μm long (Figure 6.13b).

These three examples; hinges, glue and wires have shown some of the possibilities of using DNA as a nano-building block. Researchers are

Figure 6.13

Use of DNA as a template for growing nanowires.

a) Schematic of the electrodes, the DNA template and the subsequent metal incorporation procedure.

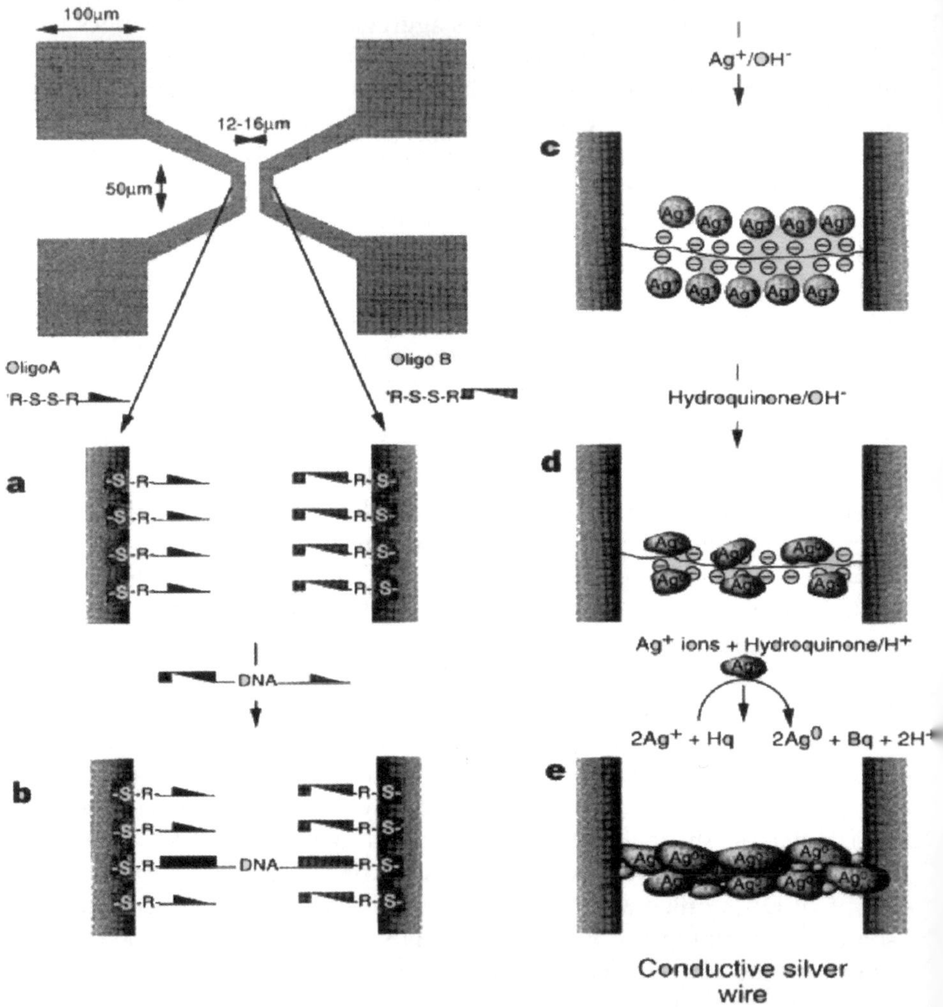

using DNA as structural elements to form squares, cubes and moveable hinges; as a molecular glue to join structures such as nanoparticles; as a template to build nanoscale wires in nanocircuits — and we haven't even mentioned the work that is going on in constructing DNA computers that can solve problems that digital computers cannot.

All of this versatility from a molecule made from just four basic subunits.

Figure 6.13

Use of DNA as a template for growing nanowires.

b) Atomic force micrograph (AFM) of the Ag wire showing the 100 nm diameter.

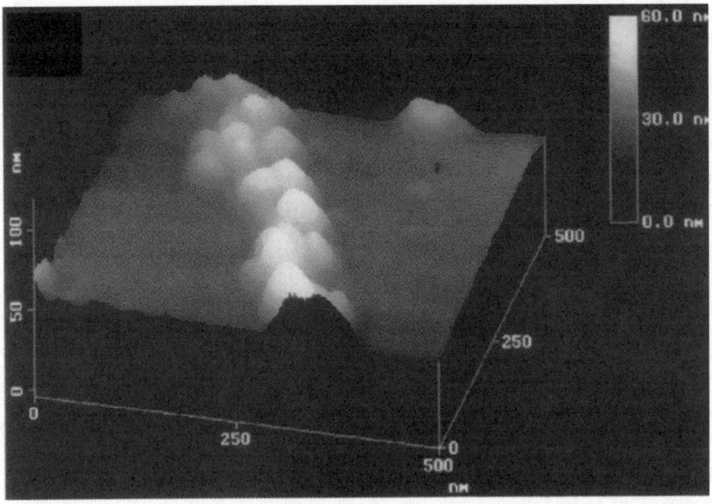

6.6 A BIOLOGICAL NANOTECHNOLOGICAL FUTURE

Clearly there are lots of lessons that we can learn from biology. We have looked at just a small sample of the sort of work and ideas that are being explored at the intersection of biology and nanotechnology. This is certainly not an exclusive list. We have not even discussed biology and nano-optics [19], biomineralisation [20], the use of DNA for computation [21], or any of a number of other areas. However, the point is not that the particular examples that we looked at are the ones that will definitely herald the new technologies. Rather, the examples are shown to provide inspiration about the versatility of biological principals and to highlight some key points.

One of the most intriguing discoveries is that with only a few basic building blocks it is possible to construct an astonishing range of materials, devices and structures. Using four DNA bases, over 20 amino acids, or the basic geometric shapes of lipids, it is possible to build vast quantities of different nanomaterials, nanostructures and scaffolds and even some functioning nanomachines.

The use of simple building blocks to produce more complex structures with new properties is the basis of nanoarchitecture. These new structures can then be used as building blocks to produce even more complex structures.

The mechanism by which this is done is self-assembly. The individual building blocks contain the information of the final structure that they will form. It is programmed into the three-dimensional molecular structure of the individual building blocks. The trick is to learn the 'programming language' of different molecules and their self-assembly.

If we now apply these principles to the materials that we find easy to produce, such as nanotubes, buckyballs, nanoparticles, metals and ceramics (described in Chapters 3 and 4), then we already have a toolbox that will enable us to develop many nanotechnologies, including perhaps biological optical devices.

Most importantly there is the lesson of looking beyond the obvious and expanding our horizons outside the traditional boundaries of the individual scientific disciplines. Perhaps biological principles can be applied to traditional manufacturing problems to develop a way of building a biomimetic system that builds the next generation of computers. These would not be produced in a billion-dollar factory, but in a beaker. Better yet, maybe we could construct a device that takes sunlight, builds itself, converts CO_2 from the atmosphere into a usable fuel, and as a bonus desalinates the soil. At the moment, these macroscopic ensembles of hierarchically assembled nanodevices are called Mangrove trees. They work superbly, but not in the high-salinity farming areas where we need them.

WHAT YOU SHOULD KNOW NOW

1 The structure of DNA, lipids and proteins.

2 How nature forms self-organising supramolecular structures.

3 How the ion channel sensor works. A biological on/off switch can be made to work using a membrane by anchoring one gramicidin half onto an electrode surface while the other half can move freely.

4 How to make a biological computer. Bacteriorhodopsin is essentially a light-driven proton pump. The proteins are lined up inside the cell wall of the bacterium and when it absorbs light it transports a proton across the lipid cell wall. By using bacteria in a cube, and by placing two sets of lasers (green and red) at right angles to each other on two sides of the cube, information can be written anywhere in the three-dimensional space of the cube. This computer operates on photons, not electricity.

5 The structure of DNA glue, hinges and cubes. You should know that it is possible to use chemical reactions to glue DNA strands in specific places.

6 That virtually every living cell is powered by a myriad of tiny rotating nanoturbines called ATPase.

6.7 EXERCISES

1 *The human genome project paper is in the February edition of* Science 2001. *Have a look at it and count the number of authors on the paper. You will understand why teamwork is necessary in big projects and when you go for a job as a scientist a selection criteria will be the capacity to work in a team. Read the introduction to this work in the editorial sections.*
2 *Find out what is meant by λ-DNA, and gramicidin.*
3 *Read papers 8–11 on molecular nanomachines.*
4 *Write an essay about the use of DNA in computing applications (reference 21).*

6.8 REFERENCES

1 Ringsdorf H, Schlarb B & Venzmer J (1988) *Angew. Chemie Int. Ed. Engl.* 27: 113–58.
2 Liu J, Feng X, Fryxell GE, Wang L-Q, Kim AY & Gong M (1998) *Adv. Mater.* 10: 161–65.
3 Langevin D (1992) *Annu. Res. Phys. Chem.* 43: 341.
4 Mann S (2000) *Angew. Chemie Int. Ed.* 39: 3392–406.
5 Soten I & Ozin GA (1999) *Curr. Opinion in Coll. & Interface Sci.* 4: 325–37.
6 Pileni MP (1998) *Supramol. Sci.* 5: 321–29.
7. Nikoobakht B, Wang ZL & El-Sayed MA (2000) *J. Chem. Phys. B.* 104: 8635–40.
8 Ullmann A (1996) *Chem. Rev.* 96: 1533–54.
9 Montemagno C & Bachand G (1999) *Nanotechnology* 10: 225–31.
10 Birge RR (1995) *Sci. Amer.* March: 66–71.
11 Cornell BA, Braach-Maksvytis VLB, King LG, Osman PDJ, Raguse B, Wieczorek L & Pace RJ (1997) *Nature* 387: 580–83.
12 Raguse B, Braach-Maksvytis VLB, Cornell BA, King LG, Osman PDJ, Pace RJ & Wieczorek L (1998) *Langmuir* 14: 648–59.
13 Bayley H & Martin CR (2000) *Chem. Rev.* 100: 2575–94.
14 Haupt K & Mosbach K (2000) *Chem. Rev.* 100: 2495–504.
15 Seeman NC (1997) *Acc. Chem. Res.* 30: 357–63.
16 Chen J & Seeman NC (1991) *Nature* 350: 631–33.
17 Storhoff JJ and Mirkin CA (1999) *Chem. Rev.* 99: 1849–62.
18 Braun E, Eichen Y, Sivan U & Ben-Yoseph G (1998) *Nature* 391: 775–78.
19 Srinivasarao M (1999) *Chem. Rev.* 99: 1935–61.
20 Ozin GA (1997) *Acc. Chem. Res.* 30: 17–27.
21 Adleman LM (1994) *Science* 266: 993 & 1021–24.

7

OPTICS, PHOTONICS AND SOLAR ENERGY

In this chapter you will learn that nanostructures behave differently in reflection, transmission, radiation and polarisation of light. You will learn about photon trapping and plasmons. You will find out why heat flow in nanomaterials is different to conventional materials and how certain nanomaterials can be used as transparent sunscreens. You will learn about evanescent waves and getting real pictures from 'ghosts' that may allow super resolution of objects.

7.1 PROPERTIES OF LIGHT AND NANOTECHNOLOGY [1]

Light and UV control are especially important for human welfare and everything that flows from our sense of well being. The levels and quality of light we experience can affect us both psychologically and physically. Nanotechnology has already produced some key improvements in lighting that can impact on our health.

Some materials allow light rays and photons to pass through without changing direction. They are called transparent materials. Other materials allow light to emerge travelling in a new direction: they are called translucent materials because we can't see through them, even though light gets through. Nanotechnology can be used to control each of these properties. When the translucent roof of Stadium Australia for the 2000 Olympics was designed, this deviation had to be measured and used in computer models of lighting from the sun. That work was done in 1997, yet if it was done today, recent developments in nanotechnology would result in quite a different translucent roofing material being used. It would provide better light while transmitting less solar heat, and it may even be cheaper. Our own skin is translucent and some emerging biophotonic technologies analyse light transport through tissue to diagnose various disorders.

Nanotechnology may revolutionise optics. To better grasp the changes that occur in small scale structures we must first revise some basic issues concerning how light and materials interact. Optics technology is about controlling the production of photons and where they travel, making use of photons, and selecting particular photons from a group of different ones. Electromagnetic radiation is often treated as waves (Chapter 1.6), with electric and magnetic fields that oscillate. That is, the local strength of the electric field oscillates in time as the wave passes through. In its particle form, electromagnetic radiation is defined as the number of photons of quantised energy in a given time. These particles travel at the speed of light, c (3×10^8 ms^{-1}) in free space, with energy given as $E = h\upsilon$ where h is Planck's constant and υ is the frequency of the wave (which also defines colour). As noted in Chapter 1, colour is due to our ability as humans to identify different wavelengths of visible light, provided there aren't too many of them mixed up together, in which case the light appears colourless, or white. These particle-wave concepts are revised in Figure 7.1a. The response of the human eye compared with wavelengths of solar radiation is shown in Figure 7.1b and a comparison of the size of a 100 nm nanoparticle with red optical radiation of 720 nm is shown in Figure 7.1c. It can be seen that they are comparable. In fact, the wavelength of visible radiation goes down to about 400 nm (Chapter 1, Figure 1.6). Optical studies are intrinsically a nanoscale science because the key length scale in optics, the wavelength of light, is in the nano domain. Particles under a wavelength in size, or arrays or patterns where spacing between the components is around a wavelength, generally have their optical response dominated by geometric parameters. The ability to engineer on the nanoscale thus has the potential to revolutionise optics and photonics, as well as lighting and colour technology.

Current valuable techniques for producing optical devices include:

1 precision nano-imprinting [2] using master stamps produced with the aid of an electron microscope or ion beams (see Chapter 8);

2 electron beam or ion beam lithography, plus chemical or ion beam etching (see also Chapter 8);

3 patterning polymers with interference fringes made from crossed beams of UV light;

4 depositing metal layers over ordered arrays of nanoparticles, then vibrating to free the particles and leave holes [3] (see also Chapter 3.6 and Figure 3.9);

5 building stacks of ceramic nanoparticles, fusing, filling the space, then etching out the originals;

6 aligning arrays of nanotubes (Chapter 3.4);

7 sculpturing thin films using oblique vacuum deposition of ceramic materials and various substrate rotation protocols [4] to give a vast array of patterns (Chapter 3.5, and Figure 3.8);

8 physical or magnetic manipulation of nanoparticles using atomic force methods or nanomagnet arrays (Chapter 2.7 and 2.8); and

9 organic surfactants as templates for hollow nanostructures (Chapter 3.8).

Note that some of these patterning techniques are suitable for mass production, so industry scale production of nanopatterned arrays can ultimately be quite cheap. Many of these nanostuctures also have quite distinct properties with respect to variations in the direction they are viewed or the direction of incident light. Many change colour spectacularly as the viewing angle changes. Some new paints have this property, where each very small paint particle has a nanocoating of controlled thickness applied using advanced coating techniques. Nano-thick coatings of clear material such as 'nano-rust' and 'nano-sand' (iron and silicon oxide layers) give colour due to the wave from the front and back surfaces of the nano-coating adding to cancel some colours and make others brighter. This is called interference. Different thicknesses give different colours, and different angles mean the light rays must travel through different thicknesses. These paints are already being used in some cars, nail polish, jewellery and other decorations. Thickness variations in thin slicks of oil form rainbows for the same reason.

REFLECTANCE OF LIGHT

The reflectance of an object depends on the surface structure. Thus, shiny structures are often flat and even and dull structures are rough. However, this is an oversimplification. A special case is when the wavelength of the incident light is similar to the distance between repetitive bumps on a surface; that is, the bumps form a pattern. Patterns of nanosize features on a surface can be used to control optical effects in striking ways. The nanobumps or nanoholes must be separated by around a wavelength to achieve these effects. Nature can do the same thing. Nature has learned to engineer colour and special reflectance effects using nanostructures for camouflage or procreation. Some examples are fish scales, butterfly wings, fur, and flowers. Some insects, such as moths, have used nanostructures to develop eyes that can harvest light from all directions, so they can see clearly at high peripheral angles. There are many important opportunities for nanotechnology in photonics engineering, in lighting, optical memory storage and holography, which indeed mimic biology.

Figure 7.1

a) Diagram illustrating the behaviour of a photon as both a particle and a wave.

b) The distribution of solar energy with wavelength at the earth's surface after coming through 1.5 air masses (one air mass is the shortest path through the atmosphere) and the component that we see. This shows the sensitivity of our eyes to different wavelengths.

c) Photons, waves and a nanoparticle relative to a wavelength.

a
)

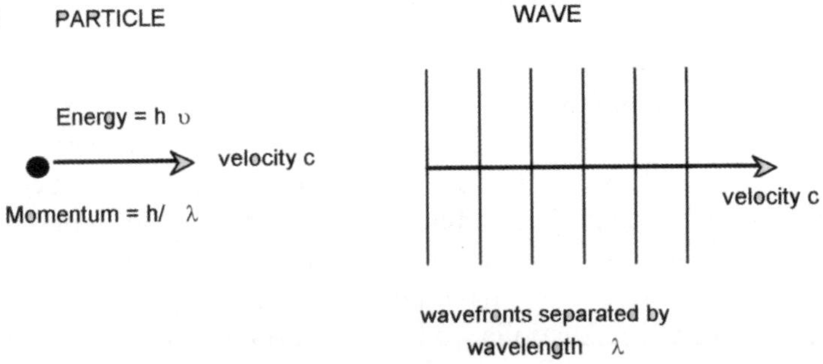

b

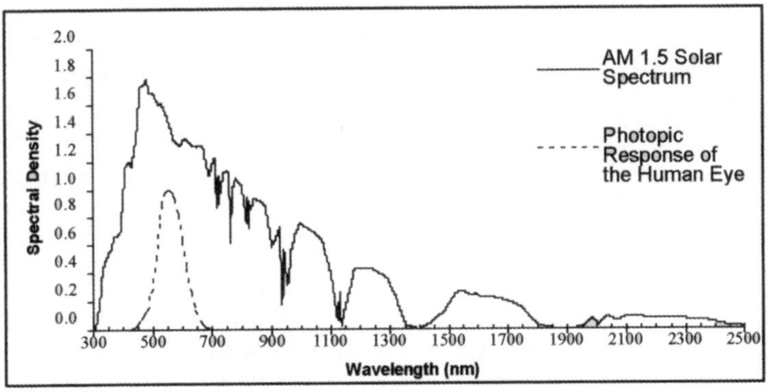

c

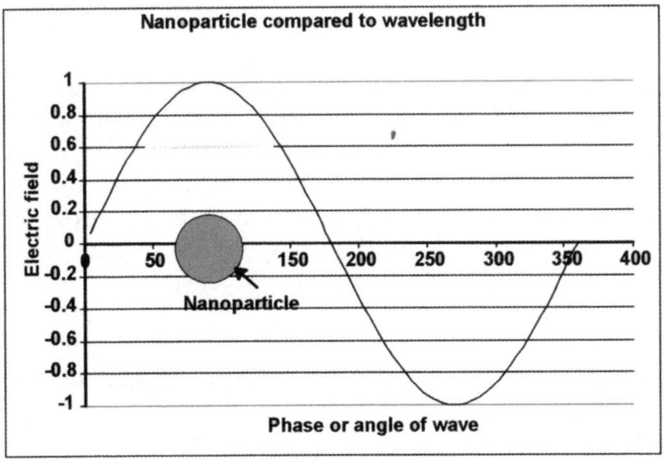

Solar radiation contains both ultraviolet and infrared as well as visible light. The UV photons are the most energetic and can most easily degrade many things by breaking chemical bonds in, for example, paint pigments and plastics, as well as our own DNA (DNA is discussed in Chapter 6). Some nanomaterials such as nanoparticulate titanium dioxide, TiO_2, actually enhance this photoactivity caused by UV and visible light by interaction, which can in turn lead to accelerated degradation. Titanium dioxide can also be put to use to encourage desirable reactions, such as breaking up pollutants by photocatalysis.

TRANSMISSION OF LIGHT

If nanostructures are aligned as thin films, they have a degree of transparency depending on the material. However, nanostructures are not like thin films of normal material. Nanostructures are not continuous — there are lots of gaps in them. They are like nanoscale venetian blinds, and just like blinds they can be oriented at different angles. Angularly selective nanostructures have been pioneered and developed over the last 12 years in Australia [5], and in Japan and Sweden. Because both the gaps and blinds can transmit light, nanoscale optical physics can be puzzling to the newcomer because it doesn't follow everyday experience. Nanoscale 'venetian blinds' are like this for some colours. Since different colours have different wavelengths, they may pass through gaps or transmit through material in different degrees. Some colours actually prefer to transmit in the direction where the 'venetians' block some of the light.

POLARISATION

Oscillating electric fields carry the energy in light waves just like oscillating air molecules carry energy in sound waves. Their wavelengths and frequencies define the periodicity or repeating patterns of fields in space and time respectively. At any instant in time the strength of an electric field changes continuously over a wavelength. After half a wavelength it also changes direction. If the directions in which the electric field can point are constrained in some way, the light is said to be polarised. Polarisation sensitivity is important for much modern optical technology, for example in CD/DVD and CD ROM players (Plate 5) because it suppresses noise. Because nanomaterials can alter the polarisation of light they can have special purposes. Nano polarising systems are already replacing expensive quartz crystals for this job because they are less than one fifth the cost.

RADIATION

The radiation given off by warm and hot surfaces is called black body radiation. Black body spectra (Figure 7.2a) show different intensity as a function of wavelength and temperature. This energy is in the infrared region, 1–30 x 10^{-6} m. Our optical communication systems all work best in the lower end of the near infra red region, around 1.3 and 1.5 μm wavelength, because that is where fibre optic cables absorb radiation the least. Many 'nasty' military lasers also work at wavelengths around 1 μm, where they can't be seen.

Figure 7.2

a) Black body spectrum for radiation loss for three temperatures at 23°C, 50°C and 100°C respectively from bottom to top.

b) A light ray being reflected, absorbed and transmitted through a material.

c) An array of nanoparticles, polarised due to an applied field. The combination of the field due to the polarised nanoparticles and the applied field as shown, is well below the applied field strength. The applied field plus that of the internal dipoles is equal to the internal field.

a

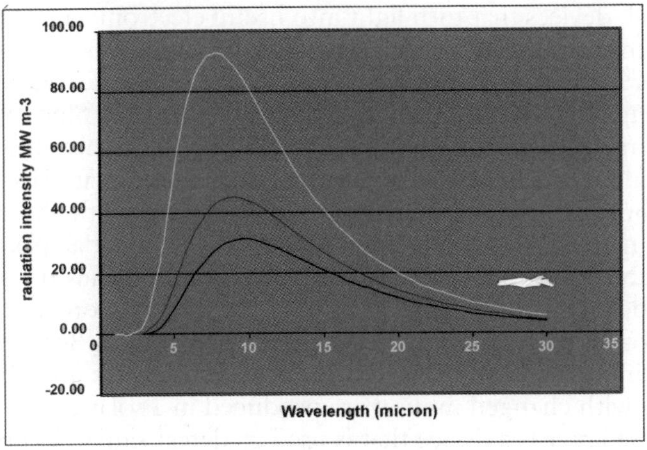

b

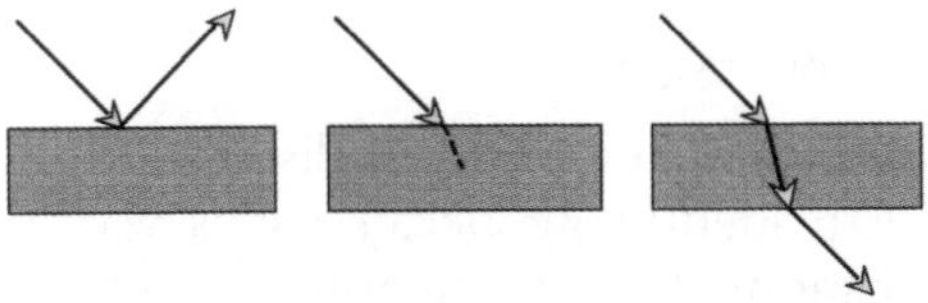

c

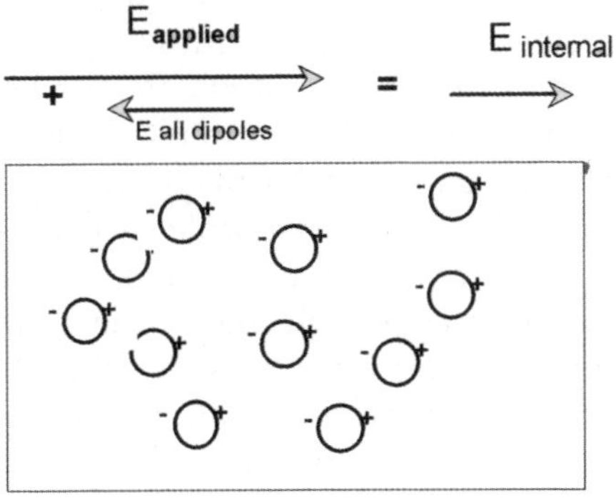

Black body radiation results from photons released during thermal transformations in materials. Since these transformations depend on molecular and atomic structure, and the wavelengths are only a magnitude larger than the size of nanoparticles, nanoparticles may affect black body radiation.

The promise of nanotechnology is thus to give us unprecedented control by producing specific radiation rather than black body spectra so as to make windows with better heat and UV control while reducing glare, and also better fibre optical communications. It may allow the production of better displays and signs, new safety lighting, new cosmetics, stunning paints and decorations, and new general lighting techniques using new lamps and light fittings.

7.2 INTERACTION OF LIGHT AND NANOTECHNOLOGY

In general, devices that turn light into useful electronic motion or electronic signals are called optoelectronic or photonic devices. Since electric fields interact with electric charge they can transfer some of their electromagnetic energy to any charge they encounter. Since all materials contain electrons and ions, light waves or photons are always affected by materials in one way or another, even if the materials are clear. When photons or electromagnetic waves interact with the charge in different materials, they can either reflect off the material, pass through it or be absorbed by it (Figure 7.2b). Absorption means the photon is destroyed and its energy appears either as heat or as a long term change in the motion of internal charge, the electrons or ions. Both effects can be used if we are clever. Heat is produced in a solar absorber, while electrons with changed motion are produced in a solar cell or a photodiode. The latter is a device that is used to detect signals in fibre optics communications.

PHOTON TRAPPING AND PLASMONS

Apart from the photon/material interactions discussed above there is a special case that is worth identifying because it is very pertinent to some amazing new nanophotonics phenomena. Photons can be trapped (that is they just bounce around) for quite long periods inside a material or structure, sometimes even in an open hole, just like sound waves can do in an open organ pipe. Lasers are the best known trapping system but there are many other examples. Lasers in effect have very shiny mirrors at either end, which only let out a small amount of the interior electromagnetic energy that strikes them. Trapping usually happens when the photon is produced inside a material or inside a cavity, such as when a fluorescent molecule in the material or cavity is excited and emits a photon that cannot exit easily. This gives a resonance, where the resonant colour is determined by the geometry of the system, for instance the spacing of the 'mirrors'. When this happens, only certain wavelengths (they are called cavity modes) can resonate and hence participate in any given laser structure. The special geometries in nanosystems open up some spectacular new opportunities for bouncing trapped photons around inside certain nanoparticles, between packed nanoparticle arrays, inside bulk nanostructures, and along surfaces with special nanostuctures. Apart from the creation of spectacular colours they also open up a whole new generation of photonics engineering in terms of spectral control and non-linear response needed to operate the latest large bandwidth communications systems.

The ability of a material to absorb electromagnetic energy and trap a photon is normally strongly dependent on a match between the frequency of the radiation and energy levels of electrons in the material. However in certain semiconductors this is not true, so that photon interaction creates additional waves called plasmons. These waves carry charge and occur at the surface of the material. Plasmons are expected to be useful in a range of new optical devices as yet undiscovered. Since the production of plasmons is surface dependent, we can expect new types of plasmons and photon trapping with nanoparticles. Thus nanotechnology may contribute to new devices. We do not know what these devices will be, but they should be as exciting as the discovery of lasers and may well be analogous to lasers but do different optical tricks.

DIELECTRIC CONSTANT AND POLARISATION

The dielectric properties of a material, which are defined by the dielectric constant, are useful in forming electronic devices. Electric fields polarise materials by slightly separating positive and negative charge within them, in much the same way as dipole forces occur in atoms and molecules (Chapter 1.5) but on a different scale. The amount of polarisation determines the value of the dielectric constant, which measures how much the internal movement of charge caused by an electric field cancels out or weakens the applied field. For example, the electric field between the plates of a capacitor is reduced when a dielectric material is placed between the plates because the charge separation produced within the dielectric material partly counteracts the applied field. The ratio of the field with no material to the reduced field with the material gives the dielectric constant. For large particles, polarisation of the material makes it appear to act as one large dipole, but for nanoparticles the behaviour is quite different. Nanoparticles appear to be individual dipoles when polarised. A collection of such particles embedded in a solid or liquid can thus change the overall polarisation substantially. This is diagrammatically shown in Figure 7.2c. Hence it can be expected that new nanomaterials will have new electronic dielectric uses.

Light also affects polarisation and hence the interaction of light with materials can affect the dielectric constant. The case for metals is particularly interesting. Mathematically, the dielectric constant is defined by an equation that contains two terms. One term contains a complex number ($i = \sqrt{-1}$), an imaginary number (Chapter 1.3) that determines photon absorption. When the equation is solved for visible and infrared frequencies, the answer can be negative. This is because something special happens at the surface. Light forms plasmons (Chapter 7.4) and the wave changes direction on the surface it is travelling on [7]. It is most marked when the surface changes direction over the scale of a wavelength or smaller, or surface features are very regularly laid out. The surface of nanoparticles change direction at distances of the order of the wavelength of light. Thus metal nanoparticles may be expected to have particularly unusual properties when producing plasmons.

Until recently, polarisation in most optical technology was simply taken to be directly proportional to the electric field, E. However, in the last ten years or so some materials have been shown to respond to the square of the field. This gives a non-linear optical response and some exciting new tricks with light then become possible. There are very few good strong non-linear materials around, and they are mainly expensive crystals. Many nanostructures are likely to give non-linear responses and therefore they are a useful new type of electronic component.

The polarisation process means charge has to move, but moving charge can produce heat due to the presence of internal friction, so photon energy degrades finally to heat and the temperature rises. Absorption of photons and the production of heat may be negligible in clear materials at visible wavelengths, but it is not so in metals and other related materials with a lot of easily moved charge. Just as they have electrical resistance for static or direct current fields, metals also have electrical resistance for high frequency optical fields. They can still be assigned a dielectric constant, but it must take account of the fact that metals produce heat when charge (current) flows at optical frequencies. How much of this absorbed energy ends up as heat depends on whether the excited charge energy can be used for anything specific before it is converted to heat. This means that action must be taken during the time that the original excitation exists. This time is usually called a relaxation time. Relaxation is surface dependent, and the surface area is large in nanoparticles. Thus the relaxation time of excited electrons is shorter in nanometals due to collisions with the particle surface, and it follows they have different heat properties to conventional metals.

REFRACTIVE INDEX [6, 7]

The refractive index is a measure of how fast a light wave travels inside a solid, its direction inside a solid, and also the amount of reflection off the surface. It thus determines whether the solid is clear, hazy or white. The path is due to refraction or direction change when light rays cross a surface separating two different materials (discontinuities). It can simply be thought of as a photon changing direction to conserve its momentum component parallel to the interface, since photon momentum is higher for higher refractive index or smaller wavelength. Nanotechnology provides greater opportunities than ever before to engineer effective refractive indices because each nanoparticle is a different surface. Hence we may be able to make light do just what we want when it strikes a material. A general rule is that nanoparticles scatter light weakly or negligibly. Thus they will produce a lot of new see-through materials.

Consider zinc oxide, ZnO. Most of us think of it as a bright white sunscreen and it is one of the best UV blockers available. Sunscreens work because zinc oxide strongly absorbs UV photons, so they don't get through to our skin. The reason it is white is because the particles of zinc oxide are micron-sized and scatter light waves. If they were to be reduced in size to 50 nm or less (see Chapter 3, Figure 3.1) the system would become clear (Figure 7.3a). The first big market for nanoparticles was thus in clear sunscreens. One such zinc oxide product originating in Australia is now on the supermarket shelves. Reducing the size of a particle does not remove its absorbing power; in fact it actually makes it more effective than the bulk material in many cases. The difference is that nanoparticles now allow all visible light through. The advantage of colourless sunscreens is mainly cosmetic.

7.3 NANOHOLES AND PHOTONS

One strange new phenomenon that has been discovered is that photons can pass through nanoholes without interaction. Early work at 200 to 300 nm hole diameters demonstrated this phenomenon, but it has recently been demonstrated that it can occur even for 20 nm holes [8]. This was thought to be impossible, since the light was expected to be diffracted, but now a whole field of 'new optics', where diffraction is cancelled or bypassed, appears to be possible *via* related physics. We do not know what these new materials will be used for but new optical devices are certain.

Nanoholes in a thin metal behave particularly remarkably and do something special with plasmons. They can cause surface plasmons to couple and behave as if they are molecules [3, 6]. They absorb and re-emit at least some light without converting it to a different form of energy. The potential applications for this light include optical lithography at finer scales than ever thought possible. Other possible applications include mass-produced atomic scale nanopatterning and electronics; tuneable optical filters; flat panel displays; insertable probes for optical imaging of nanoscale systems, including segments of DNA; and optical modulation.

Figure 7.3

a) Light rays passing mostly undeflected through an array of nanoparticles, so the array is transparent. When the particle size of the same material increases, light rays are scattered and the material becomes translucent.

b) A fixed wavelength ray carries information about the detail in an object as it leaves the surface (rays a–c).

a

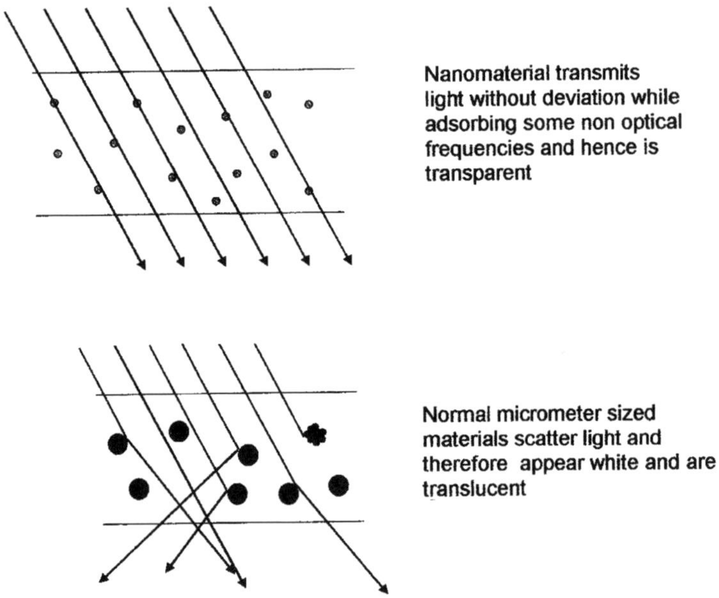

b

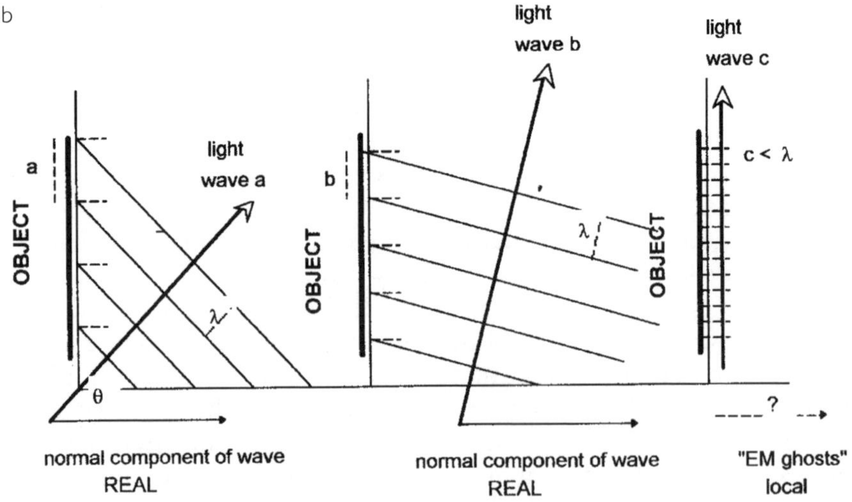

7.4 IMAGING [8]

Nanotechnology will contribute to advances in imaging technology. We can image objects at all wavelengths, using opto-electronic detectors if necessary, when the objects are too small to see with our eyes. Nanotechnology may just revolutionise the imaging process itself, not just because it can help by getting more pixels in a smaller space (which it can and will do), but because it may allow imaging with no loss of resolution at any scale using light. Ordinary images always blur at a fine level. Once neighbouring features get too close we can't differentiate between them, and if they are too small they cannot be seen at all. But new optical imaging systems may provide perfect images of objects much smaller than a wavelength. Physicists had always thought this was beyond their grasp — an immutable natural law because light is a wave. There now appears to be a way out. It involves the fact that very fine details do send out faint evanescent waves that usually fade away very quickly in empty space. If we could refocus these, we could regain our resolution.

A new class of optical materials with negative refractive index is needed to do this. These were thought to be impossible, but they have now been found [6], or rather made, for microwave radiation. Nanotechnology has the potential to provide these optical materials for visible radiation. If it can, then science and medicine could undergo a revolution as big as the one that occurred when the first optical microscope was made. Nanotechnology itself will also be a major beneficiary — an imaging capability better than current electron microscopes may be sitting on every desktop.

Geophysics has taught us a lot about waves. In this technique we measure the depth of an object by reflecting a wave from it and collecting it again. Sound waves are popular but other waves can be used. Seismological theory needs to consider that waves also bend to accommodate the differences in stratification. Depending on the material in which the wave is travelling down, the wave will reflect, and more for some wavelengths than others. That is, waves bend. Indeed, in reflection seismic surveys, the velocity contrast between shallowest and deepest reflectors ordinarily exceeds a factor of two. Thus depth variation of velocity is almost always included in the analysis of field data. Snell's law relates the angle of any of these waves in one layer with the angle in another. However, in theory, at infinite depth a wave can effectively be horizontal. Such waves are called evanescent. Evanescent waves are important because they occur in lots of systems, not just geological rocks. They affect resolution and thus they also affect imaging.

Until now, diffraction or wave effects have always limited image perfection. When we form an image of an illuminated object using a finite size lens, some detail gets lost. To image smaller objects or smaller features in an object, we have always had to use smaller wavelengths because the limit of resolution, or the point where images get fuzzy and hence cannot be seen in an image, is around a wavelength. Why is this? It is due to diffraction or wave effects from both the finer features in the object and by the lens aperture itself, such as the iris in our eyes. However, the diffracted evanescent waves emerging from a surface at different angles contain information about features separated by different amounts. These are governed by the distance between neighbouring wavefronts that emerge from the surface (Figure 7.3b). These waves are also called ghosts.

Nanotechnology offers the prospect of rebuilding something real from evanescent waves — of getting real pictures from ghosts. If a super lens can be constructed which has not only a negative dielectric constant but also a negative refractive index, then the evanescent waves could be focussed. We have learned that nanomaterials — particularly nanometals — may have these properties. A lens that works at microwave wavelengths (10^{-2} m, Chapter 1, Figure 5) has been invented [9]. Its equivalent has not yet been realised in the visible region, though it should be possible to do using special nano-arrays including metal. It should be possible to focus evanescent waves with a sheet placed close to the object, because refraction at a surface sends light as shown in Figure 7.4a. It can also give a *perfect* image with no missing detail because it can restore all components of information in the object including those normally lost, as they are evanescent. If this can be realised it could be one of the really big triumphs of nanotechnology. It could open up a new scientific era with (in principal) imaging capability down to the spacing between the fundamental internal elements in a material —perhaps even the electrons, which are rather close together in a metal. In practice, the technique will help sub-wavelength nanotechnology-based detectors and some clever image processing and collecting techniques to achieve super images.

7.5 NEW LOW COST ENERGY EFFICIENT WINDOWS AND SOLAR ABSORBERS BASED ON NANOPARTICLES

NANOMETALS

Gold and silver propagate electromagnetic waves or photons internally, but after a short, nanometre scale distance, they are destroyed by absorption. If the metal is thin enough (under 200 nm) some light can get through. That is we can see through it and daylight can pass through it. Thin films can be deposited on glass or plastic in vacuum to give transparent metal coatings. Using other special non-metal dielectric nanolayers, such as titanium dioxide TiO_2, on either side of the metal, the amount of light passing through can be increased to make the coated metal look almost clear and transparent. Since the light is transmitted rather than reflected there is no glare. If nanosized particles rather than thin films are used, they can play a similar role.

Figure 7.4

a) The effects on light emerging from a small object: comparison of a standard transparent material with a positive refractive index with that of one with a negative refractive index. It can be seen that the second produces an image of the object.

b) Spectral transmission of a laminated window in which the plastic laminate layer (poly vinyl butyral) contains 0.2% (lower trace) and 0.5% (upper trace) by weight of indium tin oxide nanoparticles. The window has very high visible transmittance (400–700 nm) while the near infrared (~1000nm), normally transmitted to 2.5 mm, is partially blocked.

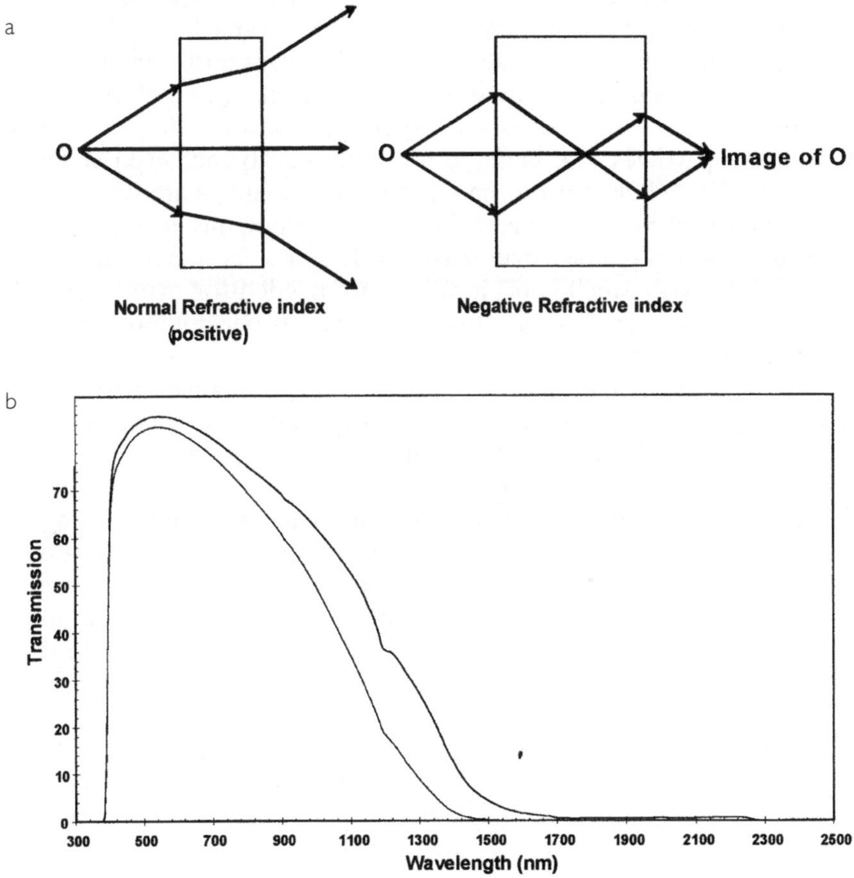

NANOTECHNOLOGY AND DAYLIGHT

Ideally for warm countries a window should provide a view and some daylight but not allow in too much solar heat or ultraviolet radiation. Nanotechnology can make it easy and cheap to do this. If the material doesn't absorb at wavelengths we see, it can transmit visible radiation, but block the other radiation that we don't want. Since nanoparticles are normally clear we just need to use one that absorbs other wavelengths, to reduce glare. Nanoparticles of these materials can be incorporated in a sheet of plastic, such as the laminate layer in a car window, to reduce solar heat gain and UV degradation. The window will still let in daylight and provide a clear view. We have already seen how nanoparticle zinc oxide can be used in good quality, transparent sunscreens, however it is not an efficient near infra red absorber. Antimony tin oxide (ATO), indium tin oxide (ITO) and other materials can absorb near infra red wavelengths and also UV but let visible light through. It is possible to use more than one type of nanoparticle material to block all but the narrow band of wavelengths required.

Thin film coatings on glass composed of several layers, including some very thin metal ones, can also provide solar control in clear windows. This has been widely used until now, especially in high rise buildings in warm climates, but is too expensive for most cars and trucks. Using nanoparticles, the same performance can be achieved at much less cost. Why? Because the nanoparticles in many materials are not very expensive; they can be added directly to a polymer using standard existing processing lines that mass produce the polymer sheet; and because only very small concentrations are needed for many materials of interest. 'Very small' means hundreds of parts per million, or a fraction of a percent by weight.

Materials such as antimony tin oxide, ATO, have a nanoparticle resonant absorption at around 2.0 μm, while indium tin oxide, ITO, has a resonant absorption around 1.5 μm. Examples of spectra for a laminated window with 0.2% and 0.5% by weight of indium tin oxide particles in a standard thin polymer laminating material is shown in Figure 7.4b. This is for totally clear glass, but if slightly tinted glass or a tinted laminate is used, even better solar performance can be achieved while still providing a high enough visible transmittance (> 70%) to be acceptable for cars. A new laminated solar control window product using nanoparticles should soon be on the market for automobiles and architectural use.

Metal nanoparticles are perfect if we want to absorb nearly all solar energy and make things really black [10]. Since metal nanoparticles are so good at absorbing, they can be used as a very thin layer on conventional metal to absorb nearly all the incident solar energy yet keep the radiative properties of the metal underneath low so the heat is trapped and doesn't radiate back out. This is the secret behind some of the best absorbers such as black chrome for domestic heaters and metal ceramic mixtures containing molybdenum nanoparticles used in high temperature power-generating solar collector systems.

SOLAR CELLS, NANOPARTICLES AND NANOSTRUCTURES

New approaches to low cost solar cells will almost certainly emerge with the help of nanotechnology. There has been little work to date, but this should change as nanoparticles of the various necessary constituents become available. New solar cells could be based on nanoparticles of semiconductors or coated or filled nanotubes, individually or as films, maybe cross-linked with conducting electrode nanowires, or specially doped titanium oxide in solid or liquid electrolyte. They may stand alone or be coated onto or doped into nanopolymers or metal nanostructures; or they may be embedded in a charge transfer medium.

Nanosystems could lower costs and give better efficiencies by enabling more efficient use of the solar spectrum. Most current technologies sample only a part of the incident solar energy and use it with reduced efficiency because they employ just one material that has poor absorption coefficients across some parts of the solar spectrum. Geometric compromises are made to get the best optics while minimising recombination (electrical) losses. It is no use if most photons go right through the device or are absorbed well away from the internal junctions that collect them; or if the photo generated electrical carriers produced can't get to the outside world because they have to traverse too large a distance in the junction itself. Some thin film solar cell systems overcome the spectral problem to some extent by putting stacks of different types of semiconductor junctions on top of each other [11]. With nanotechnology, the use of mixed materials should be easier in principle. To date some of the best cells utilise half buried layers where metal fills laser-cut grooves. Nano-techniques could go further, but an integrated approach to production of the whole cell would need to be used, including the incorporation of arrays of internal nanowires, which might be conducting carbon nanotubes, for instance.

OPTICALLY USEFUL NANOSTRUCTURED POLYMERS

It is relatively simple to make polymer particles down to 20 nm diameters using cross linking agents and solution polymerisation with special agitation techniques and other control additives to give the desired sizes on polymerisation. These particles can be easily added to polymer moulding compounds used in common mass production methods, even though the host plastic melts as it is extruded or injection moulded, because the polymer particles do not melt even if they soften. Rohm and Haas in Germany make a moulding product called Plexiglas, which can be used to produce such polymers. In many cases these particles can also be added to monomers without being dissolved and can survive the full polymerisation process.

Nanostructured polymer surfaces can be made by nano-imprinting into warmed polymer [2]. They can be used directly as anti-reflection surfaces [12] to let through more light or for computer generated holographic or diffraction effects for applications such as security. Over-coating such surfaces with very thin metal, around 20 nm coatings, can produce very sharp absorption bands and colours that vary with the size of surface features [13].

Polymer particles embedded in a very similar polymer have created a new class of translucent materials with high total transmittance and very good spread of light [14]. That is, as light is transmitted through the sample it illuminates a large area, with little angular or conical spread about the incident beam direction. The new materials may provide ideal translucency for lighting, either in skylights or in lamp fittings, giving low glare without reducing light throughput.

Nano- and micro- polymers may make their biggest impact *via* additives or modifications. For instance, placing a nanoparticle of cadmium sulfide, CdS, into a polymer particle, especially if it is hollow, has produced some remarkable new photon trapping and bright output materials called whispering gallery modes (WGMs) [15]. Photon intensity from fluorescent emission from the cadmium sulfide builds up very strongly into many very narrow (in wavelength terms) intense bands of closely spaced but distinct colours.

7.6 PHOTONIC CRYSTALS, SURFACE WAVE GUIDES AND CONTROL OF LIGHT PATHS

Photonic crystals are types of crystals that can manipulate light signals to go wherever they are required inside a solid structure. Photonic crystals are not completely new. Some decorative materials are photonic crystals and the effect can be used to produce special colours. Opals, which can be made artificially now using nanostructure techniques, are examples (Figure 7.5). However, these materials are genuine light circuits, analogous to electrical circuits. They guide light rays by keeping them out of regions of solids where the patterned structure of the particle arrays forbids particular frequency photons — these are called photonic band gaps. Particular light rays will thus stay out of the patterned region but can be guided through it with a track of defects deliberately laid out.

Related to photonic crystals are surface wave guides, where nanostructure patterns on a metal surface can guide photon beams in a defined track around a surface: even turning it at right angles seems possible [16]. The light follows a path in a track set between the regular arrays on the surface. The surface plasmons on the metal do the guiding. The world of optoelectronic devices is now a vast, important industry which nanotechnology could well severely disrupt by threatening existing techniques. However, an exciting possibility is to use these new optical materials to transport energy as photons rather than as electrons, and build optical devices, much like we have built transistors, actuators and motors for transforming electrical energy to other forms. Indeed, the days of electricity may be numbered!

Figure 7.5

A photonic band-gap nanostructure made from semiconductor material. This is a cross section through an artificial opal. There is a face-centred array of overlapping air spheres (see sodium chloride structure, Chapter 1, Figure 1.2). A nematic crystal is shown on the top centre. These fill each sphere. Only one is shown. It is made by first creating a pile of spherical nanoparticles of silica, filling the gaps between the spheres, then etching away the silica. Reproduced with permission from Kurt Busch and Sajeev John, University of Toronto.

WHAT YOU SHOULD KNOW NOW

1 The ability to engineer on the nanoscale has the potential to revolutionise optics and photonics, as well as lighting and colour technology, since light waves and nanoparticles have the same dimensions. The small size of nanoparticles makes reflection, transmission, radiation and polarisation properties unique.

2 Angular selective nanostructures can control the transmission of light according to incident direction. They are like nanoscale venetian blinds. For some colours, light actually prefers to transmit, not when it is aimed at the nano-gaps between the blades, but in the direction where the structures block some of the light.

3 The promise of nanotechnology is to give us unprecedented control by producing specific radiation rather than traditional black body radiation. Heat flow from photon excitation in nanometals is quite different, since there is rapid contact with surfaces.

4 In some semiconductors and nanometals, the photon's interaction energy does not match that of electron quantised energy levels and

some energy is released as an additional wave called a plasmon. Nanoholes can be used to cause surface plasmons to couple and absorb and re-emit light like a molecule.

5 For large particles in the presence of an electric field, polarisation of the material makes it appear to act as one large dipole, but the behaviour is quite different for nanoparticles. They appear to be individual dipoles when polarised. Many nanostructures are likely to give non-linear polarisation responses and therefore they are a useful new type of electronic component.

6 The refractive index reflects how fast a visible light wave travels inside a solid, its direction inside a solid, and also the amount of reflection off the surface. It thus determines whether the solid is clear, hazy or white. Nanoparticles interact weakly and therefore can be colourless while still absorbing ultraviolet radiation. They can therefore be good sunscreens.

7 Evanescent waves are present at surfaces and contain information about the fine features in objects. Nanomaterials may allow the construction of a super lens in which the evanescent waves could be focussed. This may give the ability to image with no loss of resolution at any scale using light. We will see everything in nanoscopic detail.

8 Photonic crystals are certain types of crystals that can manipulate light signals to go wherever they are required inside a solid structure. These are genuine light circuits, analogous to electricity circuits.

7.7 EXERCISES

1 *Web search electrochromic windows, sunscreens, Obducat, plasmons, photonic crystals, soliton, nematic crystal, whispering gallery modes.*
2 *Examine a butterfly's wings under a microscope for fine structure and note colour changes as you change the viewing angle. Do the same with some fish scales.*
3 *Pour a little oil on some concrete with a trace of water and let it dry. Observe the colours.*
4 *Find out why titanium oxide acts as a UV enhancement agent.*
5 *Write an essay discussing the different heat dissipation mechanisms in metals and semiconductors. Research the difference between recombination and relaxation pathways.*
6 *Find out more about negative refractive index and the refractive index of metals*
7 *Go to an opal shop and look at some opals.*

7.8 REFERENCES

1 Most basic optical textbooks discuss this issue. For example Ward L (1998) *The Optical Constants of Bulk Materials and Films.* Adam Hilger Series on Optics and Optoelectronics, Pike & Welford (eds), IOP Publishing, Bristol.
2 Cho SY (1997) *Science Spectra* 10: 38–43.
3 Sonnichsen C, Duch AC, Steininger G, Koch M, von Plessen G & Feldmann J (2000) *Applied Physics Letters* 76: 140–42.
4 Lakhtakia A, Messier R, Brett MJ & Robbie K (1996) *Innovations in Materials Research* 1: 165–76.
5 Smith GB, Dligatch S, Sullivan R & Hutchins MG (1998) *Solar Energy* 62: 229–44.
6 Ebbeson TW, Lezec HJ, Ghaemi HF, Thio T & Wolff PA (1998) *Nature* 391: 667–69.
7 Smith GB, Hossain AKM & Gentle A (2001) *Applied Physics Letters* 78: 2143–44
8 Mullins J (2001) *New Scientist* April 14: 35–37.
9 Shelby RA, Smith DR & Schultz S (2001) *Science* 292: 77–79.
10 Granqvist CG (1989) *SPIE Tutorial Series in Optical Engineering*, Volume TT1, The International Society for Optical Engineering Press, Bellingham, USA.
11 Hamakawa Y (1997) *Phys. Stat. Sol. B.* 194: 15–29.
12 Gombert A, Glaubitt W, Rose K, Dreibholz J, Blasi B, Heinzel A, Sporn D, Doll W & Wittwer V (2000) *Solar Energy* 68: 357–60.
13 Takei H (1998) *Proceedings of the International Society for Optical Engineering (SPIE)* 3515: 278–83.
14 Smith GB, Earp A & McCredie G (2001) *Proceedings of the International Society for Optical Engineering* July.
15 Artemyev MV, Woggon U & Wannemacher R (2001) *Applied Physics Letters* 78: 1032–34.
16 Bozhevonyi SI, Erland J, Leossen K, Kovgaard PMW & Hvam JM (2001) *Physics Review Letters* 86: 3008–3009.

8

NANOELECTRONICS

This chapter explains how microelectronics has developed into nanoelectronics in some detail. It follows the birth of electronics, the invention of the transistor, integrated circuits, optical, electron beam, atomic lithography, molecular beam epitaxy and molecular (including DNA) electronics. You will learn how a quantum computer works and how it differs from a conventional computer.

8.1 INTRODUCTION

In the late 1960s, Gordon Moore, the co-founder of Intel Corporation, made a memorable observation that has since become known as Moore's Law. He noted that the number of transistors on a chip roughly doubled every 18 months. What is remarkable is that this trend has remained true for the past four decades. A consequence of this doubling is that the individual feature sizes of the electronic components decreases every year despite the continued difficulty in fabricating smaller and smaller electronic components. Following on from this law came the prediction that by the year 2015 the feature sizes of devices will become less than 0.1 μm, where the electronic properties of the materials will change from obeying the familiar classical physics to the less familiar quantum physics (Figure 8.1a). Another consequence of Moore's Law is that as transistors get smaller they contain fewer and fewer electrons. Figure 8.1b shows how the number of electrons used to store one 'bit' of information is decreasing as miniaturisation continues. Eventually we will reach a limit of one electron per 'bit'. This is a fundamental brick wall for the microelectronics industry

— how can we store more than one bit of information on a single electron? In addition it becomes increasingly difficult to fabricate smaller and smaller devices using conventional techniques. At the end of the twentieth century it costs an incredible US$5 billion to build a new semiconductor manufacturing plant. If this trend continues, then by the year 2035 it will cost more than the gross domestic product of the entire planet!

a

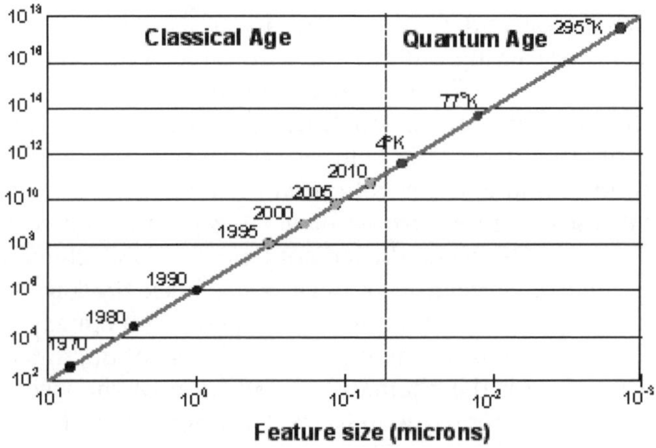

b

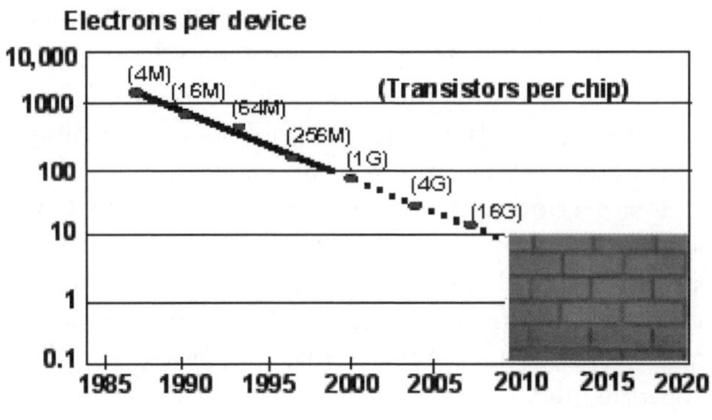

Figure 8.1

a) An illustration of Moore's Law showing the crossover from the classical to the quantum age around the year 2010.

b) The brick wall of the microelectronics industry.

Whilst quantum effects represent a fundamental limit to the miniaturisation that has been one of the key methods of increasing processor performance, a school of thought believes that these effects may be used to our advantage — if we knew how to control them. Nanoelectronics is the emerging field of building electronic devices at the atomic level to harness these small-scale 'quantum' properties of nature. This field unites physicists, chemists and biologists in order to understand how nature works at the atomic scale, and how we can control it. It is expected that the ability to manipulate individual atoms and molecules to build new kinds of quantum electronic devices will revolutionise the twenty first century in much the same way that the invention of the computer led to the Information Age (see Chapter 1.1).

This chapter is about the development of nanoelectronics and the tools required to observe and manipulate atoms. In particular, it will highlight the way in which the microelectronics industry has progressed over the past 50 years. As we noted in Chapter 1, there are essentially two different approaches to creating very small devices. Firstly there is the increasingly precise 'top-down' approach of finely machining and finishing the materials, which can be compared to a sculptor carving a statue out of marble. The sculptor uses the best and finest tools to fabricate the delicate features in the marble. However the limit of the smallest feature sizes they can create depends on the tools they use. This field of condensed matter physics (the study of solid, liquid and amorphous systems) is where most of the fundamental research in electronic device design and the creation of new fabrication tools occur. Strict control over the chemical, structural and geometric perfection of devices has been the key. Amongst numerous crystal growth techniques, molecular beam epitaxy (MBE) has allowed the fabrication of single crystals of semiconductors one atomic layer at a time, with the chemical composition of the layer controllable to an accuracy of approximately 1%. Combined with lithographic techniques it has permitted the fabrication of complex electronic devices with feature sizes down to around the 100 nm scale. However, the invention of scanning probe microscopy in the 1980s [1] provided a tool that can sense and image individual atoms (see Chapter 2). The recent adaptation of the scanning tunnelling microscope (STM) to atomic manipulation [2] now offers the possibility of extending this manufacturing technology down to the atomic scale.

The second approach briefly discussed in Chapter 1 is called the 'bottom-up' approach, where individual atoms and molecules are placed or are self-assembled precisely where they are needed. This is a close approximation to understanding how nature works. For many years chemists have been using the 'bottom-up' approach to synthesise molecules to produce millions of different molecular structures. In doing so, researchers have developed recipes and protocols for making plastics, ceramics, semiconductors, metals, glass, fabrics and many other materials of modern life. In parallel, nanotechnology researchers have been developing a set of techniques known as molecular self-assembly. Here, molecular building blocks are designed that automatically snap together in predefined ways. It is hoped that this technique will allow a new route to making everything from circuit elements to new polymers that can manipulate light in optical communications systems. We have already seen how scientists have produced nanoelectronic components, such as molecular switches made of a few molecules [3], and molecular wires [4] in order to realise the next challenge of fabricating a molecular transistor. At the same time, biologists have been rapidly learning how certain mechanisms in nature work, such as understanding how a bacterium is propelled through water using a molecular motor [5] or how electronic signals are generated across biological membranes (bio-sensors) [6]. All this is discussed in Chapter 6, but it is also relevant here. It is the combination of the understanding gained from biological self-assembly, the chemical development of new molecular structures and the physical development of new tools of nanofabrication that promise to revolutionise the electronics industry.

8.2 WHAT WILL NANOELECTRONICS DO FOR US?

It is never certain where an investment in fundamental research will lead. An example of this goes back to 1947, when physicists were trying to figure out what to do with a new kind of material called a semiconductor. This material had the novel property that its electrical characteristics could be precisely controlled from being insulating to metallic, just by adding a miniscule amount of impurities. The subsequent invention of the transistor [7] using this unusual property, not only led to the Nobel Prize in Physics in 1956 but, more importantly, it accelerated us through the Information Age. During the past 50 years this invention has changed many aspects of modern life, and all modern appliances such as radios, washing machines, computers, mobile phones, televisions and calculators utilise transistors built at the micron scale. Whilst it is not easy to predict what the next big discoveries will be, it is easy to see which technologies dominate the present. Global communications, the Internet, information and fast computation all drive today's economic and social lifestyles. We are in the midst of the Information Age, where information is all and the ability to rapidly process and interpret huge amounts of data is paramount. It is clear that nanoelectronics will assist in processing and transferring huge amounts of data. It is also certain that the computer and hence nanoelectronics will be essential in the developing Genetic Age or other forthcoming ages (see Table 1, Chapter 1).

Nanotechnology offers the possibility of building a new generation of electronic devices in which electrons are confined quantum mechanically to provide superior device performance. The high electron mobility transistor (HEMT) and the quantum well laser are just two examples where quantum mechanical confinement has led to better performance in terms of efficiency, speed, noise reduction and enhanced reliability. These devices are considered the first generation of quantum semiconductor devices that operate with quantum electronic states. They are used in the widespread commercial exploitation of communication and computational systems. As our manufacturing capabilities become increasingly advanced, we are now approaching the possibility of building electronic devices at the atomic scale. Over the past two decades discrete nanoelectronic devices have been proposed and successfully demonstrated in research laboratories, such as resonant tunnelling diodes (RTD), single electron transistors (SETs) and a broad class of devices comprised of quantum dots and molecules. However there is a gap between device physics and nanoelectronic systems integration. As the gap closes, new and exotic quantum phenomena will emerge in networks composed of entangled quantum devices.

One of the ultimate quantum electronic systems is a computer that operates purely on quantum principles using individual atoms or molecules. Even a small so-called 'quantum computer' has been predicted to be so powerful that it can perform certain calculations that all the computers on the planet linked together could not complete before the end of the universe. However, trying to construct a quantum computer at the atomic scale is far from easy. Whilst a practical quantum computer has yet to be realised, there is intense international interest in building such a device.

8.3 THE BIRTH OF ELECTRONICS

SEMICONDUCTORS

All materials have electrical properties that allow them to be organised into three broad categories: metals, insulators and semiconductors. Metals such as aluminium and copper conduct electricity. Insulators such as ceramics and plastics do not conduct electricity under normal circumstances. Semiconductors are in between — they are neither good metals nor good insulators. However, by intentionally adding just 0.01% of impurities it is possible to change the electrical resistance of a semiconductor 10 000 fold — enabling semiconductors to be engineered between metallic and insulating behaviour. This ability to switch between these two states has meant that semiconductors have formed the basis of the transistor device used in the modern electronics industry. Silicon-based materials (from group IV of the periodic table) dominate the semiconductor industry for the production of devices like computers and calculators, whilst gallium-arsenide-based materials (from groups III and V of the periodic table) are used in high-speed devices such as mobile phones, and in optoelectronic devices such as the lasers of a CD player and fibre optic communication systems.

The electrical properties of semiconductors give them an inherent flexibility. In metals we understand that the electrical current is a flow of electrons with charge $-e$. However, semiconductors are interesting because they behave as if there were two different kinds of charge carriers: electrons with charge *-e* and holes with charge *+e*. The type of charge carriers that exist in the semiconductor can be controlled by the dopant element added to it. If we consider pure silicon, all the electrons in its outer shell are used for bonding to the four nearest silicon neighbours in a diamond-like structure. There are no free electrons and therefore silicon does not intrinsically conduct electricity. However, if we add a small amount of material from group V of the periodic table, such as antimony, to a silicon crystal, we can create an n-type (negative type) semiconductor. The antimony atom provides an extra electron and this extra electron is used to carry a current. Conversely, if the silicon is doped with an element that has at least one less electron than the host material, such as Boron (group III) then a p-type (positive type) semiconductor is formed. This missing electron creates a positively charged 'hole', which is also free to carry a current. This doping usually only occurs in the range of parts per million concentrations and modern crystal growth techniques allow these precise dopant additions without disturbing the crystalline structure of the host crystal. The active parts of most semiconductor devices involve combining p-type and n-type regions together to form a p-n junction. A simple p-n junction can be used as a diode for radio reception, as a light-emitting diode for displays, and as an electronic capacitor for tuning digital radios. However, one of its most significant applications is in the transistor — the most important of all solid-state electronic devices.

THE INVENTION OF THE TRANSISTOR

Colour plate 6 highlights some of the pivotal moments during the history of electronics, from the first transistor through to integrated circuits, quantum turnstiles, nanotubes, quantum corrals and DNA-based interconnects. As the chapter progresses these devices will be discussed in detail to demonstrate how quantum mechanics has become important as device sizes shrink.

The first transistor, shown in plate 6, was about a centimetre high and made of two gold wires separated by 0.02 inches on a germanium crystal. The transistor worked by applying a potential to the 'emitter' wire, which modulated and amplified the current between the 'collector' wire and the base electrode (the germanium crystal). Bardeen and Brattain won the Nobel Prize in Physics for its invention in 1956. By today's standards the size of the original transistor is mammoth — a modern processor contains transistors that are ten thousand times smaller. We have come a long way in the last 50 years from this single, fairly bulky and unreliable transistor to the more than 42 million transistors in a modern processor, where each and every transistor has to work reliably.

Transistors can operate in two ways: bipolar junction and field effect, both of which can be used in a circuit to amplify a small voltage or current into a larger one (for example, Hi-fi amplifiers) or to function as an on-off switch (for digital computers). Figure 8.2 illustrates the operation of a field effect transistor first developed at Bell Laboratories in the 1960s. Instead of using doping to alter the concentration of electrons, an electric field on a metal gate electrode is used to control the excess number of electrons in the active channel of the device. The gate electrode is placed on top of an insulator (silicon dioxide) that is grown on the silicon semiconductor surface (hence the terms Metal-Oxide Semiconductor, MOS, and Metal-Oxide-Semiconductor-Field-Effect Transistor, MOSFET — see Figure 8.2a). With no voltage on the metal electrode there are no free electrons in the silicon to conduct and the transistor is in its 'off' state, see Figure 8.2b. By applying a positive voltage to the metal gate electrode, electrons are attracted towards the gate (since opposite charges attract) from the source and drain electrodes. The insulating layer of silicon dioxide prevents them from reaching the gate so that they form a thin, two-dimensional sheet of electrons underneath the oxide layer, as shown in Figure 8.2c. These electrons form a conducting channel between source and drain, allowing a current to flow. The transistor now conducts electricity and can be considered in its 'on' state. The transistor can be turned between on and off states repeatedly just by changing the applied voltage. Right from its discovery, Shockley recognised the transistor's huge potential for computing, stating, 'in these robot brains the transistor is the ideal nerve cell'.

Variants of MOS, such as CMOS, which were developed in the late 1960s, progressively replaced early transistor designs containing both p and n-type devices in integrated circuits. A modern 256 Mb CMOS dynamic random access memory (DRAM) chip may contain several hundred million transistors with 0.25 μm features in a 1 cm^2 area [8].

Figure 8.2

a) A schematic diagram of a MOSFET showing its switching operation,

b) with no voltage applied to the gate: OFF and

c) with a positive voltage applied to the gate: ON.

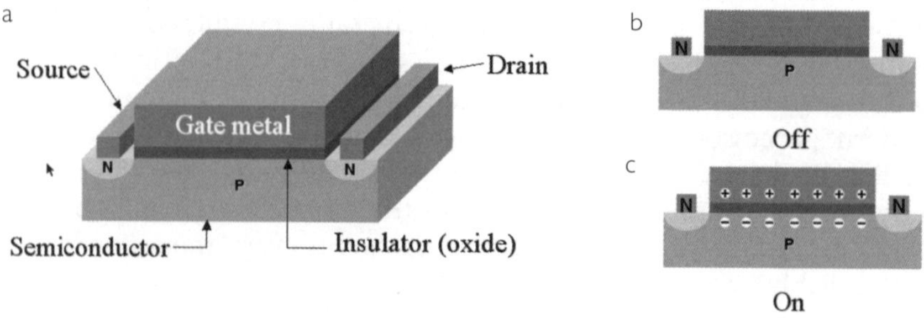

INTEGRATED CIRCUITS

Initially, electronic devices such as the transistor radio were made by painstakingly soldering all the individual resistors, capacitors, diodes and transistors onto a circuit board with interconnecting wires. However, in 1958 Jack Kilby at Texas Instruments realised that if conventional circuit elements, such as resistors and capacitors could be fabricated in silicon they could be incorporated with transistors onto a single silicon substrate. As well as miniaturising circuits by consolidating the circuit elements, and doing away with the need for interconnecting wires, this procedure also eliminated assembly (soldering) errors. Kilby used photographic techniques to transfer patterns to the silicon wafer and introduced precise concentrations of dopants to the semiconductor to specific areas to form a new integrated circuit. This first chip contained just five components. Since this first integration of all the components into a single 'chip' the miniaturisation of these devices has been an on-going process with ever more elements being packed into even smaller areas.

8.4 THE TOOLS OF MICRO- AND NANOFABRICATION

There are several different top down and bottom up fabrication methods that are used to make feature sizes in semiconductors from the micron to the sub-100 nm range.

OPTICAL LITHOGRAPHY

Lithography is the key technology of the semiconductor industry. It is the process of transferring patterns to semiconductor materials in analogy to the photographic process. In photography, a series of lenses is used to record an image onto a thin film, which contains a light sensitive silver-based emulsion. In semiconductor lithography, a thin layer of polymer (known as a resist) that is sensitive to ultraviolet radiation replaces the emulsion (Figure 8.3a). Using a mask that contains the pattern required for the transistor fabrication, radiation is projected through the mask onto the resist thereby altering the resist's chemical nature and hence solubility (Figure 8.3b). The ultraviolet light alters the resist layer chemically so that the exposed area becomes more soluble (positive resist) or less soluble (negative resist) in a developing fluid. Using conventional optics a much reduced image of the mask is thus projected onto the resist layer, which is then developed away to reveal the clean semiconductor surface in the exposed (positive resist) or unexposed (negative resist) regions (Figure 8.3c).

We can now etch away the semiconductor in this exposed area to form patterned devices in the surface, where the remaining resist acts as a protective mask for the rest of the semiconductor (Figure 8.3d–f). Alternatively, metals or dopants can be deposited through these holes to form highly conducting regions or interconnects. By dissolving the

Figure 8.3

A schematic overview of an optical lithography process:

a) a silicon wafer is covered with light sensitive resist;

b) ultraviolet light is projected through a mask;

c) the exposed region becomes more soluble in a developing fluid so that the pattern is reproduced on the resist layer;

d) the underlying semiconductor material can be etched away in the regions of the holes using an acid;

e) a solvent can be used to wash away the resist; and

f) the pattern is now transferred to the semiconductor material.

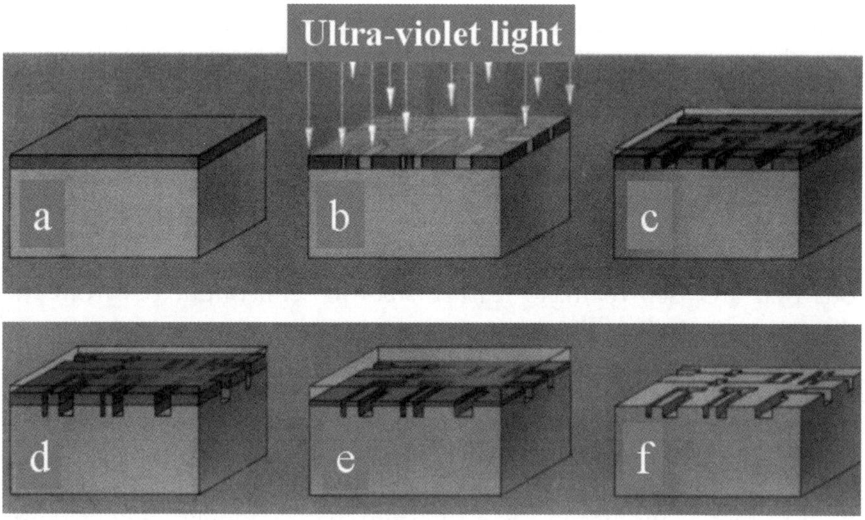

remaining resist layer, any metal on top of the resist disappears. This procedure is known as the 'lift-off' technique. A metal pattern remains on the surface in the areas where the holes used to be. By successive patterning, developing, etching and depositing many different layers it is possible to generate complex integrated circuits.

The resolution of this patterning process is determined by the wavelength of the radiation used and the numerical aperture of the focussing lens:

Minimum feature size (resolution) = C x wavelength / numerical aperture

where C is a material-dependent constant. In commercial chip manufacturing this resolution is currently around 0.15 μm and is set by the wavelength of ultraviolet radiation. In order to obtain higher resolutions, higher energy (shorter wavelength) radiation is necessary.

Using x-rays in the 1–1.5 nm range is one way to improve on these diffraction limitations. However, the lack of refractive or reflective x-ray optics does not allow the use of projection systems, so only contact and near-contact printing has been used to date. The need for higher resolution lithography commercially is currently being addressed by industry with plans to introduce extreme ultraviolet lithography, using wavelengths of 10–14 nm, within the next few years [9].

ELECTRON BEAM LITHOGRAPHY

One of the main higher energy, shorter wavelength radiation lithography systems used in research laboratories is electron-beam lithography. This can be used to fabricate devices many orders of magnitude smaller than can be made with standard optical lithography. The resolution limits in these systems are determined by a variety of factors, including the spatial distribution of the deposited energy, the granularity of the resist employed, the contrast of the resist developer and the statistical distribution of photons at each pixel. Electrons are emitted from the electron gun of a scanning electron microscope and are focussed on the sample using electron optics with a resolution of approximately 0.5 nm. However, a different chemical resist is required that is now sensitive to the wavelength of electron irradiation rather than ultraviolet light. The resist itself limits the final resolution to approximately 5 nm. During the lithographic process the whole system is kept under vacuum and a single beam of electrons is focussed at the surface of the resist-coated semiconductor wafer. The electron beam is moved across the surface under computer control using a pattern generator. Rather like writing with a pen, the pattern is carefully inscribed into the resist before the wafer is removed from the system and developed in the normal way. Whilst providing higher resolution and greater flexibility than optical lithography, this system has several disadvantages for commercial application. The main problem is the length of time to write the pattern compared to optical lithography, where a whole chip is exposed through a mask in a single, quick exposure.

Another problem is the so-called proximity effect: the high-energy electrons (5–100 kV) undergo a large degree of elastic scattering in the resist and substrate so that the region of the electron beam resist that is affected by the electrons is much bigger than the size of the finely focussed electron beam spot. This tends to blur the pattern. It is possible to correct for these 'proximity' effects but the pattern data then grows by orders of magnitude and consumes considerable computer time [10]. Nonetheless, such systems are extensively used in research laboratories to study the fabrication of ultra-small devices in which quantum effects become important.

ATOMIC LITHOGRAPHY

The ultimate limit of fabrication is to control the arrangement of individual atoms — but how can this be achieved? As we have discussed in Chapter 2, in the 1980s Gerd Binnig and Heinrich Rohrer at the IBM research Laboratories in Zurich, Switzerland, invented an instrument called the scanning tunnelling microscope (STM). The STM is capable of imaging individual atoms on the surface of a material with a resolution of a fraction of an atomic diameter, allowing the most detailed images of surfaces ever made. This invention revolutionised surface science, and earned Binnig and Rohrer the Nobel Prize in Physics in 1986.

Figure 8.4

A schematic diagram of the operation of a scanning tunnelling microscope, showing the tunnelling current measured from the tip-sample (inset circle), which is measured and fed back to control the tip-sample distance as the tip rasters across the surface. The change in tip height is recorded and displayed as a 3D topographic image of the surface (see also figure 2.3 for a different design).

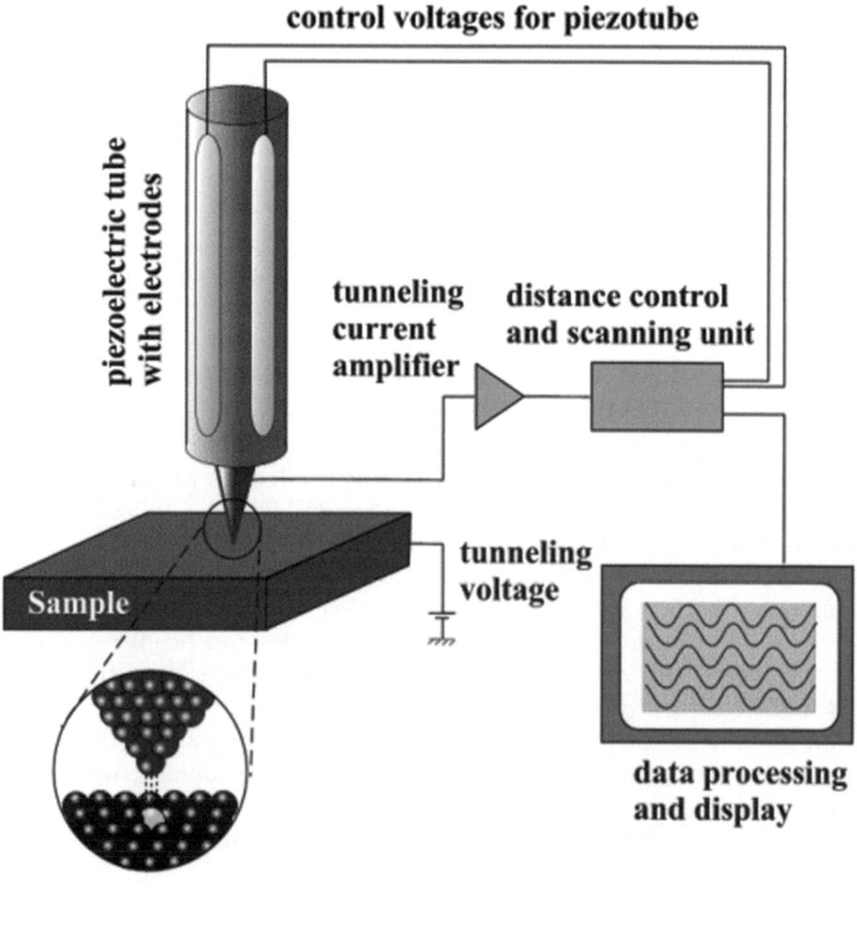

Figure 8.5

a) A close-up photograph of the sample stage of a high resolution scanning tunnelling microscope showing the sample tip (facing upwards).

b) Using an STM tip to knock off individual 'resist' atoms on a semiconductor surface for atomic scale lithography.

c) A quantum corral of iron atoms on a metal surface.

a

b

c

Recall from Chapter 2 that the STM (Figures 8.4 and 8.5a) works by using a very sharp, needle-like tip that is positioned over the electrically conducting sample to be studied (Figure 8.4). A voltage difference is then applied between the tip and the sample, causing a small number of electrons to tunnel from the tip to the sample surface. The magnitude of this tunnelling current is very sensitive (exponentially) to the separation between the tip and the nearest surface atom, as shown in Figure 2.3. As described in Chapter 2, the magnitude of the tunnel current is therefore used to monitor the distance between the tip and sample surface. In order to obtain a topological map of the surface, the tip is carefully moved a small distance across the sample surface. As the tip moves, the tunnelling current changes, reflecting the change in the tip-surface distance. This change in current is fed back to an electrical circuit that controls the up-and-down motion of the tip. If the tunnelling current decreases, the tip-surface separation must have increased, so the feedback circuit causes the tip to be lowered until the current returns to the previous value. Conversely, if the tunnelling current increases, the tip-surface distance must have decreased so the feedback circuit raises the tip. In this way the feedback circuit works to keep the tip-surface distance constant as the tip is scanned across the surface. The up and down motion of the tip is then tracked by a computer which creates a topographical map of the height of the tip over the surface, effectively plotting the height of the surface atoms.

The advent of STM has allowed us to directly observe the atoms on the surface of a crystal. The STM works best in the stringent conditions of ultra-high vacuum, where it is possible to prepare clean, electrically conducting surfaces, although over the past decade a number of other nanometre-scale imaging techniques that do not rely on ultra-high vacuum have been developed [11].

As described in Chapter 2, the STM has also been adapted as a nanometre-scale tool to move individual atoms around on a surface. To demonstrate the enormous potential of this technique researchers at IBM in the early 1990s wrote the letters 'IBM' using individual xenon atoms (see Chapter 2, Figure 2.5) — although it took nearly a day to achieve [12]. Even a molecular abacus has been formed (plate 7). Applying a few volts between the tip and surface can result in electric fields that can break local chemical bonds [13], or even initiate local chemical reactions [14]. As a result a wide variety of local manipulations and modifications can be made [15, 16, 17], ranging from atom displacement, removal and deposition of individual atoms and corral formation (Figure 8.5c).

Recently it has been shown that it is even possible to control the placement of atoms in semiconductors, where the bonds are extremely strong, to fabricate electronic devices at the nanoscale. After covering a semiconductor surface in an ultra-high vacuum compatible resist it has been possible to use the tip of the STM to knock off individual atoms from the resist surface, thereby creating a mask of atomic-scale resolution. In the case of silicon surfaces, hydrogen has been extensively studied as a lithographic mask [18, 19], since H-terminated areas are less reactive to adatoms than the dangling bonds of silicon surface atoms. Subsequent deposition of adatoms onto the surface leads to selective growth to the exposed dangling bond sites. The atomic scale extraction of hydrogen atoms from a H-terminated Si (001) surface has been extensively studied in order to develop nanoscale electronics and optical devices [20]. In particular, recent results have demonstrated the ability to controllably desorb single hydrogen atoms on a silicon surface to form an atomic scale fabrication mask [21, 22]. In addition, a single-electron transistor that works at room temperature has already been created using a STM [23].

MOLECULAR BEAM EPITAXY

Scanning probe microscopes are not the only tools that allow nanomanipulation. Another useful tool is known as molecular beam epitaxy (MBE) [24]. MBE is essentially a refined form of evaporation in which molecular beams interact on a heated crystalline substrate under ultra high vacuum (UHV) conditions to produce a single crystal film. Here it is possible to fabricate crystals one atomic layer at a time. Figure 8.6a shows a photograph of a Varian GENII MBE system that is used to grow ultra high quality GaAs/AlGaAs heterostructures. Figure 8.6b shows a schematic of this process. The system consists of a large stainless steel UHV system that is pumped using a combination of vacuum pumps, including sorption, sublimation, ion and cryopumps. Each source contains one of the constituent elements or compounds required in the grown film, be it part of the matrix crystal or one of the dopants. The evaporation rates of the source materials are chosen to give growth rates of typically one to two monolayers per second, which corresponds to a pressure of 10^{-4} Pa. In order to produce even moderately pure films it is necessary to reduce the partial pressure of unwanted species, which might stick to the film, to the 10^{-10} Pa regime — hence the need for ultra high vacuum quality. During growth the heated crystalline substrate is rotated to ensure optimum film uniformity, both of composition and thickness. Additional control over the growth process is achieved by inserting mechanical shutters between the individual sources and the substrate. The shutters prevent the molecular beams reaching the substrate and so allow the composition of the incoming molecular beams to be altered in a fraction of a second — much less than the time it takes to grow a single atomic layer. This control allows the chemical composition of the semiconductor to be precisely controlled for each atomic layer.

Figure 8.6

a) A photograph of a Varian GENII MBE research machine used for the fabrication of ultra high purity GaAs/AlGaAs HEMTs.

b) A cross-sectional diagram of an MBE machine showing the effusion sources with control shutters and the sample manipulator, housed in an ultra-high vacuum chamber with liquid nitrogen cryoshrouds.

a

b

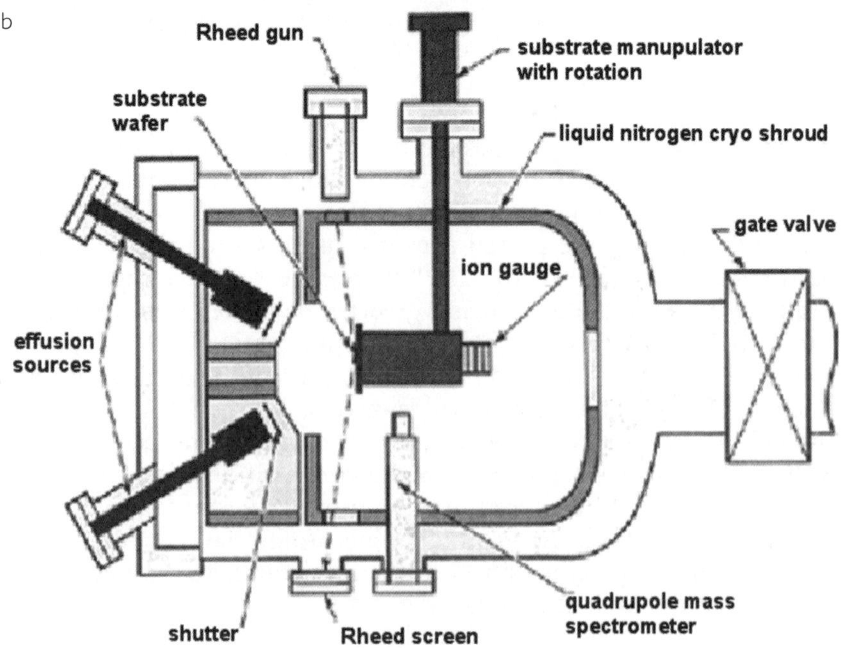

Figure 8.6 (cont.)

c) A schematic of a modulation doped GaAs-AlGaAs HEMT, with both front and back gate electrodes. The black downward-pointing arrows are the ohmic contacts to the 2D electron layer formed at the interface between the AlGaAs spacer and the GaAs buffer layer.

d) Quantum Hall effect data showing the quantisation of the Hall resistance, R_{Hall} with Shubnikov-deHaas oscillations in the longitudinal resistance, ρxx as a magnetic field B is applied perpendicular to the plane of the 2D electron system, h =Planck's constant, e = charge on electron (after Hamilton et al. 2000).

c

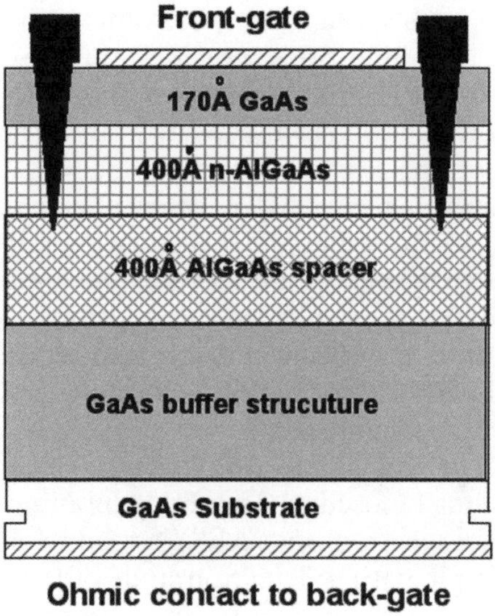

d

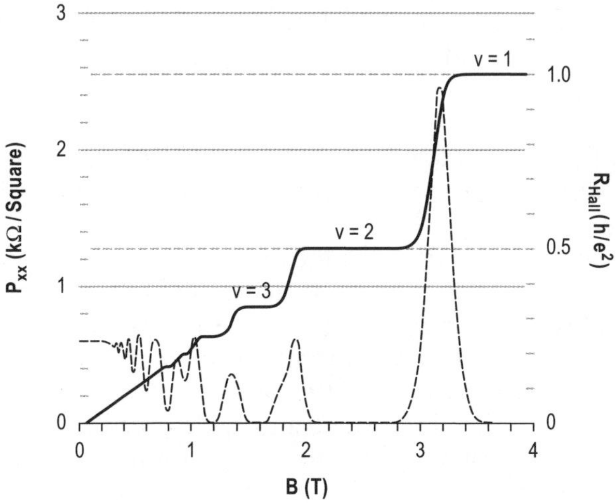

The sources and growth environment are surrounded by a liquid nitrogen cooled cryopanel (cryoshroud) to minimise the unintentional incorporation of impurities in the deposited layers from the residual background. Any unwanted impurities in the chamber stick to the cold walls and are not incorporated into the grown film. To minimise contamination of the growth chamber substrates, the substrates are initially loaded through an airlock system into a separate preparation chamber. Here they are heated or 'outgassed' to remove residual H_2O, O_2, CO_2 and CO on the surface. The clean heat-treated substrate is then transferred to the growth chamber through a gate valve before growth begins. A quadrupole mass spectrometer in the growth chamber is used to measure impurities remaining in the vacuum system. At pressures below 10^{-9} Pa the most common impurities are H_2O, N_2 and CO, most likely originating from the substrate surface or various hot filaments in the system.

One of the advantages of MBE over other crystal growth techniques is that it is performed in a UHV environment such that a whole range of surface analysis techniques may be used to monitor the growth process before, during and after deposition. An example of one of these techniques is reflection high-energy electron diffraction (RHEED), which gives information about the substrate cleanliness during thermal cleaning prior to growth [25]. RHEED also allows the growth dynamics of individual atomic layers to be studied during growth, and can thus be used to accurately monitor growth rates [26, 27]. Auger electron spectroscopy (AES) is used to analyse the topmost 1–3 nm of the sample with a detection limit of 0.1–1% of a monolayer. Here a finely focussed electron beam bombards the sample, transferring energy to a core electron and exciting it to a higher energy state. As it relaxes back down to a lower energy state it transfers its energy to another electron, which is emitted with an energy characteristic of the parent element. An energy spectrum of the detected electrons shows peaks assignable to the elements present, and the ratios of the intensities of Auger electron peaks can provide a quantitative determination of surface composition. AES is primarily used as a tool when preparing surfaces prior to growth to determine when the surface is oxygen- and carbon-free [28].

In the late 1960s MBE was introduced as a research tool to grow crystals of high quality, but it was generally believed that the cost of running an MBE system was so great it would prohibit its use for commercial production. However recent advances to increase throughput have elevated MBE to one of the fastest growing production techniques. MBE now has important and widespread applications in the semiconductor industry, where device performance depends on the precise control of dopant profiles and on the production of extremely thin crystal layers with hyper-abrupt interfaces. Examples of some of the important devices produced by MBE are described more in the next section and include light-emitting diodes, laser diodes, optical waveguides, field effect transistors, high electron mobility transistors and read/write heads for computer hard drives.

At the present time, exciting new research avenues are currently opening up by combining the atomic accuracy of STM with the atomic layer-by-layer growth of MBE. This allows truly three-dimensional quantum structures to be fabricated with nanometre-scale precision in all three spatial dimensions. However, the fabrication of electronic devices using these two complementary techniques is a challenging and time-consuming task and the processing speed is very low. To increase the processing speed, researchers have developed SPM systems with large parallel arrays of tips [29, 30]. Nonetheless significant scientific and technological effort is required before scanning probe lithography can be used in commercial production processes.

8.5 FROM CLASSICAL TO QUANTUM PHYSICS

Most commercially available devices operate under the rules of classical physics, with the notable exception of quantum well lasers [31], where the lasers emit light at well-defined wavelengths and are good for tasks such as monitoring pollution, tracking chemical reactions and optical communications. The current size of circuit components in the conventional microelectronics industry is around 0.13 μm. The crossover from classical to quantum physics in industrial chip manufacturing will occur when feature sizes get below about 0.05 μm, predicted to be within the next ten years or so. Researchers have been studying quantum effects in semiconductor devices for the past two decades and this knowledge will be invaluable if device sizes continue to shrink.

In this nanoscale regime, electrons in a solid no longer flow through electrical conductors like solid objects, but the electron's quantum mechanical nature also expresses itself as a wave. This wave behaviour makes it possible for electrons to do remarkable things, such as instantly 'tunnel' through an insulating layer that normally would have stopped it dead (rather like a ball going right through a brick wall). In order to understand the weirdness of quantum mechanics we need to recall the concept of wave-particle duality, which was introduced in Chapter 1. Wave-particle duality is where things that we normally consider as solid, such as a basketball or an atom, under some circumstances, behave like waves and vice-versa. Light is a very good example of this — it can be considered as solid particles (photons) that hit semiconductors such as solar cells and make energy; but light can also be considered as a wave-like phenomenon, exhibiting effects such as interference, which is what makes a hologram work.

Young demonstrated the wave-like nature of light in his famous two-slit experiment. Light is shone at two separate slits in a wall with a screen behind them. If the photons of light only behave as particles they would pass through the slits and hit the screen behind it at two discrete points. Wave-like entities however are diffracted as they pass through each slit, and the peaks and troughs in the waves mix to create an interference pattern. This diffraction pattern is essentially a map of the momentum distribution of the photons and occurs even if the light beam is so weak that only a single photon reaches the double slit at any one time. We can put a detector at one of the slits to work out which of the two slits this photon went through. However, the attempt to determine which of the two slits the photon passes through destroys the interference pattern. This phenomenon has been described by Nobel Laureate, the late Richard Feynman, as one of the central mysteries of quantum physics. It is summed up in Heisenberg's uncertainty principle that relates uncertainty in positional information to uncertainty in momentum, such that when an entity is constrained in space (that is, we know which slit it went through) the momentum must be randomised to a certain degree. Translated, this means we can never know the energy and position of an object simultaneously. We can only measure one of these precisely. By knowing that the photon passes through slit A rather than slit B the diffraction pattern, which is a momentum distribution map, is lost. The important point for nanoelectronics is that like light photons, we cannot know the position and energy of an electron simultaneously. Because we are dealing with less and less electrons as devices become smaller we must consider wave-particle duality and Heisenberg's uncertainty principle.

To understand how and when quantum effects come into play we must consider what happens to a semiconductor device as it becomes smaller. Obviously one thing is clear, as we reduce the size the net electron transit time through the device is shorter and hence there is an increase in the speed of the device. This in itself provides a strong incentive for making electronic devices smaller and smaller. However there are other more fundamental effects. In a bulk crystalline solid, electrons occupy effectively continuous energy bands with the occupation, width and separation of these bands determining the fundamental electronic, optical and magnetic properties of the solid. At the other end of the length scale, for individual atoms and molecules, the electronic states are discrete and quantised: electrons can only have one of a number of distinct energy values, rather like rungs on a ladder. This series of discrete energy levels is a consequence of quantum theory and is true for any system in which the electrons are confined to a finite space. The main physical difference between the bulk classical and individual quantum world is the lower dimensionality: quantum effects become observable when the separation between these energy levels becomes larger than the thermal energy that allows rapid transitions at operating temperature. As the physical dimensions of the device are reduced, the separation between the discrete energy levels increases, and quantum effects persist to higher temperatures.

Using modern crystal growth techniques, such as Molecular Beam Epitaxy, to form layered structures of different materials we are able to create artificial potential (energy) profiles in semiconductors to control the quantum mechanical motion of electrons and produce a variety of electronic properties. For example, if the electrons are trapped in a narrow quantum well of sufficiently small thickness, such as in silicon/SiO_2 MOSFETs or gallium arsenide heterostructures, their motion is only possible in the two-dimensional plane of the quantum well. This confinement leads to fundamentally new electronic properties and device applications, as described in section 8.6 below. In addition, an entire new field of mesoscopic physics has emerged in which the subsequent electrostatic confinement of these two-dimensional systems into one-dimensional wires or zero-dimensional boxes has allowed the investigation of quantum electronic transport through nanometre scale architectures of ever increasing complexity. Mesoscopic systems are defined as in the middle between 'nano' and 'micro' systems, that is, they contain between 10^3 and 10^6 atoms. Nanoscience overlaps with mesoscopics, emphasising smaller dimensions, typically 1–100 nm in either one, two or three dimensions.

8.6 QUANTUM ELECTRONIC DEVICES

HIGH ELECTRON MOBILITY TRANSISTORS

In the early 1970s an alternative to the mainstream silicon effort was introduced in the form of gallium arsenide (GaAs) transistors. Compound semiconductors like GaAs are of technological importance since, unlike silicon, they can emit light much more efficiently. As a result they are the basis of solid-state lasers and light-emitting diodes essential for fibre optic communications and compact disk players [32]. In addition the electron mobility (a measure of the speed of the electrons through the device) is greater than many other semiconductors: GaAs makes it possible to significantly improve transistor switching speed. GaAs-based transistors are now used in high speed communications such as mobile phone and satellite TV decoders.

There has been a driving force to speed up communications even further. However the very dopants that are added to make transistors slows down electron transport. Nevertheless a major advance came with the invention of the high electron mobility transistor (HEMT) (Figure 8.6c), with even faster switching speeds [33, 34]. In a conventional transistor the presence of impurities is unavoidable since they are the very dopants we have added to supply the carriers of the electron current whose mobility we want to enhance. Ideally we would like to move the carriers away from the dopants and keep them in a 'spacer' channel that is spatially separated from the doped region (see Figure 8.6c) and in which the carriers are free to move without any unwanted collisions. This ability to separate the charge carriers from their parent dopants by an undoped barrier of $Al_xGa_{1-x}As$ is called modulation doping and was introduced in 1978 by R. Dingle [34]. These devices are made by epitaxially growing silicon-doped $Al_xGa_{1-x}As$ on a GaAs substrate. In $Al_xGa_{1-x}As$ a fraction, x, of the Ga atoms are replaced by Al atoms. $Al_xGa_{1-x}As$ (where typically x=0.33) forms a perfect crystalline interface between the two materials without introducing strain or defects, which can trap electrons. For the GaAs-$Al_xGa_{1-x}As$ junction, energy considerations mean that the electrons in the $Al_xGa_{1-x}As$ transfer into the material GaAs causing the formation of a potential well at the hetero-junction between the two materials. The free electrons in the GaAs are held against the interface due to the ionised silicon dopants in the $Al_xGa_{1-x}As$. The well width is similar to the wavelength of the electron and, as with the silicon MOSFET, a thin two-dimensional sheet of electrons is formed. However, the electrons can travel much faster than those in a silicon MOSFET due to the smaller GaAs mass and the perfect crystallinity and abruptness of the GaAs-$Al_xGa_{1-x}As$ interface compared with the amorphous SiO_2/Si interface. The practical importance of this structure lies in its use as a field effect transistor operating at speeds that are orders of magnitude faster than silicon MOSFETs.

In addition to their technological importance, two-dimensional systems formed in silicon and GaAs field effect transistors are also of interest for their unusual quantum properties. The electrons are only free to move in the plane parallel to the interface, so the electron energy spectrum perpendicular to the hetero-interface is split into a series of discrete energy levels. This realisation has led to the observation of fascinating quantum phenomena. Applying a perpendicular magnetic field to the two-dimensional plane further quantises the electron motion, giving a ladder of discrete energy levels known as Landau levels. This quantisation is manifested in the electrical conductivity of the two-dimensional electron system. Large oscillations of the resistance measured at low temperatures are observed as the magnetic field is increased, which is very different to the behaviour found in three-dimensional metals and semiconductors. These oscillations, known as Shubnikov-deHaas oscillations [35], were first discovered in 1966 and are direct evidence of the quantisation of the electron's angular momentum in the two dimensional plane. The most famous discovery in this field was the Nobel Prize-winning Integer Quantum Hall effect, which was discovered in 1980 by Klaus von Klitzing, Dorda and Pepper [36]. Figure 8.6d illustrates this. A full explanation is beyond this text but, here it can be seen that a plot of the Hall resistance, R_{Hall}, as a function of magnetic field, B, is quantised by units of a constant (h/e^2), with an accuracy of one part in 10^8. Not only is this quantisation found to be in units of the fundamental constants h and e, but it has since been found to be a general property of any two-dimensional system, independent of the material. In 1982, Stormer, Tsui and Gossard were studying the Hall effect in very high quality two-dimensional systems and unexpectedly found additional plateaux with fractional values of h/e^2 at much higher magnetic fields. This discovery — the Fractional Quantum Hall effect — was a manifestation of a strange new material with properties quite different to those previously observed and coined 'a quantum liquid'. Its discoverers were awarded a Nobel Prize in 1998 [37].

QUANTUM INTERFERENCE TRANSISTOR

Electrons can be further constrained to specific geometries in two dimensions by either vertically etching into the device or by imposing electrostatic confinement using surface metal electrodes. An example of such a device is the quantum interference transistor, which is viewed from above in plate 6 and labelled a quantum dot device. Here the submicron sized metal electrodes are patterned onto the semiconductor surface using electron beam lithography. The lighter regions are the metal gate electrodes placed on the semiconductor surface, essentially forming a ring-shaped structure; and the dark regions represent the thin, two-dimensional sheet of electrons underneath. By applying a negative voltage to the gate electrodes we can deplete all the electrons directly underneath the gate. This essentially confines the electrons in the device to only travel around the central ring (less than one micron in diameter).

This device demonstrates the principle of wave-particle duality. If we apply a voltage between the source (S) and drain (D) we force the electrons to move from S to D in the picture. As the electron wave enters the ring it can travel either left (P side) or right (B side) around the dot in the middle. If the two waves (left and right) travel the same distance around the dot then as they leave the exit slit they recombine to give constructive interference and a current through the ring is detected. If the distance around one side of the ring is made larger than the other then when the two waves recombine they will be out of phase, giving rise to destructive interference, and no current will be detected. By applying a magnetic field perpendicular to the dot we can control the effective length of the two different pathways (left or right) around the dot. Thus, by changing the magnetic field and monitoring the current we observe a series of oscillations in the current as the electron waves sequentially interfere constructively and destructively. This is a direct demonstration of the wave-like behaviour of the electrons in a quantum device first demonstrated by Ford et al. [38].

We can now place a highly charge sensitive detector circuit (labelled

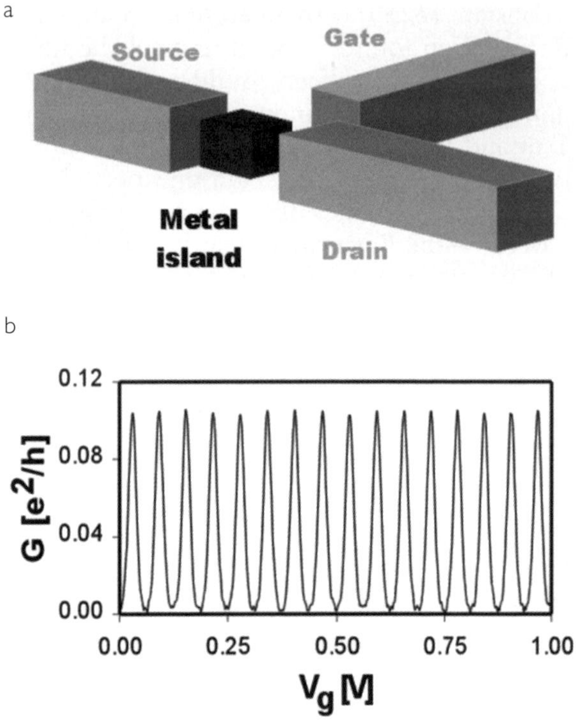

Figure 8.7

a) A simplistic diagram of a single electron transistor, illustrating the central metal island with source, drain and gate electrodes.

b) Experimental data showing conductance oscillations as a function of gate voltage for an Al/AlO_x SET taken at 20 mK by R Brenner (Dipl. Thesis 2000).

P) on the left hand side of the device to try and sense when an electron passes this side of the ring. However, as the sensitivity of this electron detection circuit is increased we learn more about the spatial location of the electron, thus destroying its wave-like nature. This then suppresses the quantum interference of the electron wave and the oscillations in the current are no longer seen when the magnetic field is changed. This electronic device, fabricated in 1998 and known as the 'which-path detector' beautifully demonstrates the principle of wave-particle duality [39].

SINGLE ELECTRON TRANSISTORS (SETS)

If we fabricate a small region or dot in a semiconductor that is no greater than 10–1000 nm in diameter, it may seem that there are a fairly large number of atoms within this area. However, all the electrons are solidly bound to the nuclei of the material, so that the number of free electrons will only be between one and 10 000. Such structures are often referred to as artificial atoms, since the small dimensions of the dot means that the energy level spacings are correspondingly large, as is found with individual atoms. The electrons occupy discrete quantum levels similar to atomic orbitals in atoms, and have a discrete excitation spectrum. In a similar way as the ionisation energy, there is a corresponding energy to add or remove an electron from the dot called the 'charging energy – E_c'.

This term arises since most of the energy comes from the Coulomb interaction required to add or remove a charge, e from the dot.

One benefit of fabricating quantum dots is the possibility of attaching current and voltage leads to probe their atomic states. This is not yet possible with individual atoms. A single-electron transistor consists of a small region or 'dot' through which electrons must tunnel to get from the source to the drain (Figure 8.7a). Such a device was proposed in 1986 [40] and realised experimentally by Fulton and Dolan in 1987 [41]. Essentially we are considering charge transport between the two electrodes (source and drain) separated by an insulating barrier with a third electrode (island) in the middle of this insulating gap. The transport of charge from the source to the island and from the island to the drain across two insulating gaps occurs via the quantum mechanical process of tunnelling. If there is not enough energy to overcome the charging energy, the transport of electrons is blockaded. This is known as 'Coulomb blockade'.

For a detailed review of single electron effects the reader is referred to a number of popular reviews [43]. The charging energy, E_c, is controlled by the capacitance C_Σ between the island and the rest of the world, $E_c=e^2/C_\Sigma$ In order to observe single charge tunnelling events this charging energy must be considerably larger than the thermal energy kT (k is Boltzmanns constant, Chapter 1.6), so that the electrons cannot be thermally excited to energies above E_c [42]. This means that the islands either have to be extremely small (to maximise e^2/C_Σ) or the temperature is very low (to minimise kT), or both. Conventional nanofabrication tools such as electron beam lithography can readily make sub-100 nm islands with $C_\Sigma \sim 10^{-16}$ F, but this requires temperatures below 1 K to satisfy $E_c >> kT$. At these low temperatures electrons can only flow through the device from source to island and then from island to drain.

Recently SETs have been fabricated using scanning tunnelling microscopy and atomic force microscopy at sub-10 nm dimensions and this has produced single electron effects at room temperature [44]. Plate 6 shows an atomic force microscope image of this device showing the source and drain made from oxidised titanium (light regions) surrounding a metallic titanium island (dark) of dimensions 30 nm^2. Here the gate is underneath the island rather than at the side (shown in Figure 8.7a). The island is so small that it can only hold a few free electrons. By applying a voltage to the underlying gate the potential barrier to an electron hopping onto the island decreases so that one electron at a time can flow through the island. Although the extra electron can still tunnel off the island, it tends to stay there as it is held in place by the positive gate voltage — so the current flow stops. No additional electrons can tunnel onto the island because of Coulomb blockade; they would need to gain an extra energy e^2/C_Σ from somewhere. If the gate voltage is further increased so that the potential on the island is lowered by E_c, then this second electron can now tunnel onto the island. Unlike the previous electron it is not tightly bound and can tunnel off again easily, so once again a current can flow.

Thus, gradually increasing the gate voltage, V_g, causes a series of periodic oscillations in the source-drain current (or conductance G (e^2/h)), with each peak corresponding to the equilibrium number of electrons on the island increasing by one. These periodic oscillations are a characteristic signature of single-electron Coulomb blockade effects (Figure 8.7b).

The ability to measure and control current at the single electron level has a number of potential uses from nanoelectronics to computing. One of the most important applications of Coulomb blockade is the detection of single electron charges. SETs are extremely sensitive to small changes in their local electrostatic environment. Semiconductor and metallic dots are therefore being proposed as local electrometers to monitor the behaviour of single electrons in circuits [45] and in nanocircuits in particular.

It is also possible to contemplate complete circuits based on single electronic devices. In principle it is possible to perform calculations using quantum dots based on either charging or quantum coherent phenomena utilising the electron's spins. The use of quantum dot systems for computing or spintronics is a presently rapidly growing field [46].

QUANTUM CORRALS IN ELECTRONICS

One of the advantages of using a STM to fabricate atomic scale features is that we can also analyse the electronic signature of the surface using the conducting tip. As discussed in Chapter 2, in 1993 researchers at IBM rearranged 48 iron adatoms on a copper surface to form a perfect circle called a 'quantum corral' of radius 7.13 nm (see plates 1 and 6, and Figures 2.6 and 2.7) [47, 48]. By confining electrons to length-scales approaching their de Broglie wavelength, their behaviour becomes dominated by quantum mechanical effects. The adatom structure confines the surface state electrons laterally because of the strong scattering that occurs between the surface state electrons and the iron adatoms. The STM can then be used to measure the differential conductivity (dI/dV) between the tip and the copper surface [49].

It is well known that it is possible to bend light waves in the atmosphere to form mirages, but the observation of such analogous wave phenomena in condensed matter physics has only been possible with recent advances in nanofabrication. As discussed in Chapter 2, one example of this is the projection of the electronic structure surrounding a magnetic cobalt (Co) atom to a remote location on the surface of copper. Researchers at IBM evaporated 0.007 of a monolayer of cobalt atoms on the surface of copper. Using the adatom sliding process [50, 51], they carefully arranged over 30 cobalt atoms to form an ellipse (Figure 2.7). An ellipse has the property that all classical paths emanating from one focus bounce specularly off the walls and converge on the second focus. By placing a cobalt atom on the surface of the copper crystal at one focus of the ellipse the electron waves scattered from the real cobalt atom are coherently re-focussed to form a spectral image or 'quantum mirage' of the cobalt atom at the other focus. Cobalt atoms display a distinctive spectroscopic signature called the many particle Kondo resonance [52, 53], arising from its magnetic moment. By positioning a cobalt atom at one focus of the ellipse a strong Kondo signature was detected, not only at the atom, but also at the empty focus. As discussed earlier, if a Kondo blip is seen at the remote location, this raises the question of whether the electronic structure of the cobalt atom and therefore the atom itself is projected from one focus of the ellipse to the other. At this stage it is difficult to draw any conclusions about remote projection but such an elegant experiment may open the way to probe atoms and molecules remotely on a surface whilst minimising the influences and interactions that occur with a nearby local probe. This is also a different way of transferring electrons. In this new 'nanoelectronics' there is transfer of electrons on the atomic scale using the wave nature of electrons instead of conventional wiring.

CARBON NANOTUBE TRANSISTORS

In Chapter 4 we explored in detail the structure of carbon nanotubes, which are essentially a sheet of graphite rolled up into a tube of less than one nanometre in diameter [54]. Nanotubes offer many exciting applications, from microscopic wires to diodes and transistors. They have been considered the dream material for building tiny circuits, since they are strong, non-reactive, tolerant to extreme temperatures and can pass current almost without any resistance. They are also smaller than any wires in today's electronics. One other valuable property is that, like semiconductors, they can be made either insulating or metallic (see Chapter 4) [55]. The picture of the carbon nanotube in plate 6 demonstrates the first nanotube transistor [56]. Here a particular type of nanotube that is a semiconductor (see Chapter 4.6 and Figure 4.8) about one nanometre in diameter is laid across two electrodes — a source and drain — on a silicon surface. Applying a voltage to the silicon substrate induces carriers onto the nanotube to turn the transistor on. Nanotubes, as the name suggests, are as narrow as the double-stranded DNA molecule that carries our genetic information. So arranging nanotubes into electronic circuitry could allow miniaturisation by a factor of about 100 over the current limit. By combining techniques from engineering, chemistry and biology researchers are already learning how to grow, cut, sort and chemically modify nanotubes in new and exciting ways, and the first purely carbon electronic circuits have been fabricated. Time will tell whether they can replace silicon. NASA is one of the major investors [57].

MOLECULAR ELECTRONICS

Another new and exciting interdisciplinary field is the area of molecular electronics, which concerns the exploitation of organic materials in electronic and optoelectronic devices. There are already many commercial applications, including specially engineered polymers that emit light for thin film displays [58], conductive polymer sensors [59] and pyroelectric plastics. Longer-term goals have been the fabrication of metallic contacts to organic molecules to allow the study of the basic transport mechanisms in these systems for the potential development of molecular-scale electronic systems [60]. As noted in Chapter 5, a major breakthrough was achieved when researchers at Hewlett-Packard and the University of California at Los Angeles announced that they had built an electronic switch consisting of several million molecules of rotaxane. By linking several switches they were able to form a rudimentary version of a device that performs a basic logic operation. However there is still much work to be done, as these initial devices only worked once before becoming inoperable [61].

How does a molecular switch compare to a large scale switch such as a transistor? The key is to be able to control the flow of electrons. Using a device that moves such as rotaxanes or catenanes is one method, provided there is some way of recording the movement. Another way to do this in a molecule is to control the overlap of electronic orbitals. For example with the right overlap it may be possible for electrons to flow but if we disturb the overlap (such as by twisting the molecule or affecting its geometry) it may be possible to block flow. A two-terminal transistor based on molecular mimics (Chapter 5) of three benzene rings, where electron orbitals overlapped throughout has been fabricated [62]. Both NO_2 and NH_2 groups were added to the central benzene ring, creating a strongly perturbed electron cloud. The asymmetric, perturbed cloud made the molecule very susceptible to the distortion by an electric field. This gave rise to an active device where every time a large voltage was applied to the molecule it twisted and stopped the flow of electrons. When the electric field was switched off the molecule sprang back to life and current flowed again. The switching characteristics were found to be far superior to any comparable solid-state device. The task of fabricating and testing such a tiny molecular device was made possible by the use of a scanning tunnelling microscope. However for any useful molecular circuit a vast number of devices need to be orderly arranged and securely affixed to a solid structure to keep them from interacting randomly with one another. Although it might be possible to use SPM techniques to do this it would take an extremely long time.

Today's chip densities are such that wires consume 70% of the real estate on a chip. This means that they account for 70% of the defects that lower chip production. In the field of nanoelectronics this trend will be exacerbated. One way round this is to build computers that rely on regular structures rather than molecular wiring using molecular mimics such as conducting fullerenes or overlapping π-electron systems such as single molecule conducting polymers like polypyrrole. In the more distant future quantum corrals might do the job

DNA-DIRECTED ASSEMBLY AND APPLICATION IN ELECTRONICS

The ultimate nano-electronic material will allow molecular recognition and self-assembly to achieve inter-element wiring and electrical interfacing to macroscopic electrodes. When it comes to self-assembly it should be no surprise that nature has long since solved this problem. DNA has the appropriate molecular recognition and mechanical properties to achieve such a feat, but unfortunately it has poor electrical characteristics that have so far limited its direct use in electrical circuits. However, as discussed in Chapter 6 researchers at the Technion Institute of Technology in Israel found that it was possible to hybridise a DNA molecule with surface-bound nucleotides on gold electrodes such that they could stretch it across the gap between two gold electrodes (see fluorescence image of DNA, Figure 8.2 and Figure 6.13 and Chapter 6.5) [63]. The DNA molecule was used as a scaffold over which to deposit silver atoms to form an electrically conducting wire, nanometres in width and several microns in length. This experiment demonstrated the possibility of using DNA as a sophisticated substrate for the targeted attachment of conducting metal wires. DNA self-assembly technology, whilst in its infancy, has a broad range of potential electronic applications from integrating with existing microelectronic device fabrication to the longer term nanofabrication of true molecular electronic circuits and devices. A detailed review of such structures was conducted by R. Bashir [64].

8.7 QUANTUM INFORMATION AND QUANTUM COMPUTERS

Over the past 40 years scientists have investigated and tried to understand unusual quantum phenomena, but is it possible that a new kind of computer can be designed based entirely on quantum principles? The notion of a quantum computer has existed since 1982, when the famous physicist Richard Feynman outlined how such a device might operate [65]. The extraordinary power of the quantum computer is a result of a phenomenon called quantum parallelism, a mechanism that enables multiple calculations to be performed simultaneously. This is in stark contrast to a classical computer which can only perform operations one at a time, albeit very quickly. The field of quantum computation has remained a largely academic one until the 1990s, when it was shown that for certain key problems quantum computers could, in principle, out-perform their classical counterparts [66, 67]. Since then research groups around the world have been racing to pioneer a practical system. However, trying to construct a quantum computer, at the atomic scale, is far from easy, since it requires the ability to manipulate and control single atoms. Several articles and books offer a detailed review of quantum computation [68].

HOW IS A QUANTUM COMPUTER DIFFERENT TO A CLASSICAL COMPUTER?

Whilst it may appear that computers can understand us, in reality they understand nothing at all. All computers can do is recognise two distinct physical states produced by either electricity, magnetic polarity or reflected light. Essentially they can understand when a switch is on or off. Indeed the 'brain' of the computer — the central processing unit — consists of several million tiny electronic switches called transistors. A computer therefore appears to understand information only because it contains so many transistors and operates at such phenomenal speeds, assembling its individual switches into patterns that are meaningful to us. Information is therefore represented by groups of on/off switches to give us data. If we consider the writing on this page, the data is just the individual letters, which taken out of context mean nothing. A computer takes this meaningless data and groups it together into useful information, such as spreadsheets, graphs and reports.

Table 8.1
Three classical bits registering numbers between 0 and 7

bit 1	bit 1	bit 3	Represented Number
0	0	0	0
0	0	1	1
0	1	0	2
0	1	1	3
1	0	0	4
1	0	1	5
1	1	0	6
1	1	1	7

In the computer world everything is a number. Letters are represented by a string of numbers. In a classical computer everything is represented by the state of the computer's electrical switches and hence there are only two possible states, on and off, giving us the binary number system. The smallest possible unit of data is stored as a 'bit' — either a 0 or a 1. A group of eight bits is called a byte, an important unit since there are enough different 8-bit combinations to represent all the characters on a keyboard, including all the letters, numbers, punctuation marks and so on. The way in which the numbers are arranged to represent the letters of the alphabet is called a text code, and very early on programmers realised they needed a standard code that everyone could agree on. This standard code allows any programmer to use the same combinations of numbers to represent individual pieces of data. The three most popular codes are EBCDIC, ASCII and Unicode.

The processing that takes place within the computer therefore involves either comparing numbers or carrying out mathematical calculations. The computer's flexibility comes from being able to establish order sequences of different operations and changing those sequences. The computer can perform two types of operations: arithmetic and logical. Arithmetic operations include addition, multiplication and division. Logical operations include comparisons, such as determining whether one number is greater than or less than another number. Data is input into the computer, a program is written to perform operations on these numbers and then the result is read out.

Table 8.2
Three quantum bits (qubits) registering numbers between 0 and 7

qubit 1	qubit 2	qubit 3	Represented Number
0	0	[0] [1]	0 and 1 simultaneously
0	1	[0] [1]	2 and 3 simultaneously
1	0	[0] [1]	4 and 5 simultaneously
1	1	[0] [1]	6 and 7 simultaneously
0	[0] [1]	[0] [1]	0, 1, 2 and 3 simultaneously
1	[0] [1]	[0] [1]	4, 5, 6 and 7 simultaneously
[0] [1]	[0] [1]	[0] [1]	0–7 simultaneously

In a quantum computer the rules are changed. Not only can a quantum bit (referred to as the 'qubit') exist in the classical 0 and 1 states, but it can also be in a 'superposition' state, where it is both 0 and 1 *at the same time*. If every qubit in a quantum computer is in a superposition state then the computer can be thought of as being in every possible state that those qubits can represent.

Another way of understanding the difference between a quantum and classical computer is if we consider a register of three classical bits. Using 0s and 1s in binary code it is possible to represent any number between 0 and 7 at any one time (see Table 8.1). Now consider a register of three qubits. If every qubit in a quantum computer is either in a 0 or 1 state then the three-qubit register stores any one of the 8 distinct possibilities and is identical to the classical computer. However if one of the qubits is in a superposition of both 0 and 1, then the register can store two distinct possibilities at the same time. If *each* qubit is in a superposition state this quantum register can be storing all of the numbers from 0 to 7 simultaneously (see Table 8.2). The beauty of quantum mechanics is that these superposition states can be created and uncoupled afterwards. A processor that can use registers of qubits will in effect be able to perform calculations using all possible values of the input registers simultaneously. This phenomenon is called quantum parallelism, and is the motivating force behind the research being carried out in quantum computing.

HOW DOES A QUANTUM COMPUTER WORK?

Information comes in discrete chunks similar to the discreteness of energy levels in an atom. As discussed, in a classical computer this information is digital and is passed on as a series of bits. A quantum computer must match this discrete character of digital information to the strange discrete character of quantum mechanics. To do this a quantum system such as an atom can be used since this has discrete energy levels that could hold bits of information similar to transistors — in one energy state it can be 0 and in another energy state it can be 1. For a cluster of atoms to work as a computer it must also be possible to load information onto the system, process that information by means of simple logical manipulations and to read out the answer. Another way of saying this is that quantum systems must be able to read, write and do arithmetic. An example of how this can be done using atomic systems is outlined below.

WRITING TO AN IDEALISED ATOMIC-QUANTUM COMPUTER

How do we write information into a quantum computer? One way is to excite atoms using laser light. We can consider the ground state of a hydrogen atom as having energy, E_0. If we want to write a 0 into this atom we do nothing. However if we want to write a 1 we can excite it from the ground state to an excited state, E_1, using a pulse of laser light with an energy of E_1-E_0. As the electron absorbs a photon it will gradually move from the ground state to the excited state. If the atom is already in the excited state the same pulse will tell it to emit a photon and go to the ground state. Therefore the pulse of light tells the atom to flip its qubit and is a method of information storage. If the frequency of the photons does not match the energy separation between energy levels nothing will happen. But what happens if we only apply the right frequency, but for only half the time needed to cause a qubit flip? In a classical computer this would lead to errors since it can only exist in a 0 or 1 state and we wouldn't be sure which state it would end up in. In the quantum world the atom is in a superposition state of both the 0 and 1 state with equal amplitudes; that is, the qubit is only flipped halfway. These half-flipped qubits are the reason for the potential power of quantum computation.

READ-OUT FROM AN IDEALISED ATOMIC-QUANTUM COMPUTER

Reading the state of the qubits in a quantum computer is very similar to the writing process. For this we need a third energy level, E_2, which is well separated from E_0 and E_1. We now apply an energy pulse, E_2-E_1 that is different to E_1-E_0, and analyse the photons emitted. If the electron is originally in the state E_1 it will absorb this photon and be excited to the energy level E_2 — a higher, less stable state. As a result it will rapidly decay, emitting a photon of energy E_2-E_1. If the electron is in the ground state nothing will happen since it is not the right energy to excite it to E_2.

Note that read-out is only possible if the qubits are in the 0 or 1 states with a high probability. If an atom is in the superposition state, it has equal probabilities of emitting a photon or not. In this way it is possible to determine the original state of each qubit.

QUANTUM COMPUTATION

Classical computers are comprised of electronic circuits that contain many different components, such as resistors, capacitors, diodes and transistors. Some of these components, such as resistors, alter individual signals, but others, such as transistors, cause more than one input signal passing through them to interact. Calculations are performed by repeating tasks, such as flipping a bit from one state to another, over and over at great speed. Flipping a bit is equivalent to the logical operation called NOT where true 1 becomes false 0 and false 0 becomes true 1. This is an example of a linear operation, one in which the final output reflects the value of a single input. An example of a non-linear operation is the AND operation where the final output depends on some interaction of multiple inputs: if two bits are identical (say both 1) the operation creates a third bit equal to itself (1); if they are different the third bit is zero. A classical computer therefore consists of a large number of linear and non-linear logic gates that perform these operations, allowing the computer to complete any arithmetical or logical task.

How does this work in a quantum computer? All we need is the ability to flip qubits and to be able to control a suitable non-linear interaction between them. Simple two-bit quantum logic operations have already been performed with particle spins back in the 1950s. The spin of a particle is an ideal qubit because it can take only one of two values — it can either be spinning in one direction with respect to a magnetic field (1), or in the opposite direction (0). In a hydrogen atom both the proton and electron have a spin, so a single hydrogen atom in a magnetic field is thus a two-qubit system. It is possible to flip the individual spins by using short bursts of high frequency radiation. The interaction between the electron and proton also makes it possible to perform non-linear operations, such as flip the proton spin only if the electron spin is 1. In this system it has been possible to demonstrate several of the main logical operations. However it is also necessary to connect between quantum logic gates using quantum 'wires'. Wiring these quantum bits together is an extremely difficult task since it requires the manipulation of electrons and protons within individual atoms without disturbing the particles' spins. These systems may need to be wired up with molecular mimics as discussed above or even using biological material [69–71].

DECOHERENCE — THE ENEMY OF QUANTUM COMPUTATION

Two individual quantum-mechanical qubits may also be correlated so that if one qubit is in the superposition state it can affect the state of the other qubit so that it too exists in the superposition state. This kind of correlation is called 'entanglement' and only the entire superposition carries information. When two superimposed quantum waves behave like one wave they are said to be coherent; the process by which two coherent waves regain their individual identities is called decoherence. For an electron in a superposition of two different energy states (or roughly two different positions within an atom) decoherence can take a long time. It may take sometimes days before a photon, say, will collide with an object as small as an electron to expose its true position.

One of the main stumbling blocks to realising a quantum computer is the problem of decoherence [72]. The superposition state of a qubit is very fragile: almost anything, such as a stray electron or photon, can cause the coherent qubit to collapse into one of two classical states. Therefore the number of calculations that a computer can do is directly related to the time that the qubits can remain coherent. This problem is compounded by the fact that even measuring a qubit can cause it to collapse. So we can't even check a qubit to see what is happening because the test process will collapse the superposition state, stopping the calculation before it has finished. These concerns have prompted serious discussions about the feasibility of a practical quantum computer [73]. In a classical computer, algorithms have been developed that correct any errors that creep into a computation before they completely upset the calculation; error correction is a routine part of modern digital communication. However, classical approaches to error correction and fault tolerance cannot easily be extended to quantum systems. In 1995 Steane [74] and Shor [75] independently discovered 'quantum error correction codes' which are discussed in detail in some more recent review articles by Preskill and Steane [76]. Following on from this work, Knill et al. show that despite the debilitating effects of decoherence and other sources of noise, a quantum computer can successfully carry out an arbitrarily long computation successfully [77]. These new principles, and recent experiments that demonstrate that some systems can maintain quantum superposition for several hours [78] have nourished hope that quantum computers will overcome the problems of decoherence.

THE POWER OF QUANTUM COMPUTATION

With only one qubit a quantum computer can already do things no classical computer can do. Take a single atom in a superposition state of 0 and 1. If we make it fluoresce to try and discover what state it is in, half the time it emits a photon (showing that it is in the one state) and the rest of the time no photon is emitted and the qubit is in a zero state. This means that the bit is a random bit — we have produced a random number generator, something a classical computer cannot create. The real power, however, of quantum computation occurs with a many qubit system. In order to understand this we need to consider how a classical computer works.

POWER OF A CLASSICAL COMPUTER

Consider the problem of having a random phone number written on a piece of paper, but we do not know who the number belongs to. If we check the phone directory, a classical computer can help speed up the problem. The computer checks each number in the directory sequentially, starting with the As and working through to the Zs until it finds the number. The power of the computer is that it checks each number very quickly. If we wanted to increase the speed of finding the answer we could add another computer, getting one to check from A–L and the other to check from M–Z. Adding another computer means we have three computers, one checking from A–I, one from J–R and the last from S–Z. So adding another computer simply increases the power of the computation by one. Thus in a classical system the power of this look-up system increases linearly with the number of computers used.

POWER OF A QUANTUM COMPUTER

Unlike a classical computer, each time we add a qubit to a quantum computer the power doubles. Consider a quantum bit or qubit as a coin. Unlike a classical coin, which can either land as heads or tails when thrown, the entangled state of the qubit means that it can be both heads (H) and tails (T) at the same time (@). In the classical situation when we add a second coin there are HH, TT, HT and TH solutions, however when we add a second qubit there are another four solutions H@, @H, @@, T@ and @T. It is this increase of computer power that drives the push for a practical quantum computer. It has been predicted that a 40-qubit computer could recreate in a little more than 100 steps, a calculation that would take a classical computer with a trillion bits several years to finish. A 100-qubit computer would be more powerful than all the computers in the world linked together.

So to recap, what can a quantum computer do with many logical operations on many qubits? If we put all the inputs in an equal superposition of 0 and 1, each of which have the same magnitude, then we essentially have a computer in an equal superposition of all possible inputs. If we now run a series of logic gate operations to perform a calculation, the result is a superposition of all possible outputs of that computation. To put it another way, the computer has performed all possible computations at once. This massive quantum parallelism holds the promise for quantum computation. The only snag is that the qubits are now in a superposition of all possible answers. The trick is to find a quantum operation that will only amplify the correct answer, and reduce the probability of all other outputs. This is the realm of the quantum algorithm.

QUANTUM ALGORITHMS

It has been shown in theory that a quantum computer will be able to perform any task that a classical computer can. However it does not mean that a quantum computer can outperform a classical computer for all tasks. In order to input problems into a computer we need to establish algorithms –a series of generalised instructions that allow it to perform a certain calculation. If we use our classical algorithms on a quantum computer it will simply perform the calculation in a similar manner to a classical computer. In order to demonstrate its superiority and exploit the phenomenon of quantum parallelism we need to establish quantum algorithms.

SHOR'S ALGORITHM

Whilst quantum algorithms are not easy to formulate they do produce spectacular results. A good example of one such algorithm is the quantum factorising algorithm created by Peter Shor of AT&T Bell laboratories [79]. The purpose of this algorithm is to break down or factorise a large number into its prime factors. Whilst it is easy to multiply large numbers such as 1237 by 3433 it is difficult to calculate the factors of large numbers, such as 4 246 621 using a classical computer. The difficulty of factorising large numbers lies at the heart of a type of data encryption, called public key or RSA (Rivest, Shamir and Adleman) encryption, one of the most trusted methods used to protect electronic bank accounts. Here encrypted data is transferred by means of a public key (that encrypts the data) and a private key (that is used to decrypt the data). The public key is obtained by multiplying two large prime numbers together. The private key is one of the prime numbers. In order to decode a message encoded with the public key we need to factorise it. This is easy for somebody who has one of the private keys but very hard if we only have the public key. As numbers get bigger and bigger the difficulty increases rapidly. However, in principle a quantum computer that is running Shor's algorithm can crack the code within seconds.

GROVER'S ALGORITHM

Another example of an important quantum algorithm is Grover's algorithm [80], which can be used to sort through large unsorted databases. Using a conventional computer on a database with N entries would take an average of N/2 inquiries to find the data needed. A quantum computer would take $\sqrt{N}$ inquiries. For example, a database holding ten million entries would take on average five million searches using a classical computer, compared to 10 000 on the quantum computer. As databases get larger and more interlinked, quantum computers would offer a significant saving in time. Along with the design of quantum algorithms, something only a quantum computer can do, has come the obvious question: what about building a quantum computer?

8.8 EXPERIMENTAL IMPLEMENTATIONS OF QUANTUM COMPUTERS

Along with the main technical difficulty of working at the single atom level, one of the most important problems in building a quantum computer is that of preventing the surrounding environment from interacting with the fragile superposition states. However, the number of experimental proposals to fabricate a quantum computer has grown rapidly over the past five years. These include ion trap systems, nuclear magnetic resonance (NMR) systems (see Chapter 1.4 and Chapter 3.9) [81], solid-state systems [82, 83], quantum optics [84], quantum dot proposals [85], superconducting qubits [86] and many others. Indeed the amazing variety of approaches reflects nearly every branch of physics. For more details about each of these proposals the reader is referred to two recent conference proceedings on experimental systems [87]. The field of quantum computation is a truly cross-disciplinary one, bringing together ideas from classical information theory, computer science, quantum physics, molecular electronics and self-assembly. A recent review article by DiVincenzo goes on to outline the five basic requirements for the physical implementation of quantum computation [88]. These include:

1. scalable physical systems with well-defined qubits that are disconnected from their surroundings to minimise decoherence;
2. the registers should be initialised to a known state before the start of the computation. We have to be able to connect to the qubits when we want so that we can input data into the computer at the beginning of a calculation;
3. the ability to read out the answer by measuring specific qubits;
4. the decoherence time for the qubits must be much longer than the gate operation time;
5. a universal set of quantum gates.

Despite these guidelines it is not possible at this early stage to predict which of the above technologies will prevail. Whilst several proposals have been considered, each one has its own difficulties.

In the case of the nuclear magnetic resonance (NMR) system, a sea of molecules in a liquid can perform all the steps in a quantum computation: loading in an initial condition, applying logical operations to entangled superpositions and reading out the final result. NMR operates on the atomic nuclei within the molecules of the fluid. When placed in a magnetic field each nuclear spin acts like a tiny bar magnet, either aligning parallel or antiparallel with the field. This corresponds to two different quantum states with different energies, which naturally constitute a qubit. In a molecular liquid there will always be a small excess of spins in one direction over the other and it is the signal from these excess molecules (perhaps just one in a million) that is detected and manipulated to perform calculations.

As discussed in Chapters 1 and 3, NMR is used to detect these spin states. Radio waves of the right frequency and duration are used to flip the nuclei from spinning in one direction to another and vice-versa. The quantum computer is therefore the molecule and the qubits are the nuclei within the molecule. Instead of using a single molecule and trying to isolate it from its environment, the NMR approach is to use a large number of liquid molecules. There are several advantages to this approach:

1 even though the liquid molecules are bumping into each other continuously, this doesn't provide enough energy to change the spin state of the nuclei they contain;

2 the decoherence time is found to be much greater than other techniques realised so far, such that it is possible to perform several thousand quantum logic operations.

A two-qubit based NMR quantum computer has been constructed using a few millilitres of chloroform ($CHCl_3$), taking advantage of the interaction between the spins of the hydrogen and ^{13}C nuclei. Using this NMR system the first practical realisation of a quantum computation was performed — the Grover search algorithm [89]. This was an extremely encouraging start for NMR systems, but scaling up to larger numbers of qubits becomes a problem. As the number of molecules grows, the signal from the excess spins in the system becomes extremely weak, making read-out harder to detect. Whilst it has been possible to demonstrate up to eight qubits in this system, advancing beyond ten in liquid NMR systems is expected to be extremely difficult [90].

Another example of a practical design that is *scalable* is the solid-state silicon-based nuclear spin quantum computer [82]. Plate 8a shows the device architecture, where the nuclear spins of single phosphorus nuclei ^{31}P embedded in a pure ^{28}Si host crystal form the quantum bits or qubits. The Si:P nuclear spin relaxation time is extremely long (10^{13} hours at -274°C) allowing many logical operations. An electrical barrier isolates the qubits from surface control gates, which are used to control and read out the nuclear spin state of individual phosphorus nuclei. For the electron wave functions of the phosphorus qubits to overlap they must be approximately 20 nm apart and form part of an ordered array within the silicon crystal.

Plate 8b demonstrates the manipulation at the atomic scale required to achieve such a device architecture. It is then necessary to encapsulate the phosphorus atoms in a pure silicon host crystal before depositing very narrow control gates on the surface. All of these steps push the limits of our present technology. Recent progress towards this goal has been achieved with the formation of an atomically precise array of phosphorus nuclei on a silicon crystal (Plate 8c) [22]. There remain many challenges ahead for this proposal too, including registration of the metal electrodes to the individual phosphorus nuclei and the growth of extremely pure silicon that does not contain any other nuclear spins or defects which could give rise to decoherence.

Initially a low defect density silicon surface is carefully prepared and then passivated with a monolayer of hydrogen. The array is fabricated using a resist technology, much like in conventional lithography, where the resist is the layer of hydrogen atoms that terminate the silicon surface. An STM tip is used to selectively desorb hydrogen, exposing silicon atoms on an atomic scale. Phosphine gas is now passed over the hydrogen patterned surface such that only one phosphine molecule adsorbs at each of the exposed sites, thus forming an array of single phosphorus-bearing molecules on the silicon surface. Low temperature silicon overgrowth then encapsulates the phosphorus atoms whilst maintaining the ordered atomic array. Finally it is necessary to register control electrodes on the surface to individual buried phosphorus atoms in the bulk of the crystal. These electrodes will control the spin states of individual nuclei and are used for read out.

Whether or not quantum computers can be realised is a hotly debated question. However, to surpass classical simulations of quantum systems requires only a few tens of qubits — a goal that is within our grasp. When the first transistor was invented, nobody could have guessed at modern computer applications such as word processors, databases, computer networks and the Internet. Quantum electronic devices, such as the quantum computer, promise a glimpse of a new wave of similarly spectacular applications in the twenty first century.

WHAT YOU SHOULD KNOW NOW

1. By adding just 0.01% of impurities it is possible to change the electrical resistance of a semiconductor 10 000 fold — enabling semiconductors to be engineered between metallic and insulating behaviour.
2. The operation of a transistor falls into mainly two types: bipolar junction and field effect, both of which can be used in a circuit to amplify a small voltage or current into a larger one (for example, Hi-fi amplifiers) or to function as an on-off switch (for digital computers).
3. Lithography is the key technology of the semiconductor industry and is the process of transferring patterns to semiconductor materials in analogy to the photographic process.
4. Electron-beam lithography has the capability to fabricate devices many orders of magnitude smaller than can be made with standard optical lithography.
5. Atomic lithography is the ultimate limit of fabrication and controls the arrangement of individual atoms.
6. Molecular beam epitaxy (MBE) is essentially a refined form of evaporation in which molecular beams interact on a heated crystalline substrate under ultra high vacuum (UHV) conditions to produce a single crystal film.

7 Electrons can be constrained to specific geometries in the two-dimensional plane by either vertically etching into a device or by imposing electrostatic confinement using surface metal electrodes.

8 Single electron transistors (SETs) are fabrications of 10–1000 nm in diameter. The electrons occupy discrete quantum levels similar to atomic orbitals in atoms, and have a discrete excitation spectrum.

9 The ultimate nanoelectronic material will allow molecular recognition and self-assembly to achieve inter-element wiring and electrical interfacing to macroscopic electrodes.

10 In a quantum computer the rules are changed from classical computing. Not only can a quantum bit (referred to as a 'qubit') exist in the classical 0 and 1 states, but it can also be in a 'superposition' state, where it is both 0 and 1 at the same time. If every qubit in a quantum computer is in a superposition state, then the computer can be thought of as being in every possible state that those qubits can represent.

8.9 EXERCISES

1 *Find out what is meant by the term 'Miller indices' and sketch structures demonstrating different faces of a crystal, labelling the different Miller indices*

2 *Find out more about the following techniques and principles:*
   a) Auger electron spectroscopy (AES);
   b) Heisenberg's uncertainty principle;
   c) Landau levels;
   d) proton-proton coupling in NMR;
   e) proton ^{13}C coupling in NMR;
   f) Kondo signature;
   g) Quantum Hall effect;
   h) Coulomb blockade effects
   i) Reflection high-energy electron diffraction (RHEED)

3 *Make a list of all Nobel Prize winners discussed in this article and list them by date. Predict what you may consider as worthy of a new Nobel Prize in 2010 and write an article on the discovery.*

4 *List all nanoelectronic devices discussed here and in previous chapters. Write short notes on what each one can do.*

5 *If you are interested in computing, web search qubit and find out all you can on quantum computing. If you do not know much about computing find out about logic operations, NOT and AND statements. Find out what is meant by EBCDIC, ASCII and Unicode and explain the differences.*

6 *MBE now has important and widespread applications in the semiconductor industry, where device performance depends on the precise control of dopant profiles and on the production of extremely thin crystal layers with hyper-abrupt interfaces. Examples of some of the important devices produced by MBE are light-emitting diodes, laser diodes, optical waveguides, field effect transistors, high electron mobility transistors and read/write heads for computer hard drives. Web search to find data to support these statements.*

8.10 REFERENCES

1 Binnig G & Rohrer H (1987) Scanning Tunneling Microscopy — from birth to adolescence. *Reviews of Modern Physics* 59: 615–25.

2 Cuberes MT, Schlittler RR & Gimzewski JK (1996) *Appl. Phys. Lett.* 69: 3016–18; Lopez M et al. (1996) *Nanotechnology* 7: 183.

3 Reed MA, Zhou C, Muller CJ, Burgin TP & Tour JM (1997) *Science* 278: 252–54.

4 Tans SJ, Devoret MH, Dai HJ, Thess A, Smalley RE, Geerligs LJ & Dekker C. (1997) *Nature* 386: 474–77; *Nature* 386: 474.

5 Schuster SC & Kahn S (1994) *Annu. Rev. Biophys. Biomol. Struct.* 23: 509–39.

6 Cornell BA, Braach-Maksvytis VLB, King LG, Osman PDJ, Raguse B, Wieczorek L & Pace RJ (1997) *Nature* 387: 580.

7 Bardeen J & Brattain WH (1948) *Phys. Rev.* 74: 230; Riordan M, Hoddeson L and Herring C (1999) *Reviews of Modern Physics* 71: S336.

8 Fowler A (1997) *Physics Today* October 50(10): 50–54.

9 OSA Trends in Optics and Photonics Vol. 4, Kubiak GD & Kania DR (eds) *Extreme Ultraviolet Lithography.* Optical Society of America, Washington DC.

10 Tenant DM (1999) Chapter 4. In Gregory Timp *Nanotechnology* p. 161ff.

11 Bustamante C & Keller D (1995) *Physics Today* 48(12): 32–38.

12 Eigler, DM & Schweizer EK (1990) *Nature* 344: 524.

13 Stroscio JA & Eigler DM (1991) *Science* 254: 1319.

14 Avouris P (1995) *Acc. Chem. Res.* 28: 95–102.

15 Stipe BC, Rezaei MA & Ho W (1998) *Science,* 279: 1907–1909.

16 Shen TC, Wang C, Abeln GC, Tucker JR, Lyding JW, Avouris P & Walkup RE (1995) *Science* 268: 1590–92.

17 Stipe BC, Rezaei MA, Ho W, Gao S, Persson M & Lundqvist BL (1997) *Phys. Rev. Lett.* 78: 4410–13.

18 Thirstrup C, Sakurai M, Nakayama T & Aono M (1998) *Surface Science* 411: 203–14.

19 Tucker JR & Shen TC (1998) *Solid State Electronics* 42: 1061–67.

20 Lyding JW, Shen TC, Hubaceck JS, Tucker JR & Abeln GC (1994) *Appl. Phys. Lett.* 64 2010–12; Sakurai M, Thirstrup C, Nakayama T & Aono M (1997) *Surface Science* 386: 154.

21 Sakurai M, Thirstrup C & Aono M (2000) *Phys. Rev. B.* 62: 16167–74.

22 O'Brien JL, Schofield SR, Simmons MY, Clark RG et al. (in press) *Phys. Rev B.*

23 Mastomoto K (1999) *Int. J. Electronics,* 86 641; (1996) *Appl. Phys. Letters,* 68: 34.
24 Cho AY (1983) *Thin solid films,* 100: 291–317; Cho AY & Arthur R (1975) *Prog. Solid State Chem.* 10: 157–91.
25 Cho AY (1970) *J. Appl. Phys.* 41: 2780–86; Laurence G, Simondet F & Saget P (1979) *Appl. Phys.* 19: 63.
26 Neave JH, Joyce BA, Dobson PJ & Norton N (1983) *Appl. Phys. A.* 31: 1–8.
27 Joyce BA, Neave JH, Zhang J & Dobson PJ (1988). In Larsen PK & Dobson PJ (eds) *Reflection High Energy Electron Diffraction and Reflection Electron Imaging of Surface.* Plenum, pp. 397–417.
28 Munoz-Yague A, Piqueras J & Fabre N (1981) *J. Electrochem. Soc.* 128: 149.
29 Minnie SC et al. *Appl. Phys. Letters,* 73: 1742.
30 Yazdani A & Lieber CM (1999) *Nature* 401: 227–30.
31 Kung P & Razeghi M (2000) *Opto. Electron. Review* 8: 201–39.
32 Kelly MJ (1995) *Low-Dimensional Semiconductors: Materials, Physics, Technology, Device.* Oxford University Press, UK.
33 Drummond TJ, Masselink WT & Morkoc H (1986) *Proc. IEEE* 74: 773–822; Melloch MR (1993) *Thin Solid Films* 231: 74.
34 Dingle R, Stormer HL, Gossard AC & Wiegmann W (1978) *Applied Physics Letters,* 33: 665–67.
35 Fowler AB, Fang FF, Howard WE & Stiles PJ (1966) *Phys. Rev. Lett.* 16: 901–903.
36 von Klitzing K, Dorda G &Pepper M (1980) *Phys. Rev. Lett.* 45: 494–97.
37 Tsui DC, Stormer HL & Gossard AC (1982) *Phys. Rev. Lett.* 48: 1559–62; Laughlin RB (1983) *Phys. Rev. Lett.* 50: 1395–1400.
38 Ford CJB, Washburn S, Buttiker M, Knoedler CM & Hong JM (1989) *Phys. Rev. Lett.* 62: 2724–27.
39 Buks E, Schuster R, Heiblum M, Mahalu D & Umansky V (1998) *Nature* 391: 871–74.
40 Averin DV & Likharev KK (1986) *J. Low Temp. Physics* 62: 345–48.
41 Fulton TA & Dolan GJ (1987) *Phys. Rev. Lett.* 59: 109–12.
42 Altmeyer S, Hamidi A, Spangenberg B & Kurz H (1996) *J. Vac. Sci. Technol. B* 14: 4034–37.
43 Reed MA (1993) *Scientific American* 268: 118–32; Kastner M (1993) *Physics Today* 46, 24; Likharev KK & Claeson T (1992) *Scientific American* 266: 50–62; Devoret MH, Esteve D & Urbina C (1992) *Nature* 360: 547–49.
44 Matsumoto K (1999) *Int. J. Electronics* 86: 641–52.
45 Dresselhaus PD, Ji L, Han SY, Lukens JE & Likharev KK (1994) *Phys. Rev. Lett.* 72: 3226–29.
46 Burkhard G, Engel HA & Los D s (2000) *Fortschr. Phys.* 9: 965–75; Ensslin K (2000) *Fortschr. Phys.* 9: 999–1003.
47 Crommie MF, Lutz CP & Eigler DM (1993) *Science* 262: 218–20.
48 Manoharan HC, Lutz CP & Eigler DM (2000) *Nature* 403: 512–15.
49 Lang ND (1986) *Phys. Rev. B.* 34: 5947–84.
50 Eigler DM & Schweizer EK (1990) *Nature* 344: 524.
51 Stroscio JA & Eigler DM (1991) *Science* 254: 1319–22.
52 Kondo J (1964) *Prog. Theor. Phys.* 32: 37–45.

53 Madhavan V, Chen W, Jamnaeala T, Crommie MF & Wingreen NS (1998) *Science* 280: 567–69.
54 Iijama S (1991) *Nature* 354: 56.
55 Farajian AA, Esfarjani K & Kawazoe Y (1999) *Phys. Rev. Lett.* 82: 5084–87; Yao Z, Postma HWCH, Balents L & Dekker C (1999) *Nature* 402: 273.
56 Tans SJ, Verscheueren ARM & Dekker C (1998) *Nature* 393: 49–52.
57 Globus A, Bailey D, Han J, Jaffe R, Levitt C, Merkle R & Shrivasatva D (1998) *The Journal of the British Interplanetary Society* 51: 145.
58 Schadt M (1997) *Annual Review of Materials Science* 27: 305.
59 Petty MC (1996) *Measurement Science and Technology* 7: 725–35.
60 Reed MA (1999) *Proceedings of the IEEE* 87: 652.
61 Heath JR (2000) *Pure and Applied Chem.* 72: 11–20.
62 Reed MA & Tour JM (2000) *Scientific American* 282: 68–79.
63 Braun E, Eichen Y, Sivan U & Benyoseph E (1998) *Nature* 391: 775.
64 Bashir R (2001) *Superlattices and Microstructures* 29: 1–10.
65 Feynman RP (1982) *Int. J. Theor. Phys.* 21: 467.
66 Shor PW (1994) Proc. 35th Annual Symp. Found. Computer Science. IEEE Press, November.
67 Grover LK (1997) *Phys. Rev. Letters* 78: 325–27.
68 Lloyd S (1995) *Scientific American* October: 44; Preskill J (1998) *Nature* 391: 631; Brown J (1994) *New Scientist* 24th September 1994: 21; Lo HK, Popescu S & Spiller T (1999) *Introduction to Quantum Computing and Information.* World Scientific Press; Mosca M, Jozsa J, Steane A & Ekert A (2000) *Phil. Trans. R. Soc. Lond. A,* 358: 261; Milburn GJ & Davies P (1999) *The Feynman Processor: Quantum Entanglement and the Computing Revolution.* Helix Book Series.
69 Caldwell WA, Nguyen JH, Pfrommer BG, Mauri F, Louie SG & Jeanloz R (1997) *Science* 277: 930–33; Sarkar P (2000) *ACM Computing Surveys* 32: 80–88.
70 Mohan RV, Tamma KK, Shires DR & Mark A (1998) *Advances in Engineering Software* 29: 249–58.
71 Birnbaum J and Williams RS *Physics Today* 53(1): 38–42.
72 Imry J (1998) *Physics Scripta*, 26: 171; Preskill J (1998) *Proc. Royal Soc. A.* 454: 469.
73 Landauer R (1995) *Phil. Trans. Royal Soc. London A.* 353: 367; Haroche S & Raimond JM (1996) *Physics Today* 49: 51; Hadley P & Mooij JE (2000) *Quantum Semiconductor Devices and Technologies* 6: 1.
74 Steane AM (1996) *Phys. Rev. Lett.* 77: 793–97.
75 PW Shor (1995) *Phys. Rev. A.* 52: R2493–96.
76 Preskill J (1998) *Royal Soc. A.* 454: 385–97; Steane AM (1998) *Phil. Trans. Royal Soc. London A.* 356: 1739–45.
77 Knill E, Laflamme R & Zurek WH (1998) *Proc. Royal Soc. A.* 454: 365–84.
78 Cirac JI & Zoller P (1995) *Phys. Rev. Lett.* 74: 4091–94.
79 Shor PW (1994) *Proc. 35th Annual Symp. Foundations of Computer Science* 1994: 124–66.
80 Grover LK (1997) *Phys. Rev. Lett.* 79: 325–29.
81 Gersenfield N & Chuang I (1997) *Science* 275: 350; Cory D, Fahmy A & Havel T (1997) *Proc. Nat. Acad. Sci.* 94: 1634–39.

82 Kane BE (1998) *Nature* 393: 133–36.

83 Vrijen R, Yablonovitch E, Wang K, Zhang HW, Balandin A, Roychowdhury V, Mor T & DiVincenzo DP (2000) *Phys. Rev A.* 6201: 2306–10; Tucker JR & Shen TC (2000) *Int. J. Circuit Theory and Applications* 28: 553–60; Berman GP, Doolen GD, Hammel PC & Tsifrinovich VI (2001) *Phys. Rev Lett.* 86: 2894–99.

84 Turchette QA, Hood CJ, Lange W, Mabuchi H & Kimble HJ (1995) *Phys. Rev. Letters* 75: 4710–15; Imamoglu A, Awshalom DD, Burkard G, DiVincenzo DP, Loss D, Sherwin M & Small A (1999) *Phys. Rev. Lett.* 83: 4204–206; Knill E, Laflamme R & Milburn GJ *Nature* 409: 46.

85 Sukhorukov EV & Loss D (2001) *Physica Stat. Solidi B.* 224: 855–60; Tanamoto T (2000) *Phys. Rev. A.* 61: 022305; Sherwin M, Imamoglu A & Montroy T (1999) *Phys. Rev. A.* 60: 3508–20.

86 Averin D (1998) *Solid Stat. Comm.* 105: 659–61; Shnirman A, Schön G & Hermin Z (1997) *Phys. Rev. Lett.* 79: 2371–74; Mooij JE, Orlando TP, Levitov L, Tian L, van der Waal CH & Lloyd S (1999) *Science* 285: 1036–39.

87 *Fortschr.der Phys.* Volume 48, 2000; *Quantum Information and Computation.* Volume 1, 2001, Rinton Press.

88 DiVincenzo DP (2000) *Fortschr. Phys.* 9: 771–77.

89 Chuang I, Vandersypen LMK, Zhou X, Leung DW & Lloyd S (1998) *Nature* 393: 143.

90 Warren WS (1997) *Science* 277: 1688; 1672.

9

FUTURE APPLICATIONS

This chapter reveals some new developments just around the corner. These include micro electromechanical machines, ageless, invisibly mended materials, new optics and new electronics. Some ways in which nanomaterials can affect the environment are discussed. These include filtration and green chemistry.

9.1 MICROELECTROMECHANICAL SYSTEMS

Over the last twenty years, microscale machines have been developed, and their uses are growing rapidly. They currently represent the lowest size limits of commercially available machines. Microelectromechanical systems (MEMS) are more recent. They have small moveable structures such as mirrors, or cantilever beams, which are typically coupled to very small motors that are actuated using electromagnetic induction or electrostatic forces. Their production is not too different from the procedures described in Chapter 8. Figure 9.1 shows a cross section of one such device [1]. In this structure, the upper diagram shows layers before etching. The structure is formed by depositing a series of layers (A to D, Figure 9.1a) on top of an insulator (E layer) and silicon substrate, and then etching them away in specific positions so that the structure in Figure 9.1b is formed. The Y-shaped structure in the middle undergoes a seesaw action in which it moves a mirror that allows light to be transferred from one place to another. Layer C is conducting, and hence actuates the Y shape and thus the mirror.

MEMS are largely based on silicon single crystal machining technology, which has evolved from the techniques developed by the microelectronics industry. These are currently used as very sensitive and accurate sensors, small, portable and fast chemical analysers, and optical switches, but their uses will expand in the future. In time they might also serve as active elements, such as flow controllers in bio-fluid systems.

Figure 9.1

Cross section of a MEMS microstructure, in which the central plate is actuated by a voltage. It is formed in silicon using microelectronics-based patterning and etching techniques. The first figure shows the steps in their manufacture. The sacrificial layers must be etched to allow the beam to move. The second figure is the final product after etching. It is the Y-shaped object in the centre that moves.

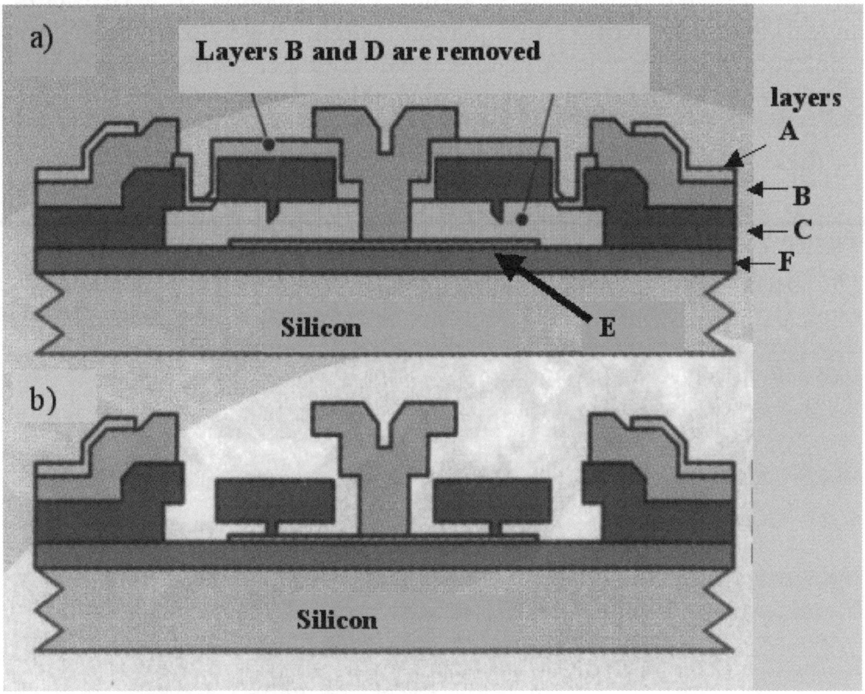

MEMS have been used as pressure sensors since the mid 90s. Until recently, they were primarily used in the management and monitoring of car engines and air conditioning systems. They are also used in accelerometers for car air bags, ink jet printers, computer hard disk drives and biomedical analysis. In the near future the use of MEMS as optical cross connect switches and as other opto-mechanics in fibre optic telecommunications networks, will equal and then eclipse the pressure sensor market. Globally, in 2000 MEMS generated US$3.4 billion in sales. This figure will grow to US$31.5 billion in 2010, mainly due to the expected upsurge in telecommunications and other optical work [2]. Cross connect optical switching systems, which will form the hub in the routing system for information and telephone traffic, are referred to as microoptoelectromechanical systems, or MOEMS. This technology means that light doesn't have to be converted into electrical signals to re-route as it does now. MOEMS technology allows switching light beams with very low loss of light signal power. Figure 9.2 shows a MOEM system where a mirror array dictates that a given incoming fibre's light goes almost exactly to the desired output fibre [3]. Arrays of MOEMS may require up to 100 million switchable micro-mirrors at an exchange, although several arrays will be needed for this, because individual arrays are unlikely to be this big. However, these machines can be packaged together and duplicated in large groups, just like transistors on a chip. As an example of what is coming, in 2001 the world's first petabit switch, based on MOEMS, was constructed by Lucent. A petabit is 10^{15} bits per second, or a million gigabits per second. This particular MEM silicon mirror array switches between 1296 ports, each of which carries 40 separate signals, and each of these 40 is carrying data at 40 gigabits per second [4].

Figure 9.2

MEMS mirror arrays for switching light at a telephone/data exchange between incoming and outgoing fibre optic arrays.

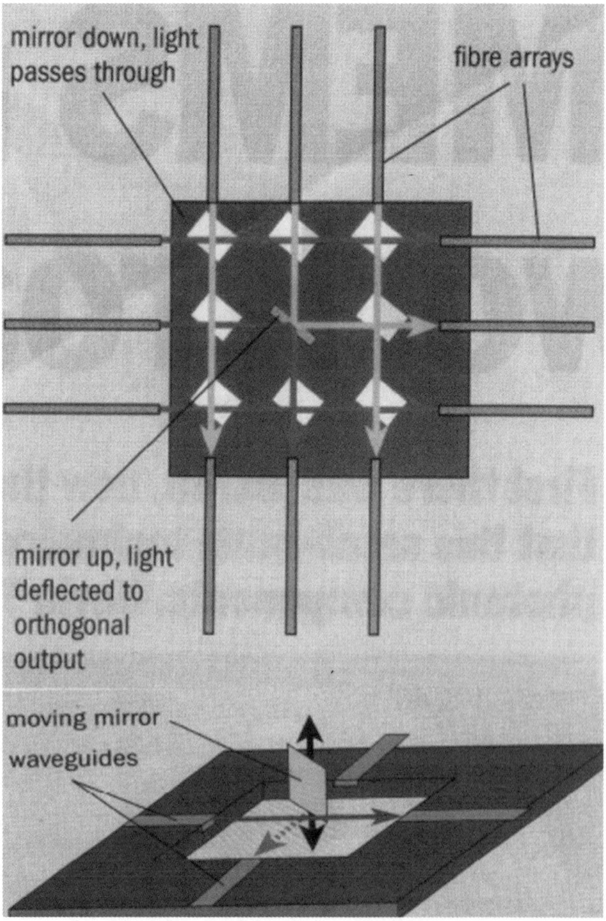

Other MEMS-based sensing includes a recently developed highly sensitive magnetic field strength device [5], where a magnetic field induces distortion and oscillations in a very small silicon bar. The distortions are optically monitored. Changes in magnetic field as small as 20 nanotesla can be detected. The earth's magnetic field is about 10^{-4} Tesla, so this MEMS device can detect changes in magnetic field more than a thousand times smaller than the earth's field. Since electrical current produces magnetic fields, currents in our brains or other body areas could be detected with such a device, which by its nature is also very small. Other potential applications of this device include heterodyning (mixing) of oscillating magnetic field signals, electromagnetic field frequency analysis, and mineral exploration. Chemical and biological sensing can use the effects of different adsorbed species through their different mass, on the vibrational properties of the tiny MEMS plate.

Lab-on-a-chip, or the μChemLab concept [6] uses MEMS and offers a compact way of performing biological and chemical analysis for environmental, military, medical or home health care with the same sensing ability as large laboratory instruments. This tiny device performs electrochromatography to separate compounds and/or provide special micro-mirror arrays for separating colours, or different fluorescence caused by different molecules. It has an integrated laser containing nanosize components, which is directed at the sample. Using this device, ten different explosives and their components have been detected at the 50 ppm level in less than a minute.

Fluid flow control is a major concern in many mechanical systems such as pumps, motors and engines, as well as in much process equipment, such as heat exchangers and mixing. It is also often useful to be able to vary these flow properties in total or locally. Creating uniform mixing in many processes can have big benefits. Arrays of MEMS might help by altering and monitoring flow characteristics in small pipes or blood vessels, in multiple feed networks or in volumes where mixing is weak.

The nanotechnology equivalent of MEMS is nanoelectromechanical systems, or NEMS. These machines already exist in the laboratory, but they are in a very early stage of development. At this stage, work with NEMS has involved using or mimicking naturally occurring biological motors (Chapter 6); nanotubes and related materials; and chemical issues that may help realise engineered nanomotors such as the molecular mimics described in Chapter 5. It may be possible to use very small NEMS biomotors to mimic natural movement. Alternatively, artificial muscles could be made using nanoscale bundles of fibres whose individual elastic or conformational properties can be altered at different sections along their length, either chemically or electrically, and possibly also optically, so that bending and gripping ensues. Conducting polymers, such as modifications to polypyrrole and nanotubes may have a role here (Chapter 5). Simple movement has already been demonstrated. Although it may be possible for all the functions of MEMS to be taken over by NEMS, some functions can only be performed at the NEMS level.

9.2 ROBOTS — HOW SMALL CAN THEY GO?

A robot is much more than a machine. Typically, it requires sensors for pressure, position, vision, temperature and movement, plus a computer to process sensor signals, decide upon an action to take, and initiate controlled action. Some industrial robots are really just computer-controlled machines, but our use of the term 'robot' implies a higher level of capability. To describe something as a robot depends on the task they are expected to do, whether they are autonomous (free moving), and in particular how much power they need and for how long. A very small robot is shown in Figure 9.3. Power source problems at present mean that free moving robots are limited to about 1.5 cm long and about 30 g in weight. The use of solar cells may not solve this problem, since the smaller the robot's size, the smaller the area available to mount the cells — and cell area controls power. This is the case even though smaller also means lighter, which means that less power is needed for movement. If the robot needs power for a task such as lifting, carrying, or electronics or materials work, then power will be a major limitation to size reduction. However, if power is required only for movement and running the robot's sensors, then very small may be possible. Just as solar cells utilise energy from the environment, very small robots will have to learn to extract energy from their surroundings. Other sources of power may be beams of light or radio frequency energy that track each robot; or the robot may be tethered or run with a contact to a powered surface or wire grid (a nano- or micro-tram). In biology, including human anatomy, energy is extracted chemically or electro- or photo-chemically from the immediate environment. Nanorobots may have to mimic nature to source their power. Fuel cells use chemical reactions to produce electric current, and micro and nano fuel cells or small ion generators should be an area of interest here. Other options are the biomotors or artificial muscles noted above, actuated by chemical reactions and ion flow.

Programming small robots will be interesting. It may be possible to use molecular electronic switches or analogue computer nanoelectronics linked to sensors. Most likely solutions will probably centre on a separate computer or dedicated processor with wireless communication for these very small robots. Thus, while we are at the dawn of the nanomachine's potential, it may be a while before we can contemplate semi-intelligent free-ranging nano- or micro-robots. Applications where they are tethered or in wireless communication with their power and processing units are more achievable, but even these will require a lot of ancillary developments in nanotechnology, including software. The microprocessor will be replaced by the nanoprocessor and Microsoft by Nanosoft. The new Bill Gates is waiting at school.

Figure 9.3

At the time of writing, this was the world's smallest free ranging robot.

9.3 AGELESS MATERIALS

Nanotechnology is going to revolutionise the very big, as well as the very small, because it allows us to create totally new types of materials in terms of strength and durability. It will be used to make revolutionary new construction materials and industrial tools, and do better than before with fewer resources, because the strength to weight ratio of the materials we use will improve, and they will have long or infinite lifetimes. These materials will help make recycling mandatory if anything is demolished, and even demolition will be a new art. Nanotechnology should help reduce waste, not just because we will use less and recycle more, but because we should be able to nano-engineer materials that are usually trashed (such as packaging) to biodegrade or environmentally degrade rapidly into harmless, even beneficial products.

Nanotechnology offers the promise of ageless materials and hence ageless structures built from these materials. This will in turn greatly help in our quest for a sustainable earth and our quest for new worlds in outer space. For studying lifetimes, we are interested in issues such as cracking and fracture, corrosion, surface wear, and hardness. Materials can fail due to mechanical and thermal effects, or because of chemical attack, such as rust in metals or UV effects on polymers. Stress is a local accumulation of elastic energy [7, 8] due to atoms being displaced from their normal positions in a material. Macroscale defects, such as a small chip in glass, can be quite bad because they cause large perturbations. In comparison, nanoscale defects are so small they can often be added without initiating any serious problems while improving strength and durability. Stress is often 'built in' when materials are manufactured and can increase in service due to various normal events: bending back and forth in the wind; pressure and load variations; thermal expansion and contraction; chemical attack; or simply due to use, which may involve a lot of friction or heavy impacts.

Problems start with localised nanoscale distortion or dislocation in materials at the atomic level, which may be a precursor to cracking and then total failure. An aim of nanosensors is to be able to detect these at a very early stage of development, when repair is easier. We can think of materials as being healthy or unhealthy, but just like human health problems the cause can be lurking for a long time before we become aware of it. By then it may be too late, even catastrophic. Just like human health problems, if detected early they can usually be fixed a lot more easily. Ease of detection is the problem and this is where nanosensors come in. While traditional distortion sensors such as strain gauges have become a lot smaller in recent years, it is usually necessary to dig a small hole or attach them to a surface. Relatively large changes are required to generate a signal. Nanosensors may be integrated into the material at the production stage without distorting it, or they can be invisibly incorporated as thin polymer layers on a surface that gives an optical signal when changes occur. One such new sensor has just been built and demonstrated. It is proposed for use on Mars habitation 'bubbles'. The one nanosensor can detect excessive heat, UV effects, and impacting chemicals [9]. This early warning system should allow time for repair before an issue becomes critical. Nanosensors may be active, that is, linked into an electrical or optical fibre circuit, or they may be passive and subject to external interrogation at periodic intervals using various types of radiation or by imaging. It should be possible to conduct all material health checks non-destructively.

Nanosensors can also be used non-intrusively to measure vibrations, forces and torques on materials while they are in use, such as metal when it is being die stamped to produce car parts; a drill; a motor; or a whole building. This also applies to humans — for instance, measuring forces and accelerations produced by athletes or swimmers will be a lot easier and less intrusive.

9.4 INVISIBLE MENDING OF ATOMIC DISLOCATIONS INSIDE DAMAGED MATERIALS

The next question is, is it easy to make repairs at the local atomic level? Advanced laser and ion beam techniques, or atomic force nanoprobes, could be used to supply energy locally to enable atoms to move back into correct, low energy positions. Alternatively, it may be possible to apply, fill in, or 'ion stitch' in, nanopatches, nanoparticles, or ions of the correct material. Bacteria may be useful for some applications, since some bacteria produce metal. They may even be genetically engineered to perform a specific task. It will be a big challenge to achieve localised repair. It will be more like delicate surgery than traditional materials repairs using big messy techniques like welding, panel beating or riveting. Just as with our own health, if bad cells or dead tissue are to be removed or replaced it is best if we leave the nearby good tissue undisturbed. It will be interesting to see whether the very sophisticated nanoproduction and imaging techniques used in high-tech laboratories can be adapted to become repair tools for use in the world at large. Inbuilt nanosensors may have to communicate with or be sensed by the repairing agent or tool to give it direction or to guide its beams. What will be achieved at an atomic scale will be like the invisible mending done by tailors when a complex cloth in a damaged fabric is restored. It must have both the look and strength of the original. It will be a while yet before atomic or nanoscale level repairs in the field become a realistic possibility, however, if it can be achieved in medicine it can also be done for materials, and vice versa.

Can we make an ageless material? The first step will be to greatly increase the lifetimes of materials, and nanotechnology has a lot to offer. The aim is to stop the formation of dislocations and cracks, or if they do form, to prevent them from moving or extending. If many internal atoms can't move in unison, you cannot indent or bend a material easily. In the last decade it has been realised that nanosize grains in metals contribute to hardness. These crystals pack together along the grain boundaries. But just using grain size has limits, as the grains can still slide along each other, so the material still deforms. Super strength materials can now be made using nanocomposites, such as nanocrystalline particles embedded in a non-crystalline matrix, as discussed in Chapter 3.9. These materials containing nanoparticles are thus almost as hard as diamond (hardness > 100 GPa).

Thin films can be coated onto cutting and drilling tools, and dies for stamping metal and plastic shapes. This method can extend their lifetime way beyond even the big advances that have occurred with special materials in the last decade. They can also provide the cutting and polishing ability of diamond, which is used nowadays for much industrial work. Another key issue here is thermal effects. Can nanotechnology help to reduce the impact of higher temperatures on deformation in common materials? Maybe we could produce materials with negative coefficients of expansion, that is, materials that shrink when things get hotter. Certainly nanosystems should help produce materials that distort very little, if at all, with large temperature changes. Thermal insulation in very fine structures also appears to be possible, and has been achieved using sol-gels (Chapter 3.5), which set as porous films composed of a fine network of solid material enclosing large voids.

Structures built at the nanoscale from nanosize components can also have interesting strength and wear properties. Fibres or long narrow materials with nanosize cross sections can be stacked closely together to produce a very strong composite, which may also have excellent thermal properties. This is the old story of 'a match is easy to break but a stack of matches side by side is very tough'. It is already used in microfibre composites, but nanofibres should do even better. Nano-thick platelets of one or more material can be stacked randomly on each other to form wear-resistant surfaces that are very hard to penetrate. The shiny shells of abalone have such structures — again nature got there first — but we are learning to copy, and recently a technique for mixing nanoscale polymers with silica nanostructures in a regular array produce a material that has been found to have similar structural properties as some sea shells [8].

The principle of pinning any defects that form at a nanoparticle can also be extended to polymers. Polymer chains can be terminated with ceramic nanoparticles, or the cross-linked polymer nanoparticles that we discussed earlier can be added to moulded or extruded parts during manufacture, to reduce internal stress or block the growth of defects.

9.5 NANOMECHANICS AND NANOELASTICITY

This is a new field of science and engineering. It concerns the unique elastic and mechanical properties of nanosize objects. They have not been studied in great detail because it is technically very challenging to do so — how do you indent, squeeze or pull a nanoparticle or a nanotube? Techniques using atomic force probes and special sample arrangements are emerging, but computer simulation is still being relied on heavily to provide answers. It seems as though materials behave quite differently on the nanoscale [7]. Graphite is very brittle, yet nanotubes look like being quite resilient after being distorted. Will we be able to split nanotubes or nanoparticles in two in a controlled fashion, or drill holes into nanoparticles mechanically, or will we need ions or photons to do the job? The answers will shape the future possibilities in nanoengineering.

Good understanding in this area is very important, since materials or coatings containing a high density of nanotubes or nanoparticles will also have different elastic properties to their atomically continuous counterparts. How do nanoparticles interact with each other in an array? When pushed, will they roll over or slide past each other, or will they remain in a fixed position? How do the surfaces of a nanoparticle interact with other surfaces? Clearly their own surfaces and any matrix or solution they are in will play a role, but we have a lot to learn.

Expect new effects and as a result new engineering opportunities. One of these is friction modification, which we examine below.

9.6 NANOPARTICLE COATINGS — SPECIAL NEW EFFECTS

Thin layers of nanoparticles can be directly applied to materials to give special effects. The layers can be applied very simply using a solution dip or a squeegee or roller. The nanoparticles can form either compact arrays, where the particles are in direct contact with each other, or arrays with small spacings between each particle. The layers are typically one or two particles thick. These layers can reproduce some properties of vacuum deposited thin films, such as those used for solar and UV control. Sumitomo and 'ShowaCoat' in Japan already offer such nanoparticle products, but new properties can also be achieved. For instance, a layer or two of nanoparticle silica can make a surface very hydrophilic. That is, water drops do not form but spread out uniformly on a surface so that it doesn't appear to fog and also dries more easily. This property is measured in terms of a wetting angle, or the angle made by the tangent to the edge of a surface drop with the supporting surface. If a big drop forms, this angle is high, but if drops barely form at all it is very low. Outdoor surfaces may also stay cleaner when coated with such materials. It is thus possible to produce consumer antifogging products for bathrooms and kitchens, and external self-cleaning surfaces in which rain will keep surfaces very clean. They are also useful in water collection and distillation technology to maximise collection efficiency.

Another issue that is worthy of study but has received little attention to date is the ability to control surface friction with nanoparticle layers formed directly as a coating or via additives to lubricants. It may even be possible to eliminate lubricants in some cases. Friction is a major cause of energy consumption, wear and breakdown. Extra complexity in much engineering is needed to dissipate the heat resulting from friction. Motor, generator and engine parts that wear less and do not heat up in use would have reduced maintenance requirements.

9.7 NANOELECTRONIC AND MAGNETIC DEVICES AND NEW COMPUTING SYSTEMS

The shrinking of microelectronics has reached the point where it should be correctly called nanoelectronics. If your computer is less than two years old, the circuits were developed with nanotechnology. As noted in Chapter 8, for over 30 years electronics has followed a Law called 'Moores Law', which says that chip memory capacity doubles every two years as devices get smaller. Will it stop soon or will nanotechnology allow us to go on for another 30 years? Electronics is used for information processing, but our information transfer capability is building at an even faster rate. It is needed to cope with Internet traffic flow. The available bandwidth, that is, the amount of information per second we can transfer, is doubling every nine months and the cost is tumbling. It is even more dramatic than electronics. Just in 2001 the cost of renting 150 Mbits per second for a year has fallen typically by 25 to 50 per cent, depending where you are in the world. What will nanotechnology contribute to this? The fact is, it is doing it already, and several new nanotechnologies are about to hit the fibre networks of the world. Bandwidth will get much larger and costs will shrink much more. It will be the limitations of electronics that will cause the bottleneck!

Microelectronics will soon be a word from the past for the majority of integrated circuits — it should in fact already be so for computer chips, as device dimensions have already moved well and truly into the domain of nanotechnology. As discussed in Chapter 8, computers are based on electronic switches that use transistors. It is quite simple: the transistor is either on or off, and conducts current when it is on. Commercial amplifier and computer transistors are based on silicon technology and there are various types of silicon-based transistor devices. Many other non-silicon transistors have also been studied over the years and transistor action does not even require semiconductors. However, some type of special junction between materials or parts of materials is usually required, with the junction itself often having semiconductor-like electrical properties. For instance, conducting polymers, nanotubes and even modified DNA can be used to make transistors.

Shrinking today's silicon systems even further into the nanodomain will produce the next few generations of chips. The average device feature size in state-of-the-art chips is now around 130 nm. Feature sizes are expected to shrink to 70 nm by 2005. What then? It is by no means the end of the ride! Further size reductions will be nanotechnology-based, but of several contenders, which nanotechnology will win the race to take over at the 50 nm and lower level?

Traditional optical lithography for patterning will not suffice. X–rays or electron and ion beams will do the job as described in Chapter 8, but they are not yet amenable to low cost mass production of complex patterns. Non-traditional optical lithography may emerge from the new nano-optical physics noted in Chapter 7; while self assembly, using the inbuilt properties of molecules or modified nanoparticles to self-organise into desired patterns, is another contender. In the early days, silicon microcircuits became the sole basis of nearly all integrated circuits and subsequently computer chips. There was very little competition. There are many in the race for post-2005 electronic switches, and some contenders have a head start, having already been researched for ten years.

It is clear that the building blocks of new nanoelectronics and switching devices may be chosen from the following systems [9], and may also involve a hybrid of several of them:

- carbon nanotubes
- molecular electronics, including molecular mimics, biomolecules and DNA
- single electron transistors (SETs)
- giant magnetoresistance layers (GMRs)
- quantum entangled atoms or nuclei
- quantum dots
- counting devices based on the abacus (plate 7).

Except for GMRs, most of these structures have been dealt with in Chapters 4, 5 and 6. GMR devices have been in use since 1998 in disk heads, which are sometimes called spin valves. They have allowed us to have hard disks with 20 Gbit capacity. Sensors and possibly RAM (random access memory) may follow. These devices involve magnetic coupling between an antiferromagnetic and a ferromagnetic layer with a very thin 2 nm layer of ordinary metal between them. The device structure is shown in Figure 9.4. Note the very thin layers — they are only around 2 nm thick. Precision coating techniques are needed for such layers. Their electrical properties can be switched with a small, very localised magnetic field.

Figure 9.4

A layered structure of different types of magnetic materials and a simple metal as used in giant magnetoresistance (GMR) devices, including hard disk read heads.

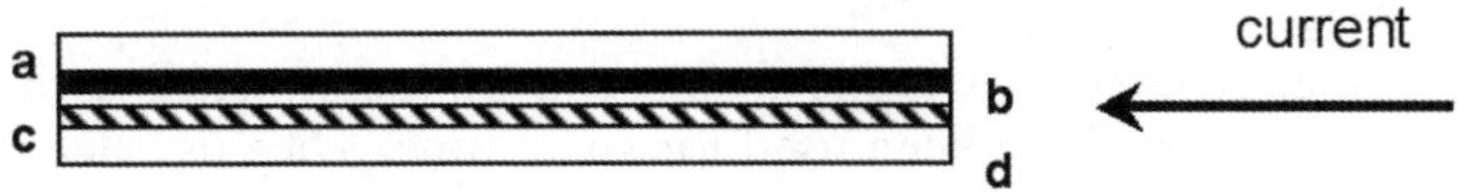

a = antiferromagnetic film (CoO) which "pins" magnetism in the ferromagnetic Co layer b which is just 2.5 nm thick
b = ferromagnetic Co layer b
c = copper metal spacer which is 2 nm thick
d = NiFe layer

We must not forget that we may even move away from using electrons or charge to do our switching and signal processing. Light may be used instead, not in free space but inside special nanostructures with non-linear optical properties. Light is used to switch light and thus achieve effectively a three-terminal device. That is, a control light beam changes the refractive properties of a material, and the state of the material determines what happens to another light beam, which is usually a different colour or frequency to the original beam. This is called an optical gate.

Circuit structures and layouts themselves may eventually change, for instance if we learn to mimic the dynamic and complex neural circuitry in our brains. However, for the foreseeable future most developments will probably maintain the ordered geometries we are used to, just on a smaller scale.

9.8 OPTOELECTRONIC DEVICES

LIGHT EMITTING DIODES

A new generation of electric power source is emerging from nanotechnology. It is based on special semiconductor light sources called light emitting diodes (LEDs), which emit light when current is passed through them. With nanosize domains of different size it is possible to engineer most colours, since the size of a nano-piece of semiconductor changes output colour according to its exact dimensions. One device can produce three colours with segment dimensions 2 nm, 5 nm and 10 nm. This new generation of LEDs has the capability to reduce the power used for lighting around the world by 90%, because they are just so efficient in converting electric power to light energy. White light can be produced either by using phosphors in combination with blue LEDs, or by using multiple colours. LEDs are mainly used in traffic signals and signs at present, but they will be in general use soon.

THERMIONIC SOLAR POWER

Thermoelectric coolers made from semiconductors such as bismuth telluride, Bi_2Te_3, have been used for some time. They use electric current to cool, or can generate power when a thermal gradient appears across them, as used in deep space probes, where heat is generated with small nuclear radiation sources. Problems with the thermoelectric coolers in some early satellites caused them to come crashing back to earth! Thermoelectric coolers are relatively inefficient in terms of heat removed over electrical power put in. Despite this, they are of growing interest because modern dense nanoelectronic chips also need compact cooling devices

Nanotechnology can make a very big impact in compact cooling devices, because there is another way to use the flow of electric current to achieve cooling and to do it with as much as 80% of the maximum possible efficiency. This is called thermionics. It uses the motion of charge induced by an electric field between hot and cold electrodes to carry heat from cold to hot regions. The basic system of potential barriers and how they provide power and cooling is shown in Figure 9.5 [10]. As expected, with no external influence hot electrons are driven over the potential barrier more than cold electrons, so a voltage is generated to produce power. With a high voltage bias, cold electrons more easily traverse the potential barrier.

Figure 9.5

The three modes of operation of a nanostructure thermionic cooling/power device in terms of potential or voltage barriers and applied voltages. Since the two metals, one hot and one cold, either side of the nanosize vacuum layer are different, a potential barrier exists even with no applied bias. This is called a contact potential. More electrons climb over this barrier from hot to cold than the reverse in an open circuit situation. They will generate a voltage that will eventually cancel the inbuilt voltage if enough heat is supplied, unless the circuit is closed to supply external power.

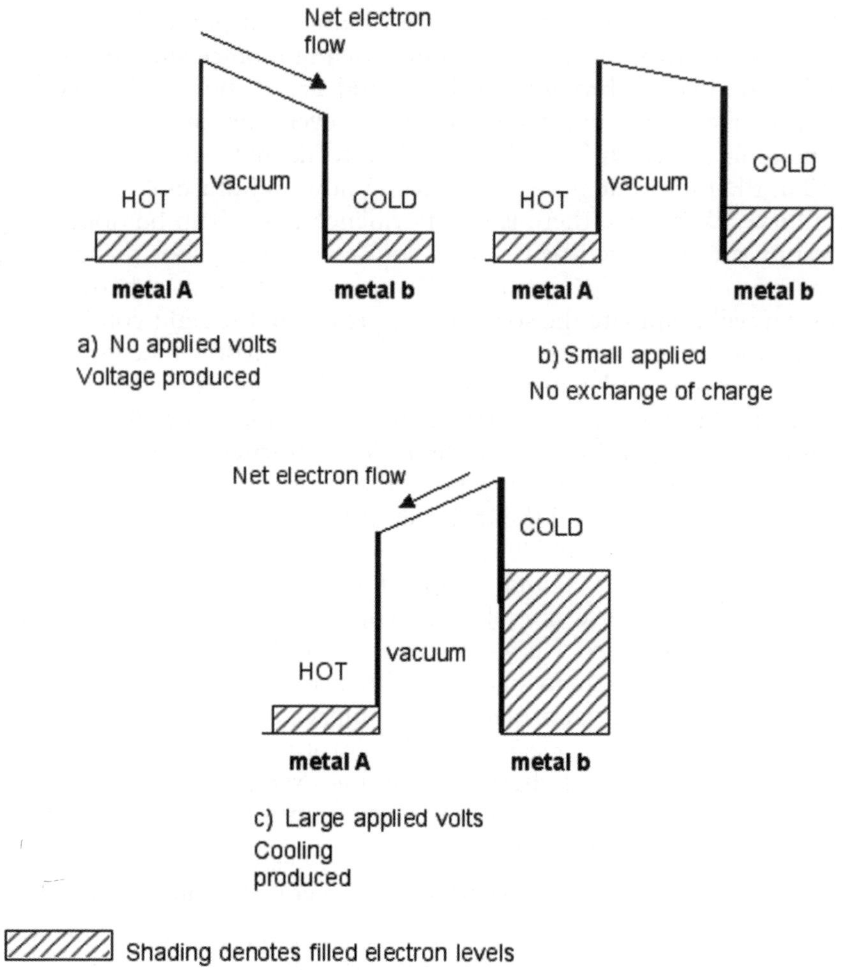

The proposed thermionic devices are all nanosystems. They use either thin multilayers (each layer around 50 nm thick) or structures that allow some conventional transfer and some parallel solid-state electron tunnelling, which assists in the process. Such devices have been estimated to be capable of providing cooling of 100 W/cm^2 [11]. If realised we could see a whole new refrigeration and air conditioning technology emerge and cooling with solar power in developing countries might even become an attractive proposition. It might even be possible to make an integrated nano solar cell thermionic cooler on a nanorobot, with a tiny cell junction itself driving the cooling electrons across a thermionic gap.

9.9 ENVIRONMENTAL APPLICATIONS

There is no question that the greatest threat to the future welfare of mankind and to future political stability is our own impact on the environment coupled with the depletion of natural resources. Most of these activities are essential to our economic and physical well-being. Some problems are hangovers of our earlier ignorance and carelessness, and some persist through selfishness and greed, but in the main we have a dilemma: improving the life of everyone on earth without degrading the earth itself. Is it possible to do both at affordable cost and in the time scale needed — starting now and finishing the job this century? We believe there is good evidence that it can be done using existing and expected new science and engineering. For this purpose alone nanotechnology could not have come along at a better time, since it will dominate the solutions by providing the right combination of economics and technical capability. The main cost will be investment in the necessary and appropriate research, development, education and training. Business opportunities in these technologies will be vast — nanotechnology will be one of the major industries of the 21st century, since cleaning up existing pollution problems and stopping new ones occurring is one of the major tasks facing humankind in the next few years. The quality of life can improve dramatically for many billions if this can be done well and cheaply. Life will also noticeably improve (or stop getting worse) for all those who are currently living in acceptable environments, so everyone on earth stands to benefit. The problem is the scale of activity needed and hence the cost.

Nanotechnology can make a big impact because it can perform many of the necessary clean up and control tasks well at much lower cost than conventional alternatives. Some examples already exist and others are on the way. Areas that need attention and that nanotechnology can address include:

- biohazards due to the contamination of water by humans and animals;
- pollutants, such as carcinogenic chemicals and other chemical hazards in the air from combustion and chemical processes;

- particulates in the air from burning diesel and other fuels;
- rising salinity in farm, grazing land and country towns due to land clearing and over-irrigation;
- sewerage and waste disposal.

As average wealth grows, so the average person on earth will use more water and energy, drive more and consume more. Add to this a population that will continue to increase for some time yet, and the size of the task is daunting.

Why will nanotechnology revolutionise our ability to improve the environment? Many of the hazards that have to be removed are microorganisms and other particulates that are larger than the nanoscale in size, so they can be physically filtered out using nanofilters. Electrical and chemical activation can be used to attract the contaminants to the walls of a filter.

Another area where nanotechnology excels is its ability to discriminate between species of molecules or dissolved solids and hence help deal with select species. For instance, nanotechnology is used in special activated filters based on a capacitor with aligned carbon nanotubes that act as electrodes and hang densely into the channels through which the water flows. This can remove salt or other contaminants from water to produce clean water quite cheaply. The main competition for desalination, based on cost, will be solar thermal energy for distillation. However, in many locations burning liquid fuels and gas for distillation is still cheaper at present.

Lack of water and the related problem of massive land degradation is one of the world's major environmental issues. Opening up other cost-effective water sources will also be of inestimable benefit, whether sea water or groundwater. With nanotechnology we will be able to filter ions easily and cheaply to purify and more easily recycle the water we use. Recycling is going to become essential in many world locations. We will need to start using sea water on a grand scale.

In the field of green chemistry, nanoparticle catalysts can be used strategically to break down airborne pollutants, such as combustion products from fuels, more efficiently than ever before. Photocatalytic effects are especially useful in breaking down polluting chemicals, sometimes into other valuable compounds. When UV light falls on a molecule that is in contact with the catalyst, the molecule can lose a bonding electron to the catalyst. UV radiation is very good but visible light can be used in some circumstances. Titanium dioxide coatings are already used beside busy highways in Japan using solar UV to break down hydrocarbons in the air. Simple nanotechnology that has emerged from research in Australia (Chapter 7) and in Toyota laboratories in Japan into the nanoengineering of sculptured thin films for optical work, has been adapted to provide better catalysts for TiO_2.

The rate of pollutant conversion doubles relative to a smooth film. Examples of such films appear in Figure 9.6. This system is also very cheap. Materials other than TiO_2 may work better if they can absorb more of the visible solar spectrum.

Figure 9.6

Nanosculptured thin films of titanium dioxide as used for more efficiently breaking down airborne pollutants. Courtesy of M Suzuki, Toyota Central R&D Labs, Inc.

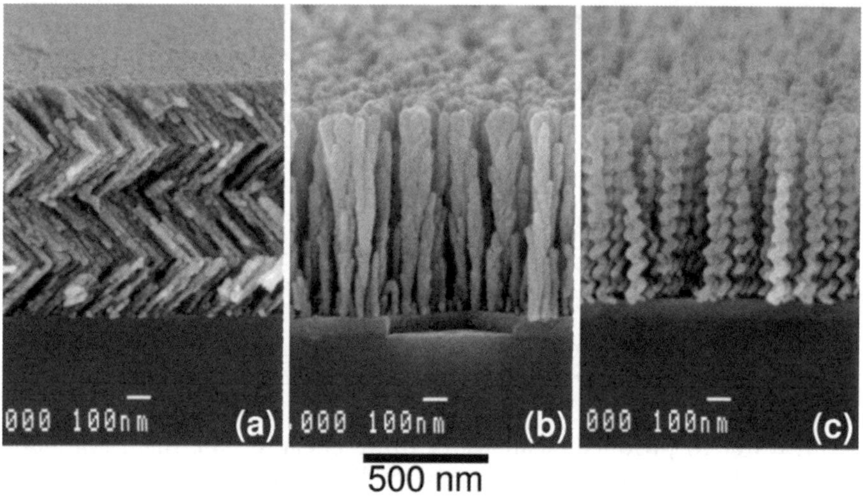

WHAT YOU SHOULD KNOW NOW

1 Micromechanical systems will be replaced by nanomechanical systems. Apart from nanomotors based on biomotors, artificial muscles using nanoscale bundles of fibres will allow bending and gripping at the nanoscale.

2 Nanosensors may be integrated into a material at the production stage without distorting it, or they may be incorporated as invisible thin polymer layers on a surface, giving an optical signal when changes occur. They are proposed for use on Mars habitation 'bubbles', where the one nanosensor can detect excessive heat, UV effects, and impacting chemicals so that corrections can be made before a tragedy occurs. Nanosensors may be active, that is, linked into an electrical or optical fibre circuit, or they may be passive and subject to external interrogation at periodic intervals by various types of radiation, or just by imaging. It should be possible to perform all material health checks non-destructively.

3 Nanoelectronics will replace microelectronics. The components of these electronics will include molecular mimics, biomolecules and

DNA, single electron transistors (SETs), giant magnetoresistance layers (GMRs) and quantum entangled atoms or nuclei quantum dots. Energy may eventually be transmitted in circuits via photons rather than electrons.

4 Nanotechnology may produce new light emitting diodes and new types of thermionic power. Electron tunnelling across nanomaterials may produce new refrigeration devices.

5 Nanofilters will assist in cleaning up pollution and reducing wasteful by-products. New catalysts will create more efficient industrial processes.

9.10 EXERCISES

1 *Find an electronics text and look at a few basic machines, such as electric motors, amplifiers, dynamos, transformers and other devices. Now draw these with parts you would use in a nano-equivalent.*
2 *Find out more about nanotechnological applications to Light Emitting Diodes.*
3 *Find out more about defects in solids and how dislocations occur when materials are stressed. Visit material science web sites and compose a library of structures that show dislocations (Figure 3.9, Chapter 3, is a good place to start)*
4 *Find out all you can on the web about nanoelectromechanical machines and labs-on-chips, including DNA on chips.*
5 *See what you can find out about how to program micro- or nano-based machines and what software will be required.*
6 *Investigate the concepts of single electron transistors (SETs) and giant magnetoresistance layers (GMRs) in more detail.*

9.11 REFERENCES

1 Bishop D, Giles R & Aksyuk V (2001) *oemagazine* (SPIE) May: 24–26.
2 Lewotsky K (2001) *oemagazine* (SPIE) May: 22–23.
3 Yallop K (2001) *FiberSystems International* (IOP UK) April: 65–69.
4 Graydon O (2001) *FiberSystems International* (IOP UK) May: 65–69.
5 Oslander R, Champion J, Kistenmacher T, Wicckenden D & Miragliotta J (2001) *oemagazine* (SPIE) March: 29–31.
6 Williams D, Picraux T & Romig A (2001) *oemagazine* (SPIE) May: 27–29.
7 Yakobsen BI, Campbell MP, Brabec CJ and Bernhole J (1997) *Computational Materials Science* 8: 341–48.
8 Ibn-Elhaj M & Schadt M (2001) *Nature* 410: 796–99.
9 Goronkin H, von Allmen P, Tsui K & Zhu TX (1999) Functional nanoscale devices. In First, Siegel, Hu & Roco (eds) *Nanostructure Science and Technology*. Kluwer Academic Publishers.
10 Lawless JL & Lam SH (1986) *J. Appl. Phys.* 59: 1875–89.
11 Hishinuma Y, Geballe TH, Moyzhes B & Kenny TW (2001) *Appl. Phys. Lett.* 78: 2572–74.

10

INTO THE REALMS OF IMAGINATION

In this chapter we contemplate 'what if?' What will happen when we no longer need to communicate by writing? What will happen when materials can repair themselves? What will happen when we have control of our own health? What are the social implications?

10.1 INTRODUCTION

If we believe the prophets of nanotechnology then we are reaching a period when we will be able to build any detailed object completely at the atomic level, consistent with the laws of chemistry and physics. Since we can now see atoms it may well be that much of the brilliant predictive thought that composes the basis of chemistry and physics is now redundant. Feynman said [1]: 'The problems of chemistry and biology can be greatly helped if our ability to see what we are doing, and to do things on an atomic level, is ultimately developed — a development which I think cannot be avoided.'

Drexler [2] has provided the extrapolations required to make us think and, in his own way, has revitalised the enabling disciplines of physics and chemistry to provide the excitement that must have been generated by the discoveries of the nineteenth century and in the first half of the twentieth century. Nevertheless, it takes vision to see what the new science will create. The last chapter dealt with the likely. Here we will try to step out further into the future. History can help because, as we began this book with the challenges of humankind, we know these same challenges will continue. Stone Age humans were concerned with the same endeavours as modern humans. That is, bettering their lot through communication, manufacturing, health, society and religion [3, 4]. Nothing has changed. These activities have been with us since the species began.

10.2 COMMUNICATION

The development of writing followed the development of language. While the development of writing did not kill off language, it limited its use. No longer were the members of a tribe required to remember history through stories passed from generation to generation. They could be written down. We all know that computers have meant that a lot of typists lost their jobs. Oral communication with a computer will soon mean writing is less important. Indeed it is difficult to think of a use for it other than slowing down the learning process to a speed that the brain can assimilate. In the future, this book will be a curiosity, like the hieroglyphic stone castings of the Pharaohs. However, things will go further. Miniaturisation has brought about considerable progress in the capacity and speed of computers. This will continue despite the concern about the brick wall effect due to the inability to miniaturise lithography, which was discussed in Chapter 8. This is not unlike the concerns that were voiced before the invention of the light microscope and then the electron microscope.

What does this mean? Who, fifty years ago, would have believed the sort of information we can now put on a CD ROM in the last few years? Some of the authors remember the huge heavy disks used in the 1970s to store 1 MB of data. At that time, 650 MB would have filled the lab. On paper, this would have filled the building! The miniaturisation of data by computers has caused the information revolution. There is no doubt that there will be a time in the future when we will be able to equip every individual with all the knowledge in the world, including their own genetic code, on the equivalent of a single floppy disk. The challenge will be updating the knowledge at the pace at which it improves. The easiest way would be to have this biologically connected directly to the brain, a sort of information archive attachment which is updated from some remote source. This would resemble the automatic updating of data such as virus information that we now do on our computer through the Internet. There may be some resistance to the idea of a communication chip implanted in the brain that can be used for updating computer data, but people have rushed to purchase any new external portable device for similar purposes, such as laptop computers and mobile phones. In the latter case they are now almost a fashion statement. In the future, such a device will track all of us, and allow us to track everyone else, but probably as an easily physically disconnectable device. It will be possible to turn it off (at least in democratic societies), but we may not want to. A constant update on our medical condition will be desirable. Your life connector might report that 'There are an estimated 300 000 flu viruses about to approach your left nostril; it's time you ate your daily anti-ageing pill. Also, please email your oven — it is trying to cook your dinner'.

10.3 MANUFACTURING

The assemblers that Drexler talked about no longer seem far-fetched. Nanofabrication is difficult. As discussed in Chapters 2, 3 and 6, there are three possible methods. In the first method lithography may be used. As discussed in Chapter 8, ultraviolet methods are standard but typical 'best practice' microelectronics are made by a printing process called electron beam lithography, which resembles conventional lithography but relies on the wavelength of the electron rather than UV radiation. One method of achieving smaller manipulations will be through nanolithography, based on the same masking principles but using something with a shorter wavelength than the electron. That is, the material you do not want coated or changed is protected by masking and then the mask is removed. This method appears to be the easiest to achieve, but it is also the most cumbersome. The technique involves deposition of a whole range of layers, like a club sandwich, and then selectively removing parts of each layer so that the desired landscape is achieved. Ultimately we need a beam of particles that can cut the substrate atom-by-atom to design objects.

The second method involves scanning probe microscopy and may be used with or without masking techniques. It allows structures to be deposited through apertures within the cantilever tip of the scanning tunnelling microscope. Some examples of how this can be done using dip pen technology were presented in Chapter 3. The future will involve predefined plans for deposition of the sample using computer control and the spraying of atoms into structures as if they were sticky ping-pong balls being blown from an air gun. Complex patterns, such as rings and intersecting lines, are readily produced. With this method, the material composition of the as-deposited line can be varied, allowing for the formation of junctions within a single layer. Two-dimensional structures can be fabricated with different material compositions. The sample will be moved underneath a cantilever by actuators using a pre-programmed sequence of two-dimensional lateral movements. To date, line widths of 70 nm with gaps of 40 nm have been produced by this method but atomic lines are also possible and whole books could be written in atoms if they are arranged as letters of the alphabet on a surface. Indeed, Feynman's original paper has already been reproduced. Using the letters of the alphabet is unlikely to be a practical way of storing information, although sequencing atoms like DNA bases or as consonants is a very sensible way of writing. Crossed wires have been produced with shadow mask writing and, through optimisation of translation speed and/or deposition rates, it will be further possible to tailor the shapes, widths and thicknesses of the deposited lines. Initial experiments with C_{60} have produced nanometre-scale patterning of fullerenes. With an array of such apertures on a single cantilever, it is possible to deposit hundreds or possibly more identically spaced and aligned structures simultaneously. By using a shutter system for the incoming material beam and multiple sources of material, whole swathes of multi-layer devices can be fabricated. Many of the synthetic mimics may be applied. We could coat the surface with a layer of rotaxane switches and use them for counting. This scheme has considerable potential to revolutionise the rapid fabrication and testing of prototype nanoelectronics. In short, the dip pen technology discussed in Chapter 2 has only just started to be explored. Using fullerenes, (Chapter 4) we are in effect writing a dotted line of full stops, but we should be able to write with a lot of other shapes.

The third technology involves biological material. This was discussed in Chapter 6. A phospholipid membrane is a very useful binding surface on which structures can be built. In a particular application, enzymes, receptors and other biological material may be assembled in a different or the same way as they are in nature. Indeed, other biological material may be used to connect these circuits and to act as a building platform for the molecular mimics described in Chapter 5.

A real problem, however, is the wear factor. Single atoms are easily removed from surfaces by friction, evaporation and contact. One approach is to attach structures to metal surfaces by means of chemical covalent bonds, as described in Chapter 5. Gold is a good surface, although for nanoelectronics it must be masked because it is conductive. However, as noted before, encapsulating metal atoms in fullerenes and polymerising the carbon cages into a rigid framework — thus trapping the metal atoms in place — makes sense. It seems that nanoelectronics may not be far off.

ACTIVE MATERIALS AND SWARMS [5–8]

The concept of nanoassemblers of materials — whether they be optical, electronic or for building purposes — can still be considered only imaginative thinking. This area includes the concepts of active materials and swarms. Even though they are currently only figments of our imagination, they should not be dismissed lightly. As discussed in Chapter 9, active materials are materials that repair themselves. To make active materials, a material might be filled with nanoscale sensors, computers, and actuators so that it can probe its environment, compute a response, and act. Although this document is concerned with relatively simple artificial systems, living tissue may be thought of as an active material. Living tissue is filled with protein machines that give living tissue its properties (such as adaptability, growth and self-repair).

These properties are unimaginable in conventional materials. Living organisms are active materials!

Active materials can theoretically be made entirely of machines. These are sometimes called swarms, since they consist of large numbers of identical simple machines that grasp and release each other and exchange power and information to achieve complex goals. Swarms change shape and exert force on their environment under software control. Although some physical prototypes have been built, at least one patent issued, and many simulations run, swarm potential capabilities are not well analysed or understood. The nearest analogy of a swarm is an ant colony or a human city, but these are not nanosized.

10.4 NANOMEDICINE [9, 10]

Medicine is concerned with diagnosis and cure. A cure may only need to be short term, such as during the process of an operation or until natural healing can take place. Nanomedicine may have some potential in each of these areas, but in the short term it may be used primarily for diagnosis. This will be done using nanoreceptors such as those discussed in Chapter 6. Instruments that are used in nanomedicine could well be biological material because of the obvious congruency. Nanoreceptors will need to identify trace amounts of biological material that are specific to the presence of a particular virus or bacterium, or which detect a particular body malfunction, such as brain abnormality or organ stress. These could be based on electrical devices built by tethering a molecular detector to a membrane so that a change in ion concentration creates a detectable electrical current, as described in Chapter 6. Hundreds of thousands of these devices could be present in a single probe, each with a specific electrical signature so that a read-out of pathogen versus concentration could be determined. Any particular body fluid could be studied. In the doctor's surgery or a forensic laboratory it would be saliva, sweat, urine, blood, sperm or vaginal fluid.

Many of the potential advantages in diagnosis can be achieved by using pre-formed biological material. Many of these materials were discussed in Chapter 6. Membranes in particular offer great advantages, however there are lots of other useful concepts. Understanding the mechanism that makes human cells stick to other surfaces could help to produce better medical devices. Nanostructures could form novel biomaterial coatings, either to prevent cells from binding to surfaces (for instance, to prevent blood clotting), or to improve their adhesion and biocompatibility for repairing skin and other damaged tissues. The living cells of the human body are very complex, and many of the effects of interactions or coupling of cells are presently unknown. In such biological systems, cells are able to stick to other surfaces. The degree of adhesion depends on both the structure of the material's surface at the nanoscale and on the characteristics of the cells. Biomaterial coatings would offer better biocompatibility for applications in the field of biomedical devices used for common medical procedures and treatments (such as the transport of blood and removal of other body fluids via catheters); and for the production of improved biomaterials used in tissue engineering and wound healing. Nanotechnology is of particular value for biomedical applications since devising ways to prevent cells from binding to a surface is an important feature of many medical devices. A low adhesion component is required for hollow devices used to transport biological fluids such as blood and urine. The low adhesion properties prevent the blood clotting as it passes through the tube, and stop the tube becoming overgrown with cells. Such properties also help to prevent bacterial contamination. The surfaces of membranes, which are now altered by drugs, could be altered by nanomanipulation of the membranes by adding nanocoatings. Damaged tissues often stick to other tissue surfaces. This can cause post-operative pain when the tissues that are stuck together are torn apart by the patient's movements. The risk of such pain can be reduced significantly if tissue adhesion is restricted by applying a biodegradable surface coating to the damaged tissue or organ. Nanocoatings are also invaluable in situations where high adhesion is required to help tissues stick together during healing. Examples include skin, muscle and bone repair.

The use of nanotechnology in diagnosis could also be achieved by mechanical nanorobots. However Chapter 9 has shown us that these are still far off. The most likely scenario is chemical robots. These are effectively smart drugs which, when injected, will rewrite DNA sequences and repair brain or other important organ damage. As we understand more biochemistry, larger self-assemblies of molecules (this defines the field of supramolecular chemistry) will undertake this function. In effect these are machines. One part will be used for penetration, another for action and others for removal or destruction. These self-assemblies of molecules may be delivered inside giant fullerene or nanotube capsules that are soluble in biological fluid.

Freitas [9] has discussed a respirocyte built by nanotechnology. This is an artificial red blood cell about the size of a bacterium, which delivers more oxygen than a normal one, and would be useful for treating people with poor lung capacity. The physical size of the device is easy to specify because respirocytes must have ready access to all tissues by blood capillaries. They must be no larger than 8 μm in diameter, but they may be as small as natural red blood cells, which are 7.82 μm x 2.58 μm in diameter. This is not nanosized. However, if built from molecular parts these artificial cells would be comprised of a number of nanomachines.

Respirocytes may be fuelled by glucose in the blood. Binding sites for glucose are common in nature. For example, cellular energy metabolism starts with the conversion of the 6-carbon glucose to two 3-

carbon fragments (pyruvate or lactate), the first step in glycolysis. This is catalysed by the enzyme hexokinase, which has binding sites for both glucose and adenosine triphosphate (ATP). Another common cellular mechanism is the glucose transporter molecule, which carries glucose across cell membranes and contains several binding sites. Freitas suggests these could be used in membrane structure to form a respirocyte.

The biochemistry of respiratory gas transport in the blood is well understood. In brief, oxygen and carbon dioxide are carried between the lungs and the other tissues, mostly within the red blood cells (erythrocytes). Haemoglobin, the principal protein in the red blood cell, combines reversibly with oxygen, forming oxyhaemoglobin. According to Freitas, at human body temperature, the haemoglobin in one litre of blood holds 200 cm^3 of oxygen.

The key to successful respirocyte function is an active means of conveying gas molecules into and out of pressurised microvessels. Molecular sorting rotors have been proposed that would be ideal for this task, however, they may look more like a nanofilter (Chapter 3) and a biological nanopump (Chapter 6). Molecular sorting rotors can be designed from structures that are generated by producing nanosized holes in plastics (Chapter 3). Instead of filling them with gold or other materials, they would be used as filters. In the model proposed by Freitas however, each rotor has binding site 'pockets' along the rim, exposed alternately to the blood plasma and interior chamber by the rotation of a disk, which composes the rotating part of the nanomachine. Each pocket selectively binds a specific molecule when exposed to the oxygen or carbon dioxide source. Once the binding site rotates to expose it to the interior chamber, the bound molecules are forcibly ejected.

Freitas points out [9] that artificial reversible oxygen-binding molecules have been studied by a number of scientists trying to make artificial blood, including cobalt-based porphyrins such as coboglobin (a cobalt-based analogue to haemoglobin) and cobaltodihistidine. Other candidates for this application have included other metallic porphyrins; simple iron-indigo compounds; iridium complexes, such as chloro-carbonyl-*bis*(triphenylphosphine)-iridium; a simple cobalt/ammonia complex; zeolite-bound divalent chromium; and non-porphyrin iron complexes. Unlike haemoglobin, haemocyanin, haemerythrin and coboglobin are not poisoned by carbon monoxide. It is likely therefore that the technology for respirocytes will be structured so they cannot be poisoned.

It is possible that respirocytes might transfer oxygen or carbon dioxide or both. Many proteins and enzymes have binding sites for carbon dioxide. For example, haemoglobin reversibly binds carbon dioxide, forming carbamino haemoglobin. A zinc enzyme present in red blood cells, carbonic anhydrase, catalyses the hydration of dissolved carbon dioxide to bicarbonate ion. In photosynthesis, carbon dioxide is added to a five-carbon sugar, ribulose biphosphate carboxylase, using chlorophyll. This is probably the greatest potential nanoecological advance: it may help to reduce greenhouse problems. The photosynthesis machine could reproduce the effect of all the green leaves on Earth. All we need to do is extract the biological machine intact, place it on a surface, reproduce it many, many times and let it work. It will need to be regulated, or it may starve natural plants of carbon dioxide.

A built-in computer or related control device is necessary to provide precise control of respiratory gas loading and unloading and other functions in respirocytes or photosynthesis machines. This must be accommodated in the μm^3-sized device. Freitas [9, 10] believes a 10^4 bit/sec computer can probably meet all computational requirements, given the simplicity of analogous chemical process control systems in factory settings. This is not much, considering that even the early PCs typically had access to 10^6 bits (0.1 megabyte) of external floppy drive memory. We have seen in Chapter 8 how quantum computers could do this with ease, but we are a long way from miniaturising such a device. A chemical computer (Chapter 5) would be sufficiently miniature.

Some studies have already investigated the biocompatibility of medical nanorobots [11, 12]. In particular Freitas [9,10] has discussed whether they will be rejected or not. The most common cause of rejection for material that does not need to be incorporated in body tissue, such as transplant organs, is an effect on body temperature. Heating blood above 47°C produces damaged erythrocytes. High temperatures also affect other organs, such as the brain. People have been known to 'go troppo' when they are suffering from hyperthermia. Hyperthermia is a heating effect that may develop during periods of intense physical exertion, dehydration, and immersion in hot fluids. It may be due to heat thrown out by respirocytes.

Another cause of overheating is when the body deliberately increases its temperature. This is called fever. Fever is a natural self-defence mechanism intended to make the host less hospitable to microscopic invaders. The intact control mechanisms of thermoregulation act to raise body temperature up to the new set point, then maintain the elevated systemic temperature. Fever is triggered by the release of fever-producing substances from cells of the immune system into the bloodstream.

Can nanorobots act as pyrogens, inducing fever? The evidence is variable. Some materials are satisfactory and some are not. For example, latex particles do not induce fever and the simple building materials of the anticipated nanorobots, such as diamond, fullerenes, or graphite, are inert. Carbon powder has been used in nasal provocation tests without eliciting fever and it has also been used with tellurium for radioactive mapping of the lungs. With rare exceptions, bulk

Teflon appears non-pyrogenic, although perfluorocarbon emulsion can cause cutaneous flushing and fever at low doses and 'polymer fume fever' or 'Teflon fever' results when Teflon combustion products are inhaled.

Other particulates are less inert. Metal fume fever (due to zinc oxide inhalation) is well known: excess trace elements such as copper and zinc can induce fever. Silica crystals and various low-solubility substances that crystallise in the human body can trigger fever. For example, monosodium urate monohydrate crystals, which are deposited during gout, cause fever. Fever has been reported from kidney stones, from calcium oxalate bladder stones, from calcified lymph-node stones in broncholithiasis, and from calcified salivary gland stones. Cholesterol crystals deposited as gallstones may cause fever, as may cholesterol crystals in the blood.

Even if this is a problem, research into new materials will solve it. Moreover, there are many medical areas where the possibilities of rejection are not as crucial, such as nanodentistry and nanocosmetics.

There are other applications for nanomachines, such as in nitrogen fixation. It is well known that legume bacteria and soil microorganisms, such as *Azotobacter* and algae (*Anabaena cylindrica*) can fix nitrogen. All we need to do is decouple this nanomachine from the organism and put it to use in the same way as the photosynthetic machine, and we can use the product to make all the proteins we need.

10.5 SOCIETY AND ETHICS

Living longer and healthier will have serious implications for the world's population. The ability to quickly and easily copy information on to a device that everyone has will make protection of intellectual property difficult. Vast databases, containing information on everything from your heartbeat to your choice of consumer products, will be assembled on everyone. Everyone will be continually monitored. The issue will be protecting access to personal information. As machines become as human as humans, are we going to kill them when we don't want them? Chip-humanoids will eventually become reality as we replace failing brain parts with chips. We will be able to have X-ray or infrared vision. Personal diagnostic and healing devices will be readily available. All of this will happen slowly, as with the genetic engineering revolution. At first it will happen for worthy causes, which we will agree with. No-one will oppose the introduction of a chip that can heal the severely mentally retarded after a road accident, or cure schizophrenia. Machines that clean the blood of those who have had heart attacks will be welcome. Machines that clean the house will be a boon. Machines that can assist thinking so I can pass a nanotechnology examination at University? This is not far removed from a calculator really, is it?

10.6 RELIGION AND MAKING EVERYTHING FROM EVERYTHING ELSE

If we are able to understand completely the structure of all matter it should be possible to manufacture all products, including biological materials. Probably the first synthetic 100% beef patty on a sesame seed bun with a Scottish-sounding name, that will not involve the death of a cow, will come from tissue culture; but according to Drexler it should also in principle be possible with nanotechnology. Lamb chops will be made from green grass but they will not arrive on our plates via lambs.

Even those who find eating meat morally unethical eat vegetables. It might be argued that this is less or more acceptable, depending on what we define as life and whether we believe that the vegetable suffers. It would not be the aim of a plant seed to be eaten. Vegetables consist of intricate molecular nanomachines involving tens of thousands of genes, proteins, and other molecular components. They also reproduce, and given enough time, they grow. It is not unreasonable therefore that we should consider building self replicating nanomachines that are vegetables without offending a group in society that call themselves vegetarians. *Mycoplasma genitalium* is the simplest natural living system that can survive on a well-defined chemical medium. Its genomic complexity of 1 160 140 bits is about one tenth of a typical floppy disk. Thus, designs for self-replicating systems both exist and are well within current capabilities.

These machines do not necessarily have to mimic nature. Indeed we have gone from analogue to digital, whereas nature prefers to use analogue. Likewise, cars have wheels but nature did not make a cow with wheels. A motor vehicle can move much faster on a flat field, even faster than a jaguar. Cars, however, need roads on which to travel, do not self repair and are unable to cope with a complex environment. In the same way, the artificial self-replicating systems will at first not be very versatile.

This raises the question: is nanotechnology dangerous? Imagination raises the question of carnivore nanomachines that eat other nanomachines or even people. One of the most far-sighted views is the 'grey goo' problem [1–4]. If assemblers can replicate themselves *ad infinitum* they could consume everything in their path as they reproduce and multiply. Is this not just like the behaviour of humans?

The grey goo problem does seem far fetched, since it ignores the laws of chemistry. A self replicating assembler will need to have a molecular structure itself and it is unlikely to find the correct diet for its replication all in one place. A self replicating machine stuck in an iron ore deposit will have all the iron and oxygen it needs but it will be short of carbon. Just like animals without the correct diet, assemblers will die unless they can hibernate. Such machines would need artificial intelligence to survive. Humans will need to install this intelligence, or devise a method by which it can evolve. That is, grey goo would have to be capable of its own evolution to be a threat. All this sounds more and more like a new life form — a new race of intelligent beings. Maybe that is our destiny: to hand over to a new intelligent life form, which may keep a few of us in some utopian zoo, perhaps a garden of Eden (again?). This brings us to the ultimate question of religion. Who is God for these new intelligent life forms — us or the real God? Will they be atheists and believe they came from the evolution of nanotechnology?

The real problems of nanotechnology are similar to those in all forms of science. It is important to understand and use what we have created safely. There is nothing wrong with nanotechnology, but we should not use it for warfare or against the service of humankind in other ways. Over the centuries, the acquisition of new knowledge, such as the discovery that the Earth revolves around the sun, or the Darwinian theory of evolution, has challenged current religious teaching. Yet it has not destroyed religion and it never will.

10.7 THANKS FOR ALL THE FISH

Finally, we should note that the science fiction television program *Red Dwarf* has been inspirational in highlighting the possibilities of nanotechnology for students. Much like the epic of a previous generation, who thrived on the lateral thinking in *The Hitch Hiker's Guide to the Galaxy* it reveals the extraordinarily blinkered and predictable way we go about our lives. This may be a genetically inherent trait, a relic of our hunter gathering days designed to ensure we keep looking for food rather than wasting time. In Chapter 1 we saw how the first use of tools took us on a path away from this pattern, and we should all stop to think 'what if?' This might be why many of us find Drexler's ideas so fascinating. If we shrink from a micrometre to a thousandth of that, a nanometre, we've reached the scale where atoms are tangible objects. A one-nanometre cube of diamond has 176 atoms in it. Designing at this scale is working in a world where physics, chemistry, electrical engineering, and mechanical engineering become unified into an integrated field.

There is an important message in *So Long and Thanks for all the Fish* (the fourth book in the Hitch Hiker series). Just before the earth is destroyed to make way for a new hyperspace bypass, mankind realises that all the dolphins have left the planet. The dolphins reveal their superior intellect by leaving a message in the sky: 'So long, and thanks for all the fish'. This comes as a huge surprise to the human race. This statement represents humankind's ignorance of what is really going on. In *The Hitch Hiker's Guide to the Galaxy*, the ultimate shock for humans was to find that laboratory mice are here to study us, not the other way around. We must use this type of lateral thinking to consider where nanotechnology is going, and this chapter gives license to such prophesies.

Imagine a research organisation functioning 200–300 years ago, which was funded in the same way that ours are funded today. Let's imagine that it was based in a land that we will call Plod, and that the main form of travel was by stagecoach. Plod had a Ploddy research organisation, which had several branches, including the Ploddy Food Research Institute and the Ploddy Energy Research Institute. In the Ploddy Energy Research Institute the scientists had to raise most of their research funding from industry, and hence most of their work was of a 'practical' nature. There were research projects aimed at breeding stronger horses, research projects to producing better axle grease and research projects into reigns and even better feed for the horses. Environmental groups were looking at ways to remove horse dung from roads, and producing odourless dung. Some people were working on advanced types of whips, because this was the Director's pet subject from his PhD days.

Meanwhile in the Food Research Institute some scientists were looking at how kettles boil, but the experiments kept failing because the steam in the kettles kept blowing the lids off. There was no use for kettles with flying lids in the food industry and the scientists were made redundant.

This story tells us why the steam engine could not be discovered in the land of Plod. It is also a lesson about the type of research we spend our time on today. Virtually nothing that was important in research in the land of Plod is useful today. So should we be doing research into different forms of energy, such as coal combustion, batteries, wind energy and nuclear power now? Are they really a side issue to the main story? What other fields of research are in reality a complete waste of huge amounts of money and time? Perhaps 75 per cent of our efforts are worthless: the remainder is necessary, but of that only 1 per cent may be life changing.

Surprisingly, science keeps throwing up a number of fundamental concepts in the Universe that turn up everywhere. Things like matter and quanta are understood. As discussed in Chapter 8, the duality of matter as particles and waves, and uncertainty and probability are less comprehensible. Perhaps we are in the land of Plod; perhaps to really understand we must waste a lot of effort on the wrong thing: if we know what we are doing we are less likely to be doing the right thing! All this smacks of a variant of an uncertainty principle!

Finally, as we conquer nanotechnology and move on to the world of the electron and smaller to research the fields of pico (10^{-12}), fento (10^{-15}) and attotechnology (10^{-18}), we are going to run in to the problem of time. It has been shown that an atomic clock at the North Pole actually keeps different time than one at the equator because as things move faster time slows down. Recall Einstein's experiment on a tram: looking at a clock as the tram moved away, he postulated that if the tram went as fast as the speed of light, the light from the clock would not change, because new light would be unable to reach the tram. The clock would permanently read the same time and time would stand still. We know also that an electron falling into a nucleus goes so fast that time also slows down, and we must take this into account in our calculations. This fourth dimension has yet to be explored and it seems that nanotechnology will be a route to understanding it. Just as quantum phenomena are necessary to understand nanotechnology, as we study smaller things in picotechnology, time relativity becomes more important. Perhaps there are some new rules of physics to be found. If we make our pico, fento and attomachines go very fast, time will actually slow down so that in effect they are going slower and eventually they may be going in reverse. In fact in a pico world the end of this book may be the beginning. Science fiction yes, but who knows?

WHAT YOU WILL *NEVER* KNOW UNLESS NANOTECHNOLOGY LETS YOU LIVE FOR A HUNDRED YEARS

1 Whether we will stop communicating by written methods in the future. Whether books will be considered as ancient relics. Whether we may even be able to relay thought patterns. Whether new hybrid humans are possible, incorporating new nanomachines.

2 How new materials will be indestructible. Whether we will only have to make what we eat and not kill plants or animals.

3 Future thinking on nanomedicine. Eventually, will we be able to replace everything and become indestructible? Only a few more generations will die if we can fix up the genetic code and treat environmentally derived diseases with nanomachines. Will you be one of them?

4 How we might deal with the social issues that arise from nanotechnology. Will we actually get so bored with immortality that we will choose death? Terminal boredom may become the new cancer.

5 Whether there are smaller worlds of pico, fento and attotechnologies, and whether study of these fields will lead to a capacity to alter time.

10.8 EXERCISES

1 *Compare and contrast how a swarm of nanomachines might resemble an ant colony.*

2 *Web search quantum mirages, relativity, Einstein, nanomedicine, respirocyte, nanoecology, social impacts of nanotechnology.*

3 *Select one of the references given in an area of interest and read it thoroughly.*
4 *Write an essay on one aspect of life in the year 2050. Theme it on an occupation that has lasted for over 1000 years, such as medical doctor, farmer or politician. A good way to project this is to compare and contrast life in 2000 BC with that in AD 1000 and AD 2000 and project again. Send it off to a newspaper or magazine for possible publication. Think big, not nano- in your life!*

10.9 REFERENCES

1 Drexler E (1990) *Engines of Creation.* Fourth Estate, London.
2 Drexler E (1992) *Nanosystems: Molecular Machinery, Manufacturing, and Computation.* John Wiley & Sons Inc., New York.
3 Drexler E, Peterson C & Pergami G (1991) *Unbounding the Future.* William Morrow and Company Inc., New York.
4 Drexler E (1992) *Nanosystems: Molecular Machinery, Manufacturing, and Computation.* John Wiley & Sons Inc., New York.
5 Storrs Hall J (1996) Utility fog: The stuff that dreams are made of. In BC Crandall (ed.) *Nanotechnology: Molecular Speculations on Global Abundance.* MIT Press, Cambridge MA, USA.
6 Wowk B (1996) Phased array optics. In BC Crandall (ed.) *Nanotechnology: Molecular Speculations on Global Abundance.* MIT Press, Cambridge MA.
7 Joseph Michael, UK Patent α94004227.2.
8 Merkle RC (1996) *Nanotechnology* 7: 210–15.
9 Freitas RA (2000) *The Sciences* July–August: 26–31.
10 Freitas Jr RA (1999) *Nanomedicine, Volume I: Basic Capabilities.* Landes Bioscience, Georgetown TX.
11 Shusterman DJ (1993) *Occup. Med.* 8: 519–31.
12 Catelas I, Petit A, Marchand R, Zukor DJ, L'Hocine Yahia OL & Huk J. *Bone Joint Surg. Br.* 81: 516–21.

INDEX